Sugar-cane

TROPICAL AGRICULTURE SERIES

The Tropical Agriculture Series, of which this volume
forms part, is published under the editorship of
Gordon Wrigley

ALREADY PUBLISHED

Tobacco   *B. C. Akehurst*
Sugar-cane   *Frank Blackburn*
Tropical Grassland Husbandry   *L. V. Crowder and H. R. Chheda*
Sorghum   *H. Doggett*
Tea   *T. Eden*
Rice   *D. H. Grist*
The Oil Palm   *C. W. S. Hartley*
Cattle Production in the Tropics Volume 1 *W. J. A. Payne*
Spices Vols 1 & 2   *J. W. Purseglove* et al.
Tropical Fruits   *J. A. Samson*
Bananas *N. W. Simmonds*
Tropical Pulses   *J. Smartt*
Agriculture in the Tropics   *C. C. Webster and P. N. Wilson*
Tropical Oilseed Crops   *E. A. Weiss*
An Introduction to Animal Husbandry in the Tropics
   G. Williamson and W. J. A. Payne
Cocoa   *G. A. R. Wood*

Sugar-cane

Frank Blackburn

Longman
London and New York

Longman Group UK Limited
Longman House, Burnt Mill, Harlow
Essex CM20 2JE, England
and Associated Companies throughout the world

*Published in the United States of America
by Longman Inc., New York*

First published 1984
Reprinted 1991

British Library Cataloguing in Publication Data
Blackburn, Frank
   Sugar-cane. – (Tropical agriculture series)
   1. Sugar-cane
   I. Title   II. Series
   633.6′1   SB231

ISBN 0-582-46028-X

Library of Congress Cataloging in Publication Data
Blackburn, Frank, 1920–1980.
   Sugar-cane.
   (Tropical agriculture series)
   Bibliography: p.
   Includes index.
   1. Sugar-cane.   I. Title.   II. Series.
SB231.B65 1984   633.6′1   83-18766
ISBN 0-582-46028-X

Produced by Longman Singapore Publishers Pte Ltd
Printed in Singapore

Contents

Acknowledgements

When Frank Blackburn died on 30 June 1980, the first draft of this book had been completed and many of the sections and chapters checked by specialists. Among the many who gave Frank their opinions and comments included

D. I. T. Walker and N. W. Simmonds – botany and breeding.

H. A. Thompson – mechanization, harvesting and skirmish plots.

C. J. Coote – irrigation.

D. W. Fewkes – pests of sugar-cane.

Ir. Boedijono (Indonesia) – top borer and its control.

I. Clarke, T. Carr and L. Donnawa – latest herbicide and insecticide use in Trinidad.

A. Johnston, P. Holliday and J. M. Waller of CMI – diseases of sugar-cane.

S. J. E. Winn – manufacture.

B. P. Baker – co-products and sucrochemistry.

W. N. L. Davies – general help and advice on many sections.

The following kindly provided illustrations:

P. E. A. Campbell, H. F. E. Deacon, J. & L. Honiton Engineering (Louisiana); J. C. Hudson, Massey Ferguson (Aust.) Ltd; J. R. Metcalf, M. Moffat, K. Mohan Naidu, B. J. Osborn, G. C. Stevenson, H. A. Thompson, Toft Bros Industries (Queensland); J. R. Williams, and, in particular, Dr A. L. Down (retired Executive Director of Texaco).

Sir John Saint, who has long associations with the sugar industry of Barbados, gave much critical advice. Tate & Lyle supported and encouraged the project from the start, and J. F. P. Tate kindly gave permission to publish data from private memoranda.

In writing a technical book a good library is essential, and the librarians at the Department of Applied Biology, Cambridge University, provided this support with their usual helpfulness.

To those who knew Frank, suffice to say he would not have accepted all the points made, particularly those that did not seem logical to him. To those whose help is not acknowledged my sincere apologies; it is simply his records are incomplete. He would have been the last person not to acknowledge their help.

When I undertook the final editing, though the book is an incomparable history of the sugar-cane industry full of practical sense, a number of developments had taken place since the text was written; for example, Zimbabwe and its sugar quota did not exist. It was necessary, therefore, to ask for the assistance of very busy specialists, and not one hesitated to offer all the help they could give.

Among those who contributed, particular thanks are due to Hugh Thompson who read the complete book and added sections to Chapters 5, 8, 9 and 10 and also wrote the section on flowering. Hugh's knowledge of Africa, which Frank never visited, introduced into the book important factors from these new sugar-producing countries, and taught me a lot in the process. R. A. Yates contributed much detail to Chapter 3, and N. W. Simmonds and D. I. T. Walker a highly technical section to Chapter 4. Patricia Cassidy volunteered to bring the complicated marketing arrangements in Chapter 1 up to date. Dr Ken Parker wrote the final section on sucrochemistry, condensing a long specialized scientific paper into a very readable full summary. Dr G. D. Thompson, Director of Mount Edgecombe Sugar Experiment Station in Natal, South Africa, was very generous in permitting the use of much material from the publications of the Experiment Station, particularly relating to weed control, diseases and *Eldana*. Similarly Dr Antoine, the Director of the Sugar Research Institute, Reduit, Mauritius, answered many questions in detail and agreed to the use of material published by the Institute.

In the editing of this book, very little has been changed. Frank was a very meticulous fellow both over factual details and the Queen's English. A number of errors inevitably occur between the author and the bookseller and I must accept full responsibility for allowing this. Frank's wife, Margaret, who has a lifetime of family association with cane growing, typed and retyped the manuscript many times, and prepared the index, will see this book as Frank's final contribution to his beloved 'World of Sugar'.

Gordon Wrigley

We are indebted to the following for permission to reproduce copyright material: Australian Journal of Biological Sciences for fig 3.6 from 'Physiology of Sugar Cane VII' by K. T. Glasziou, T. A. Bull, M. D. Hatch & P. C. Whiteman, *Australian Journal of Biological Sciences*; Commonwealth Agricultural Bureau for fig 6.6 from an article by G. A. Norton and D. E. Evans in *Bulletin of Entomological Research* 63; Elsevier Science Publishers for figs 3.9 (Shaw 1954), 6.9 (King et al 1965), 7.8 (Ricaud & Sullivan 1974), & table 4.2 (Bremer 1932); Hawaiian Sugar Planters Association for fig 3.10 (Mongelard 1973); Methuen & Co. Ltd for fig 3.4 (Miller 1931).

Whilst every effort has been made we are unable to trace the copyright holder of table 4.1 by E. K. Janaki Amal (1936) *Indian Journal of Agricultural Science*, and would appreciate any information which would enable us to do so.

Foreword

An Honours graduate from Hull University, Frank Blackburn gained selection to the then Colonial Agricultural Service and spent two years at the Imperial College of Tropical Agriculture in Trinidad, taking the College Associate course. He was then posted to Barbados as Agricultural Chemist. There he worked for Dr S. J. Saint, later knighted for services to the Sugar Industry and the Government of Barbados, under whose guidance Frank developed a keen interest in the agronomy of sugar-cane. Four years later he left the Colonial Service and took up an appointment as Research Chemist with Caroni Ltd. in Trinidad.

Caroni was a subsidary of Tate and Lyle, London, and one of the two largest sugar companies operating in Trinidad at that time. His duties included the planning and supervision of field trials but he soon became deeply involved in efforts to improve both the management and economics of cane production. Frank soon realized that one of the most important factors limiting the yield of sugar locally, was the serious and unpredictable loss caused by the Trinidad Sugar-Cane Froghopper. At worst the pest could cut cane yields by 70% and, despite many years of intensive research under the guidance of the Froghopper Investigation Committee, no satisfactory control measures had been developed.

Frank Blackburn made the vital breakthrough, using the then new organo-chlorine insecticides, and developed control measures which related the life cycle of the froghopper to precise predictions of brood emergence, adult populations and anticipated levels of damage.

This advance made a major contribution to the economics of sugar production in Trinidad and demonstrated the successful use of applied technology to a practical and serious problem. His name will always be associated with this work.

Following his time as Research Chemist, Frank was appointed Agricultural Superintendent at the early age of 30, and he subsequently became General Manager and eventually Managing Director of Caroni Ltd and Chairman of the Trinidad Sugar Manufacturers Association.

During the period of his direct involvement in Management the size of Caroni increased materially until, with the acquisition of Usine St Madeleine, it became the dominant company in the Trinidad sugar industry.

Sugar production more than doubled, reaching a total of over 200,000 tonnes from 20,000 hectares of company land and 14,000 small independent growers.

The Caroni labour force, second only in size to that of the Government, was highly unionized and greatly influenced by the proximity of the specialized and affluent oil industry. In addition, a strongly developed nationalistic spirit affected every aspect of life in a country developing from a Crown Colony to an independent Republic. During a period, when industrial relations presented singularly difficult problems, Frank's judgement and sense of humour maintained amicable relations with both his workforce and some of the more politically motivated union leaders. At the same time, he always recognized the benefits which would accrue to the Company and its employees through the deliberate promotion of organized sport. Frank not only worked hard to encourage such ideas, but personally took a most active part therein.

Under Company sponsorship many recreation grounds were established in the closely mixed rural and village community of central Trinidad, with cricket being the most popular and competitive sport. Caroni's 1st XI, "The Wanderers", was captained by "Skipper" Blackburn and occupied a dominant position in Trinidad cricket circles. All this did a great deal, not only to develop a close identification between the Company and its workers, but also to increase goodwill amongst the many friends and relations supporting the Company teams.

One way and another, Frank Blackburn's enthusiasm, coupled with his wide experience of so many aspects of sugar production, made him the right man at the right time in the right place during the development of the Trinidad sugar industry.

I first met Frank Blackburn in 1949 when, as a trainee, he helped to introduce me to the science of sugar-cane agriculture. I learned much from Frank and soon realized we had two great common interests – cane and cricket. Over the next 20 years, or so, when I was a Director of Tate and Lyle and of Caroni, we continued to have many fascinating discussions which I shall always treasure and I, and many others, are delighted that he committed to paper the thoughts and ideas resulting from his long experience of sugar-cane growing.

September 1983   John O. Lyle

Chapter 1

The origin and history of the sugar-cane

Sugar-cane varieties are species and hybrids of the genus *Saccharum* which, in turn, is of the family Gramineae in the tribe Andropogoneae. For centuries *S. sinense* has been grown in China and *S. barberi* in India; but it was the increased planting of the noble cane, *S. officinarum*, which caused the sugar industry to spread throughout the tropics and subtropics. The origin and history of the sugar-cane were described in great detail by Deerr (1949). It is thought that *S. officinarum* originated in the South Pacific area, probably in New Guinea, and was dispersed by three routes in different epochs. The first, starting approximately in the year 8000 BC, was to the Solomon Island, the New Hebrides and New Caledonia; the second, beginning about 6000 BC, was by way of the Philippines, Borneo, Java, Malaya and Burma to India; and the third, between the years AD 500 and AD 1100 was from Fiji to Tonga, Tahiti, the Marquesas and Hawaii. These migrations were described by Brandes (1958) and are illustrated in Fig. 1.1.

The Mediterranean sugar industry

The Islamic conquests of the period AD 600 to AD 800 gave impetus to the westward movement of *S. officinarum* by the second route. Its cultivation spread from India to Iran, then to Syria, and eventually to several other Mediterranean countries. Raw sugar made in Egypt, Cyprus, Sicily and southern Spain was sent to Venice to be refined. Indeed by the end of the sixteenth century Venice had become the international centre of the industry. The methods of manufacture were crude: stone-edge runner mills and screw presses, the latter still used to express oil from olives, were not replaced until 1449 when the first vertical roller mill was built in Sicily. The new invention comprised three wooden rollers, encased in iron bands and geared together, with motion imparted to the middle roller. There was little development in other aspects of manufacture. Water was still evaporated from the juice in open vessels and the sugar was

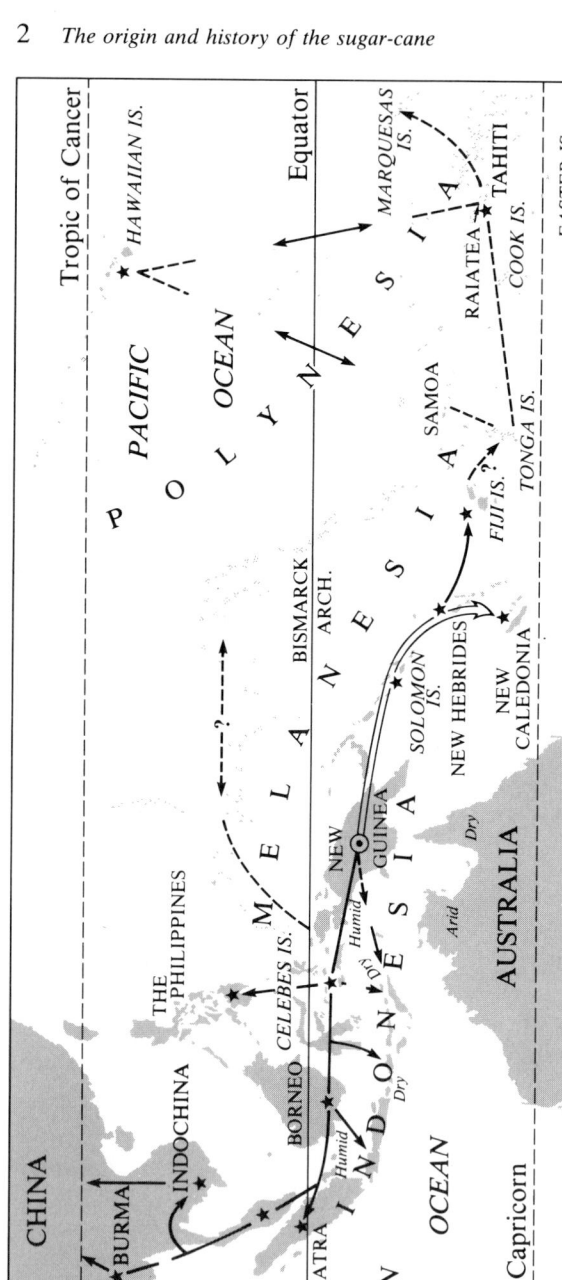

Fig. 1.1 The origin and migration of *S. officinarum* (After Brandes 1958)

⊙   Origin of *Saccharum officinarum*, derived from *S. robustum*, which occurred 8000 to 15000 BC

═══   First migration of *S. officinarum*, beginning about 8000 BC

─────   Second migration, beginning about 6000 BC

─ ─ ─   Third migration, about AD 500 to 1100

★   Satellite centres of diversity along tracks of migration

crystallized in moulds before being packed in bags for shipment. The industry survives in this area only in southern Spain and, having been revived in the nineteenth century, in Egypt.

The years of Portuguese dominance

For roughly 150 years, from the beginning of the fifteenth century to the middle of the seventeenth, international trade in sugar was dominated by Portugal. The cultivation of the cane continued its westward movement: first to Madeira and Sao Thomé, islands off the Atlantic coast of Africa; then to Fernando Po and Angola; and finally across the ocean to Brazil. Meanwhile the sometimes unseemly scramble between England (Britain from 1603), France, Holland and Spain for possession of the West Indies and the Guianas was taking place; and when the Dutch settlers, who controlled the sugar industry of Brazil, were expelled by the Portuguese in 1654, they emigrated to the new colonies and helped develop rival sugar industries in them. For example Ligon (1657) acknowledged the tremendous contribution which strangers from Brazil made to improve planting, cultivation and manufacture in Barbados. Indeed the original seed-pieces had been imported from Fernambock (Pernambuco).

The heyday of the plantocracy in the Caribbean

Political and social development

The demand for sugar by the metropolitan powers caused rapid and lucrative expansion of production in the Caribbean area during the late seventeenth and the eighteenth centuries. Flourishing industries arose in the Spanish possessions: Cuba, Hispaniola (now the Dominican Republic) and Puerto Rico; the French; St Domingue (now Haiti), Guadeloupe and Martinique; the British: Barbados, Jamaica and many of the Leeward and Windward Islands; the Dutch: St Eustatius and Surinam; and the Danish: St Croix. These were turbulent years. France and Britain disputed the possession of many islands, especially St Lucia and Martinique; Surinam was ceded by Britain to Holland, and in later years Berbice, Demerara and Essequibo by Holland to Britain. Memories of those years of strife in Tobago persist in such place names as Bloody Bay and Man of War Bay.

The expansion of the sugar industry in the Caribbean, as in Brazil, was made possible by the labour of slaves brought from

Africa. The horrors of the middle passage between Africa and America, and the human degradation inherent in the institution of slavery, are well documented and do not need repetition. At that time life for many people was brutish and short: recruits to the Royal Navy were pressed into service, without notice and against their wishes, to live in conditions but marginally better than those of the slaves, to suffer appalling punishment for trifling offences, and to fight the enemy; sheep-stealing was a capital offence; and children were hanged for the theft of pitifully small sums of money. Be that as it may, some of the slaves had been leaders of their people and, although captured in inter-tribal skirmishes, were unwilling to accept their loss of freedom with meekness. The possibility of insurrection was never far from the minds of slave-owners, whose reactions approached hysteria when the French Revolution started in 1789. Such fears were well founded: an unsuccessful rebellion by free men of colour, led by James Ogé in St Domingue in 1790, was followed by an uprising of slaves under Toussaint L'Ouverture during the period 1791–1794. L'Ouverture surrendered to the French in 1802 and died in captivity in 1803; but the struggle for freedom continued under Dessalines (also a former slave) until 1 January 1804 when the independent state of Haiti was born. Crops and buildings had been burnt down, their owners and families massacred; barbarous atrocities had been perpetrated by both sides, and the most flourishing sugar industry in the West Indies had been destroyed. Many of these events were described by Bryan Edwards (1801) (sometimes called 'the planters' historian') who, serving as an emissary of the Government of Jamaica, visited St Domingue in 1791. Conditional recognition of Haiti was given by France in 1825, and full recognition in 1838.

Meanwhile all was not well in Jamaica. When Penn and Venables reduced that island in 1655, most of the Spaniards emigrated to Cuba. Their slaves were unwilling either to go with them or to accept new masters. Consequently many (estimated by Edwards to number fifteen hundred) ran away and occupied the rugged Cockpit Country. They were called Maroons and their numbers were increased by slaves absconding from the newly formed British plantations. After several military expeditions to pacify them had met with only partial success, a treaty giving some autonomy to the Maroons was eventually ratified in 1798.

Developments in technology

Few advances in technology took place during this period. Everywhere the vertical, hollow three-roller mill, with a dumb turner to control the passage of bagasse from roller to roller, replaced the two-roller mill, with some consequent improvement in extraction. There was also greater understanding of clarification. Ligon (1657) wrote: '. . . as it (the liquor) boyles, there is thrown into the four last coppers, a liquor made of water and ashes which they call Temper, without which, the Sugar would continue a Clammy substance and never Kerne'.

More than a century later Edwards (1793) described the process as follows: 'The stream then from the receiver having filled the clarifier with fresh liquor, and the fire being lighted, the temper, which is commonly white-lime in powder, is stirred into it. One great intention of this is to neutralize the superabundant acid, and which to get properly rid of, is the great difficulty of sugar-making.' He recommended that one half pint (0.28 litre) of Bristol lime be added to every 100 gallons (45.5 litres) of liquor. The fire under the clarifier was extinguished for an hour and: '. . . the feculencies and impurities will attract each other, and rise in the scum. The liquor is now carefully drawn off, either by a syphon, which draws up a pure defecated stream through the scum, or by means of a cock at the bottom.' Thereafter the clarified liquor was ladled from one copper to the next, with intermittent skimming, until it reached the smallest vessel, the *taiche*.

Two types of sugar were produced: muscovado and clayed. They differed in the method of curing. To make muscovado sugar the mass from the last copper, having been cooled, was placed in barrels called hogsheads or *tierces*. Eight or ten holes were bored in the bottom of each hogshead, and plantain (*Musa paradisiaca* L.) leaves thrust through them. After three weeks the sugar was regarded as cured, and was then exported to be made into loaf or refined cube sugar. The substance which had drained through the partially filled holes was choice molasses. Clayed sugar, on the other hand, was made by filling conical pots (called *formes* by the French), pointed downwards, each having a hole 38.1 mm (1.5 in) in diameter at the bottom, closed with a plug. After about twelve hours, when the sugar had cooled and formed a 'fixed body', indicated by the middle of the top falling in, the plug was removed to allow molasses to drain away. This continued for twelve to twenty-four hours, until molasses ceased to be exuded. Then a layer of moistened clay (white marl, if possible) was spread over the top of the sugar. The moisture from the clay washed the crystals, which therefore were of higher quality, though some sugar was lost in the process. Clayed sugar was favoured by the French, especially the manufacturers in St Domingue, and muscovado by the British.

Depending on cane quality, the yield of muscovado sugar was one hogshead (16 cwt or 0.79 tonne) from between 1,300 and 2,600 gallons (5,910 and 11,820 litres) of juice.

The nineteenth century: Abolition, duties, free trade and beet sugar

Political and social development

Slavery and apprenticeship

The institution of slavery was attacked by John Locke, the philosopher, as long ago as 1689; and the movement for its abolition gained increasing momentum, especially in England, during the next 150 years. It was a subject of great controversy: eloquent abolitionists were faced by a stubborn lobby representing the interests of

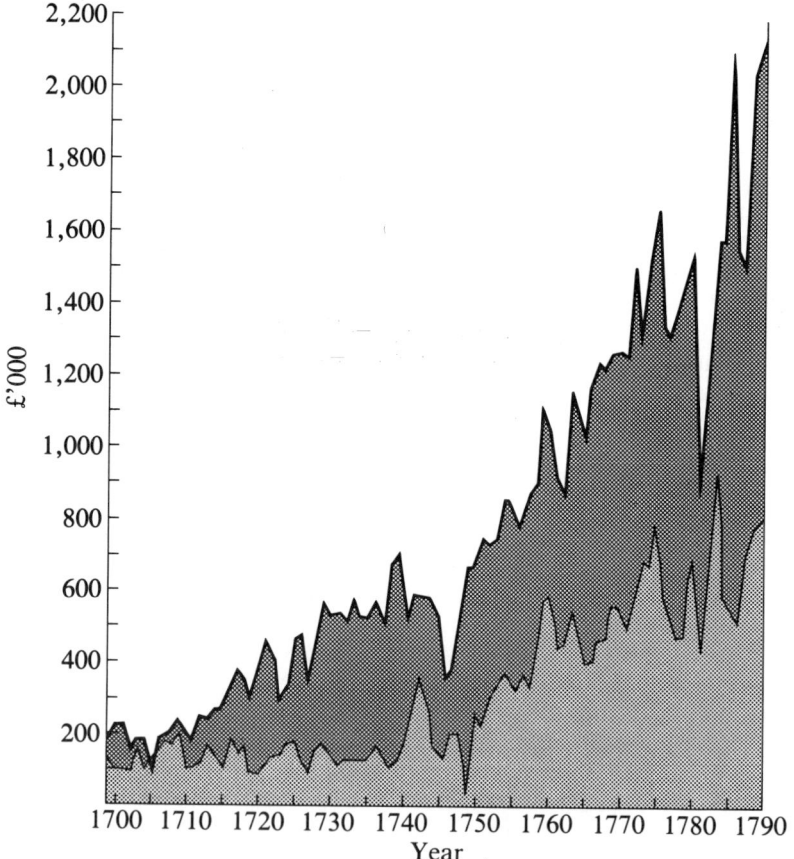

Fig. 1.2 The balance of trade in thousands of pounds sterling between England and Jamaica, 1699–1789. The bold line represents the balance in favour of Jamaica. Only in 1705 was the balance in favour of England. Based on Custom House Accounts, Ledgers of Imports and Exports (After Pitman 1917)

plantation owners. Although many books and pamphlets were published, and stormy debates took place in the House of Commons, the outcome was inevitable. Indeed a distinguished historian, who later became the first Prime Minister of Trinidad and Tobago, came to the conclusion that slavery, having provided the money to finance the Industrial Revolution in the UK, was itself destroyed by the mature capitalism which it had nurtured (Williams 1964). Certainly in the eighteenth century the balance of trade with the sugar-producing colonies was heavily in favour of the metropolitan country (Fig. 1.2), because of the value of this scarce product.

Whatever the motive, the slave-trade was made illegal by the UK (Great Britain became the United Kingdom by the Act of Union of 1803) in 1807, and by Denmark, Holland, Sweden and the USA during the next seven years. Then slavery itself was abolished by the UK on 1 August 1833 and during the next thirty-five years also by Argentina, Peru and Uruguay; by the colonial powers France, Holland and Portugal; and by the USA. In the reformed House of Commons (elected in December 1832) which passed this law there were only sixteen members with West Indian interests, whereas in 1780 there had been seventy-five.

Emancipation in the UK colonies did not bring immediate and absolute freedom to the slaves. For six years from 1834 (when most colonial legislatures, though not that of Antigua, complied with this special requirement of the Act of Parliament) they were to work as apprentices for their former masters, but would receive wages for their services. This was described rather harshly by Craton & Walvin (1970) as: 'a pious fraud, a transitional expedient designed to convert the Negro work-force from the unwieldy coercion of slavery to a system, possibly more effective but not necessarily more voluntary, of free wage labour'. It was a transitional measure by which it was hoped that the change from slavery to freedom might be achieved with minimum disruption of the economy. Nevertheless the imposition of apprenticeship was resented, met with little success, and was abandoned in Jamaica in 1834 and in Barbados and Trinidad in 1838. Where land was available the former slaves became small holders and worked irregularly, if at all, on the plantations. Consequently there was a disastrous fall in the production of sugar in Jamaica, British Guiana (Guyana) and Tobago; some reduction in St Lucia and St Vincent; but little change in Antigua, Barbados, Dominica and St Kitts. The effect of abolition in British Guiana and Jamaica (where land was available to the newly freed slaves) and Barbados (where all the land was in cultivation) is illustrated in Fig. 1.3. Because of the scarcity of labour and lower production many of the planters suffered severe financial loss and were so heavily in debt that they were unable either to work their estates or to sell

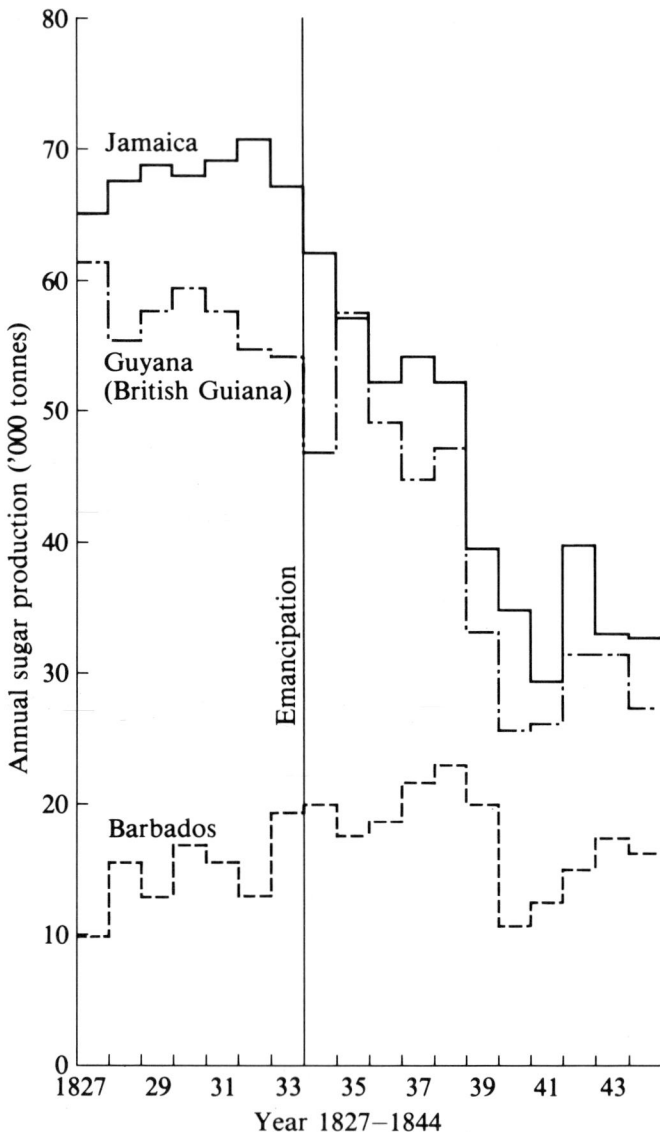

Fig. 1.3 The effect of the abolition of slavery on sugar production in Jamaica, Guyana and Barbados

them. To overcome this impasse the West Indian Incumbered Estates Act was passed in 1854, but could be operated only after the Legislature of a colony had addressed the Crown, asking for an Order in Council to that effect. Incumbered Estates Courts were formed in six colonies, with a Central Commission in London. Their purpose, to allow bankrupt estates to change hands, was achieved. The dates of establishment of Courts in the colonies, with an example of a judgement made in London (From Cust, 1865) are given.

NAMES OF COLONIES

In which the West Indian Incumbered Estates Acts are in force, with the dates of the Orders in Council by which they were brought into operation, and of the Addresses of the Colonial Legislatures upon which such orders were founded :—

NAMES.	DATE OF ADDRESS.	DATE OF ORDER.
ST. VINCENT. .	15 July, 1856	2 February, 1857.
TOBAGO . . .	22 December, 1857	31 July, 1858.
VIRGIN ISLANDS	28 December, 1859	7 March, 1860.
ST. CHRISTOPHER	December, 1859	26 March, 1860.
JAMAICA . . .	4 March, 1861	26 June, 1861.
ANTIGUA . . .	October, 1864	1 November, 1864.

WEST INDIAN INCUMBERED ESTATES COURT,

8, Park-street, Westminster, Feb. 4, 1862.

(Before HENRY JAMES STONOR, Esq., Chief Commissioner.)

Re PARKER,
Exparte WILLIAM RICKETTS PARKER.[1]

Practice—Objections.

THIS was a motion on behalf of the petitioner, Colonel Dawkins, to make absolute a conditional order for the sale of five estates in the island of Jamaica, called Hill-side, Brazaletto, Chesterfield, Bourkesfield, and Coles Penn. An objection had been filed, as to the two last named estates, on the part of William Ricketts Parker, who claimed to be interested as owner, or first incumbrancer.

Mr. *Cutler*, solicitor, appeared in support of the objection.

Mr. *Frederick Smith*, solicitor, in support of the motion.

The Chief Commissioner held that W. R. Parker had failed to prove himself the owner, and, therefore, under the 8th section of the West Indian Incumbered Estates Act, 1858, had no *locus standi* on the present occasion; but that he had made out a *primá facie* title as incumbrancer, and would be entitled to apply for a postponement of the sale, or for the carriage of the order for sale, or to get credit for his incumbrance in the biddings on the sale.

The objection was consequently disallowed, and the order for sale made absolute as to all the estates.

1 6 Sol. Journ., 250.

There were no abandoned estates in Barbados, therefore the Act was declined; and in St Lucia, Trinidad and British Guiana it was considered that sufficient remedy could be obtained under the prevailing laws, which were based respectively on the French, Spanish and Dutch Systems.

Duties, expansion and beet sugar

Scarcity of labour was not the only reason for the decline of the industry in the British West Indian colonies (the name British West Indies (BWI) was retained until independence was achieved). By the Sugar Duties Act of 1846 the difference between the duties charged on foreign and colonial sugar imported into the United Kingdom was reduced, and provision was made for equalization in 1851. By a subsequent Act equalization was deferred until 1854 when the duties levied on imported sugar, whatever its origin, were as follows:

Refined sugar	12s. 2d. per cwt
Equal to white clayed	11s. 8d. per cwt
Not equal to white clayed	10s. 6d. per cwt
Equal to brown muscovado	9s. 4d. per cwt
Not equal to brown muscovado	8s. 2d. per cwt

Successive reductions were made until 1874, when all UK sugar duties were removed. Previously both vacuum pan sugar, which attracted the 10s. 6d. duty, and muscovado sugar, which came in at the lower rates, both manufactured in the BWI, had an advantage over the clayed and partially refined sugars from Cuba, Brazil, Mauritius and elsewhere. For this reason produce from the rising sugar industries of Mauritius, Fiji and the East Indies replaced that from the declining BWI in the UK market.

There was a large increase in the manufacture of both cane and beet sugar throughout the world. In Brazil, Cuba and Puerto Rico cane sugar was still being produced by slave labour; while in continental Europe, especially in France, Austria and Germany, the beet-sugar industry expanded enormously. Exports of beet sugar were so heavily subsidized that they were sold at less than the cost of production. Imports of bounty-fed beet sugar into the UK increased from 78,747 tonnes (80,027 tons) in 1865 to 300,120 tonnes (305,000 tons) in 1878 and 600,000 tonnes in 1888.

In an attempt to alleviate the shortage of labour in the BWI, indentured immigrants were brought from Madeira, Germany, China and India to work on the sugar plantations. Of these the Indians were by far the most effective. They came to British Guiana, Jamaica and Trinidad (taken by Britain from Spain in 1797); and there were similar movements to Fiji, Mauritius (taken by Britain from France in 1810) and Natal, South Africa. This influx met with success everywhere except in Jamaica, where the Indians made little impact.

Nevertheless the British West Indian producers were in a parlous position. To the loss of long-established preferential markets and the competition of bounty-fed beet sugar was added another misfortune: the devastation of the Bourbon variety by a complex of diseases. Apart from making small quantities of sugar for local consumption, the industry declined in Montserrat, Grenada, St Vincent and Dominica; and there was complete closure in Tobago. The century ended with the appointment in 1898 of a Royal Commission to examine the economic affairs of the BWI; and though their fortunes were to improve in later years, the dominant position which they had held for so long in the world of sugar had been lost for ever. Indeed, calamities apart, the BWI could not have satisfied the rising demand for sugar made by the UK's rapidly increasing population and consumption (Table 1.1). Accounts of these trials and tribulations have been given by Beachey (1957) and Wood (1968).

The USA was exerting an increasing influence on world markets. For example in 1892 the UK concluded a short-lived treaty with the USA for reciprocal trade with Barbados, British Guiana and Trinidad; but much more important were the special commercial arrangements made between the USA and Cuba and Puerto Rico (still colonies of Spain), Brazil and Hawaii.

Table 1.1 *Annual consumption per head of sugar in the UK*

Year	lb	(kg)
1840	15.20	(6.90)
1850	24.79	(11.25)
1860	34.14	(15.50)
1870	47.23	(21.44)
1880	63.68	(28.91)

There had been early recognition of the economies of scale in Cuba where the first great central factory, Alava, was established in 1831. Many others were built both in Cuba and in Puerto Rico; and considerable expansion of production also took place in Argentina, Mexico and Peru, where modern centrals worked cheek by jowl with old muscovado plants. The abolition of slavery had little effect on the Cuban industry which, by 1894, was the largest (more than 1 m. tonnes per annum) in the world. With the cessation of the Spanish–American War in 1898, Cuba became a protectorate *de facto* of the USA, and from 1903 its sugar was imported at a preferential rate of 20 per cent. The Philippines, ceded to the USA by Spain in 1899, were also granted increasingly favourable duties on sugar imports into the metropolitan country. In the Dominican Republic, however, although the sugar industry expanded with the aid of American capital, exports were without subsidy, bounty or protection.

Hawaii became an integral part of the USA by mutual consent in 1898, and sugar production rose rapidly from 150,000 tonnes in 1894 to 250,000 tonnes in 1899

At the turn of the century the annual production of South Africa was 20,000 tonnes, of Australia 100,000 tonnes, of Réunion 40,000 tonnes and of Mauritius 150,000 tonnes; but Java, with roughly 750,000 tonnes, was second in importance only to Cuba.

Developments in technology

Organized cane breeding started towards the end of the nineteenth century, and from that time to the present day has played an important part in the development of the industry. Be that as it may, the most significant advances made during this period were in factory rather than field technology.

Horizontal mills replaced vertical mills; solid rollers, with hydraulic attachments to give equal compression whatever the thickness of the feed, succeeded rigid (sometimes also hollow) rollers; and steam-engines superseded animals and water-wheels as the providers of motive power. Consequently, juice extraction was much improved. The vacuum pan was invented by Howard in 1813; and the concept of triple-effect evaporation was conceived and developed by Rillieux during the period 1830–60: thereby syrup was boiled at a lower temperature, inversion was greatly reduced and the loss of sugar by being burnt on the side of the last copper, which took place in the old works, was eliminated. Triple-effect evaporation and the installation of multitubular boilers also led to vastly improved fuel efficiency. In the early years of the century sugar exported in hogsheads had been only partially purged of molasses, and losses in weight of between 5 and 16 per cent were incurred by further drainage during transit from the place of manufacture to the refinery. In 1837, however, the centrifugal was invented and provided much drier crystals.

Chemical control of liming was practised here and there, especially in the new central factories, but in general the owners of muscovado and clayed sugar plants still relied on guesswork for process control.

Not all of these improvements were adopted immediately or by all producers; indeed a Montserrat planter, giving evidence before a Royal Commission in 1884, stated that: 'we have no steam engines or any of that botheration here'. Nevertheless the new technology was gradually accepted, and its application and further advancement allowed the industry to satisfy a rapidly increasing world demand for sugar and, above all, to reduce the cost of production.

Browne (1939) described these developments in the continental USA in an address given to the Sixth Congress of the International Society of Sugar Cane Technologists (ISSCT).

The first half of the twentieth century: Commissions of Enquiry, subsidies and the years of Cuban dominance

Political and social development

At the Brussels Convention of 1898 it was agreed that beet-sugar bounties would be removed, effective from 1903. Thereafter the market was dominated by Cuba, where production increased from 1 m. tonnes in 1903 to 1.8 m. tonnes in 1910 and 3 m. tonnes in 1916. With a large and lucrative outlet in the USA, Cuba was able to flood the 'free' world market with cheap residual sugar.

Predictably the production of beet sugar in Europe fell sharply during the First World War. There were corresponding increases in both the output and the price of cane sugar. The boom was short-lived and was followed by a period of low world prices which lasted from 1921 until the outbreak of the Second World War (Fig. 1.4). During this great depression imports were not allowed into Germany, Russia and Italy; and the USA market was reserved for domestic production, for imports regulated by quotas from her over-seas territories and from Cuba and the Philippines. Similarly, France's requirements were met in full by home-grown beet sugar and cane sugar from her colonies. The so-called world 'free' market was limited to countries, of which the UK was by far the most important, which imported sugar on non-preferential terms and accounted for perhaps 12 per cent of the world production of centrifugal sugar. Since 1921 the UK to safeguard supplies in time

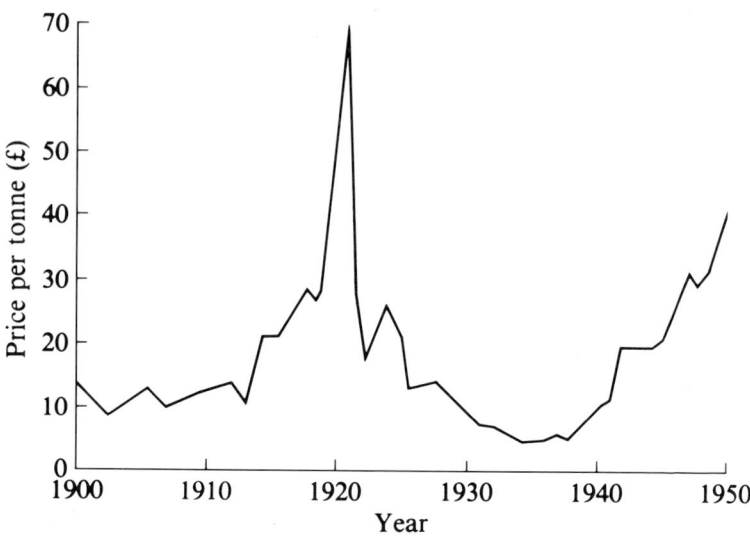

Fig. 1.4 The world 'free' market price of raw sugar Average price c.i.f. UK (£ per tonne)

of war, had also adopted the continental policy of subsidizing a beet-sugar industry. Statistics relating to production and imports during the period 1934–38 are given in Table 1.2. Conditions were particularly difficult in the BWI and Mauritius which, unlike many other cane-sugar producers, depended heavily on export sales: they did not have large and relatively lucrative home markets to cushion the effect of disastrously low 'free' market prices. Consequently, there was a succession of Royal Commissions of Enquiry and official statements on the industry in those islands. Reports were made by the Parliamentary Under-Secretary of State for the Colonies (Wood 1922) and by two Royal Commissions (Olivier 1930; Moyne 1945), publication of the latter being delayed until 1945. Numerous other governmental documents were also published.

Despite British imperial preferences of £3.15.0 per ton and Canadian preference of £4.13.0 per ton, during the 1920s the prices received were well below the cost of production in the BWI. In 1930 the Olivier Commission recommended that a stable price of £15 per ton (£14.73 per tonne) be paid; instead Colonial Sugar Certificates of £1 per ton were issued. These subsidies comprised 40 per cent of the £11.10.0 per ton, c.i.f. UK, received by BWI producers. In comparison the USA and France paid the much higher preferential prices of roughly £16 and £18 per ton respectively to their colonial, Commonwealth and protected producers: therefore their industries flourished (Table 1.3). In Cuba, also, times were good and between

Table 1.2 *United Kingdom production and imports of sugar 1934–38*

	Tons (tonnes)				
	1934	*1935*	*1936*	*1937*	*1938*
Production (beet sugar)	615,000 (625,000)	487,000 (495,000)	537,000 (545,500)	391,000 (397,000)	298,000 (303,000)
Imports (cane sugar) from foreign countries	969,000 (984,500)	1,113,000 (1,131,000)	1,161,000 (1,179,500)	891,000 (905,000)	1,179,000 (1,198,000)
from the British Empire	946,500 (961,500)	802,500 (815,500)	1,051,000 (1,067,000)	1,325,500 (1,346,500)	1,198,500 (1,218,000)

Table 1.3   *Expansion of production by suppliers to the USA and France*

	Tons (tonnes)	
	1922	1939
Hawaii	479,500 (487,000)	909,500 (924,000)
Puerto Rico	344,000 (349,500)	961,500 (977,000)
Philippines	436,500 (443,500)	961,500 (977,000)
Martinique	18,500 (19,000)	59,500 (60,500)
Guadeloupe	25,000 (25,500)	51,500 (52,500)
Réunion (exports only)	56,500 (57,500)	73,500 (74,500)

1925 and 1930 the annual production was 5 m. tonnes. Thereafter the effects of world over-supply caused a temporary fall to 2 m. tonnes in 1933. By 1946, however, output had recovered to 5 m. tonnes. The position in the 1930s was aptly summarized in the report of the United Kingdom Sugar Industry Committee (Greene 1935) as follows: 'Today, therefore, practically the only countries which are producing sugar without state assistance in one form or another are Java, Peru and Santo Domingo, and these together provide not more than 5 per cent of the world's current output.' It is not surprising, therefore, that this desperate condition was described as causing near-starvation in countries such as Java, which sold nearly all of their sugar at world 'free' market prices.

In spite of these difficulties, production in the BWI expanded from 393,100 tonnes (367,200 tons) in 1928 to 630,000 tonnes (620,000 tons) in 1939. However, even more significant developments were taking place in the Southern hemisphere. After a period of relative stagnation Australia became self-supporting in sugar in 1924, and thereafter developed a large export trade, especially with the UK. Similar expansion took place in Fiji and South Africa (Table 1.4).

The story of the emergence of flourishing sugar industries in Australia and Fiji has been told in *South Pacific Enterprise* (1956),

Table 1.4   *Expansion of production in Australia, Fiji and South Africa*

	'000 tonnes raw value crop/year			
	1920/21	1921/22	1922/23	1939/40
Australia*	186	307	314	948
Fiji	47	36	65	116
South Africa	130	135	145	540

* Of the total, the following tonnage was beet: 1920/21 – 1; 1921/22 – 2; 1922/23 – 3.
Source: 1920/21–1922/23, ISC; 1939/40, Licht.

a book written in celebration of the centenary of the Colonial Sugar Refining Co. Ltd (CSR). It is of special interest because the expansion was achieved not by developing plantations but by encouraging the settlement of small farmers, of European descent in Australia and mainly of Indian origin in Fiji, and by providing a comprehensive technical service for them. The social stability fostered by this enlightened approach is in sharp contrast with the political and personnel problems which so frequently bedevil industrial relations elsewhere. The Australian industry, located in Queensland and northern New South Wales, was stimulated by subsidies granted by the Hughes (Australian) Government in 1920 and substantial exports of 200,000 tonnes in 1925 had risen to more than 500,000 tonnes in 1939. The preferential home market compensated for low export prices.

In Fiji the CSR built factories and established plantations at the turn of the century, but in the 1920s the plantations were split up and leased to tenant farmers who by 1934 cultivated 53 per cent of all the land under cane. Of the remainder, 41 per cent was worked by independent small holders and only 6 per cent by CSR.

Thus, in spite of depressed prices because supply exceeded demand, by and large production continued to expand throughout the world. After several abortive attempts to control this trend had failed, an International Sugar Agreement (ISA), to be of five years' duration, was reached in 1937. By its terms further expansion of sugar exports from each signatory producer was limited to its proportionate share of any increase in the world 'free' market. Because of the UK's dominant position in that market, the ISA ceased to operate when the Second World War started in 1939. The war had more important effects than the abandonment of the ISA: the great industries of Java, the Philippines and Taiwan were abandoned after invasion by the Japanese.

Developments in technology

Great and widespread advances in both factory and field technology took place during the first half of the twentieth century. Prestigious research institutes and field stations were founded or expanded in many sugar-producing countries: in Australia (Queensland Bureau of Experiment Stations) the BWI (Central Sugar Cane Breeding Station, Barbados), Hawaii (Hawaiian Sugar Planters' Association, Honolulu), Java (Proefstation voor de Java Suikerindustrie, Pasuruan), Mauritius (Mauritius Sugar Industry Research Institute, Réduit), South Africa (Mount Edgecombe, Natal) and continental USA (Baton Rouge, Louisiana and Canal Point, Florida). The list is by no means exhaustive. The results of innovations, experiments and surveys carried out between 1938 and 1950 were reported in more than 100 different publications. Honig (1951) gave a catalogue of these journals, with explanatory notes concerning their content, as an appendix to a general survey of developments in production.

Equally important was the formation of technologists' societies at both local and international level. The International Society of Sugar Cane Technologists (ISSCT) was formed in 1923 and the first of many triennial congresses was held in Hawaii in 1924.

The movement towards the centralization of factories continued with increasing momentum; strict chemical control of all phases of manufacture became the rule rather than the exception; and, spurred by the shortages of jute bags caused by strife in the Indian subcontinent, methods of handling sugar in bulk were devised. Developments in field technology were equally impressive. Hybrid varieties able to withstand the ravages of disease, to give higher yields and to ratoon for longer periods, were released year after year from the breeding stations. Cane nutrition was better understood and greater use made of artificial fertilizers. The biological control of some insect pests was achieved with spectacular success, particularly in Hawaii; and elsewhere the recently developed synthetic insecticides DDT and HCH were used with equal success, but at a cost, in Australia, Louisiana and Trinidad. Towards the end of this period herbicides were being used on a large scale to replace hand and mechanical weeding. Saint (1953) quantified the contribution made by each relevant factor to increased production during the period 1855 to 1953 in Barbados, where for more than two centuries all the available land had been in cultivation. Data adapted from this paper, relating to the period under review, are given in Fig. 1.5.

Development since 1950: changing patterns of trade

Political and economic development

The most significant influences on the cane sugar industry during the immediate post-war years were three international marketing agreements: the Commonwealth Sugar Agreement, the USA Sugar Act and in less degree, the International Sugar Agreement. All of these sought, in one way or another, to stabilize the price of sugar, and to keep supply and demand in reasonable balance.

The Commonwealth Sugar Agreement

The UK, already committed to buy all Commonwealth export sugar up to the end of 1952, was anxious to obtain additional supplies from sterling sources for two reasons: to ease the drain of dollars spent on imports from Cuba and to end rationing at home. Therefore,

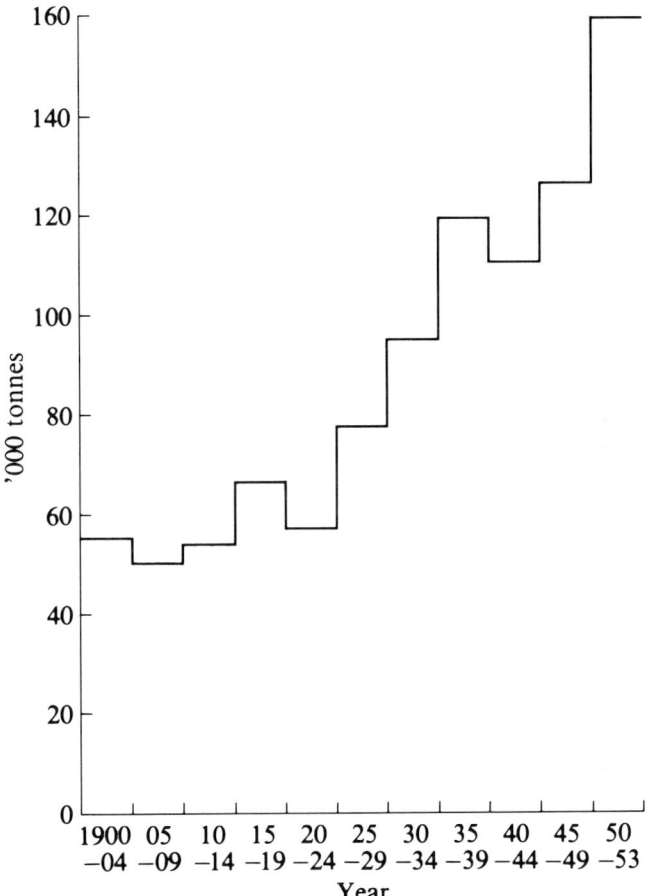

Fig. 1.5 Average annual sugar production in Barbados (in thousands of tonnes)

after three years of negotiations, the Commonwealth Sugar Agreement (CSA), originally of eight years' duration, was signed in 1951. The signatories to the contract were the UK Government as the importer, the sugar industries in a number of Commonwealth countries as the exporters. Under the CSA the exporters agreed to restrict their total exports to 2,375,000 tons *tel quel* allocated as shown in Table 1.5.

More importantly, of these overall quotas (OAQ) substantial quantities (negotiated price quotas, NPQ) were to be sold at an annually negotiated price, based on average commonwealth production costs, which would be 'reasonably remunerative to efficient producers' (Table 1.6).

Table 1.5   *The original allocation of overall CSA quotas (1951)*

	Tons tel quel
West Indies and Guyana	900,000
Australia	600,000
Mauritius	470,000
South Africa (inc. Swaziland)	200,000
Fiji	170,000
East Africa	10,000
British Honduras	25,000
Total	2,375,000

Table 1.6   *The allocation of CSA negotiated price quotas (NPQ) in 1951 and 1965 (tonnes, tel quel)*

	Basic 1951	Basic 1965	Consolidated 1965
West Indies and Guyana	640,000	641,050*	725,000
Australia	300,000	300,000	335,000
Mauritius	335,000	335,000	380,000
South Africa	150,000	—†	—†
Fiji	120,000	120,000	140,000
British Honduras (Belize)	18,000‡	18,000	20,500
East Africa	5,000§	5,000	7,000
India	—	25,000	25,000
Rhodesia (in suspense)	—	25,000	25,000‖
Swaziland	—	75,000‖	85,000
	1,568,000	1,544,050	1,742,500

* Including 1,050 tons as tonnes on account of St Vincent.
† Withdrawn from 1962.
‡ Acceded 1954, provisional NPQ until 1960.
§ Acceded 1959, quota operative from 1960.
‖ Suspended with effect from 18 November 1965.

The rest was to be sold at the 'free' world price plus UK and Canadian preference. One of the stipulations of the CSA was that 'Any Overall Agreement Quota or part thereof, which cannot be used by the territory to which it has been allocated, may be re-allocated to other territories'.

The CSA was an immediate success, and was described by colonial producers as their 'charter of existence'. The effect on production is shown in Table 1.7.

Several important changes took place during the next twenty years. South Africa withdrew from the Commonwealth in 1961 and ceased to be a party to the CSA. Swaziland (previously under the aegis of South Africa) and Rhodesia were new participants, but

Table 1.7 *The immediate effect of the CSA on sugar production (total exports)*

	Calendar year	'000 metric tonnes raw value			
	CSA NPQ 1949/50*	1951	1952	1953	1954
West Indies/Guyana	640	752.8	759	800	861	964
Australia	300	384	285	262	752	651
Mauritius	335	414	505	468	499	483
South Africa	150	59	59	10	85	204
Fiji	120	122	71	135	178	144
British Honduras	18	—	—	0.45	0.83	0.92
British E Africa	5	2	17	19	71	25

Note * 1949/50 figure only available on a September/August basis; 1951–54 'provided on a calendar year basis'.
Source: ISO Yearbooks.

Rhodesia's quota was suspended after UDI and transferred to India. Meanwhile the total OAQ requirement increased by roughly 400,000 tonnes.

The USA Sugar Act

The annual consumption of sugar in the USA rose from 8 m. tonnes in the 1950s to 9 m. tonnes in 1962 and 10 m. tonnes in 1968. Slightly more than half of this requirement was met by home-produced beet and cane sugar (including 1 m. tonnes each from Hawaii and Puerto Rico). The remainder was supplied by Cuba (more than 2 m. tonnes), the Philippines (1 m. tonnes), Peru, the Dominican Republic, Brazil, Mexico and many other foreign countries.

A complicated structure for marketing was provided by the US Sugar Act (USSA) which was administered by the US Department of Agriculture. In brief the market was 'managed': foreign suppliers were given quotas and the effective levels of these quotas were varied from time to time to ensure that the total supply to the US market achieved the price objectives of the USSA. In most years the price received by foreign suppliers was considerably higher than the NPQ price received under the CSA (Table 1.8). Quotas were set for a fixed number of years, and each time that the USSA was renegotiated there was intense competition between foreign producers to be included in the new Act, and between existing suppliers of both beet and cane sugar to have their quotas increased. Indeed, when the Act of July 1962 was signed, each new participant had been represented by a lobbyist to plead its cause in Washington. Imports into the USA from Cuba, more than 2 m. tonnes in 1960, were suspended later in that year because of strained political relations between the governments of the two countries. The deficit was met by distributing Cuba's USSA quota between members of the

Table 1.8   *The prices received for sugar exports from, say, Trinidad (£ per tonne, 96° basis, f.o.b.)*

Year	UK (NPQ)*	USA[†] (USSA quota)	Canada (world 'free' price plus preference)
1953	£42.6.8	5.79 cents/lb	3.41 cents/lb[‡]
1963	£46.0.10	7.56 cents/lb	8.29 cents/lb[‡]
1968	£47.10.0	6.89	1.98
1969	£47.10.0	7.12	3.37

[*] From 1950 to 1964 all prices were for bagged sugar 96° polarization basis c.i.f. pre-war freight and insurance. From 1965 onwards all prices are for bulk sugar 96° f.o.b.
[†] c.i.f. New York ex duty.
[‡] f.a.s. Cuba.

Table 1.9   *Changes in the source of imports into the USA between 1960 and 1963 (in tonnes, raw value)*

	1960	1961	1962	1963
Cuba	2,229,075	—	—	—
Brazil	92,624	294,020	360,947	426,226
Colombia	9	43,356	61,065	40,849
Dominican Rep.	437,592	349,937	803,434	535,244
Ecuador		32,666	60,663	51,244
El Salvador	5,360	10,473	17,369	17,207
Mexico	375,683		360,686	344,160
Australia	—	81,772	145,980	202,840
BWI and Guyana	84,064	241,249	164,805	128,239
Fiji	—	—	14,301	44,065
French WI	—	68,725	55,405	85,559
India	—	144,991	133,608	107,927
Mauritius	—	—	12,223	60,434
South Africa	—	—	90,155	120,004
Taiwan	9,480	102,581	102,587	64,657
Total imports	4,637,414	3,992,650	4,248,391	4,168,403

Note: Hawaii, Puerto Rico and the Philippines continued to be large suppliers, and the mainland production of Louisiana and Florida doubled from 0.5 m. tonnes in 1960 to 1 m. tonnes in 1963.
Source: ISO Yearbooks.

Organization of American States and other countries. The more important sources of new or increased supply are shown in Table 1.9. The changed destination of much of Cuba's export sugar, mainly to communist countries, is indicated in Table 1.10. Production in Cuba declined as attempts were made to diversify the island's agriculture. However, as has happened in many places, it was eventually realized that there was no adequate replacement for

Table 1.10 *Changes in the destination of Cuban exports between 1960, 1963 and 1980 (in tonnes, raw value)*

	1960	1961	1962	1963	1980
USA	1,948,574	—	—	—	—
Bulgaria	—	57,258	117,796	56,177	234,11
China	476,537	1,032,136	937,893	500,928	512,095
Czechoslovakia	8,988	25,322	155,680	150,105	98,775
E. Germany	61,867	111,910	179,343	244,490	209,900
W. Germany	101,924	41,231	3,800	—	—
Japan	204,559	423,256	431,482	160,771	267,082
Korea	—	32,491	14,038(N)	20,000(N)	10,897
Morocco	160,986	157,287	265,124	285,028	—
Spain	33,247	53,208	58,312	102,737	—
USSR	1,577,683	3,302,865	2,112,245	973,423	2,726,339
Egypt					138,088
Iraq					277,840
Mexico					401,122
Portugal					131,377
Syria					133,999
Production	5,861,800	6,767,034	4,815,234	3,821,070	6,805,235
Exports	5,634,513	6,413,561	5,130,940	3,520,505	6,191,074

the sugar-cane and eventually the industry was restored to its former eminence.

International Sugar Agreements

The ISA of 1937, which was an attempt to ensure equitable and stable prices for sugar sold on the world 'free' market by increasing or decreasing agreed quotas as the price rose or fell, did not achieve its objective. Having been in abeyance since 1939, a new ISA was signed in 1953, effective from 1 January 1954. The CSA and USSA quotas were excluded from the Agreement, but once more the world price could not be maintained above the minimum limit. Another agreement was reached in 1958, but was equally ineffective. In 1968, however, an ISA of five years' duration, covering 90 per cent of the 'free' market, came into being. This, at last, seemed to bring about the desired results. The world price moved steadily upwards to reach and pass the minimum limit, but the agreement lapsed at the end of 1973. There is every reason to believe that the same trend would have taken place in the absence of an ISA, but perhaps the price movements would have been more erratic.

Not only was the ISA not renewed in 1974, but the stabilizing influences of the CSA and the USSA ended in December of that year. Whereas 7.3 m. tonnes had been sold at world prices in 1961, 10.7 m. tonnes were sold in 1974 and, with the disappearance of the two major preferential markets, 15.8 m. tonnes in 1975.

At first the three agreements were not badly missed: an imbalance between supply and demand caused the London daily price (LDP) to reach unprecedented heights, as much as £650 per tonne, in 1974 and 1975. Because of these inflated prices some industrial users of sugar in the USA turned to alternative sources of sweeteners, especially high-fructose corn syrup, and sugar imports into that country were reduced by 2 m. tonnes in 1975. It was predictable that the LDP would fall as sharply as it had risen; and so it did. Consequently, from 1976 there was increasing clamour for yet another ISA to be negotiated. This was achieved at Geneva in October 1977, effective from 1 January 1978. Basic annual export quotas, 15,905,000 tonnes in total, were established for the medium and large exporters, and an entitlement was given to each of twenty-one small exporters to sell 70,000 tonnes on the world market. These were separate from and additional to quotas agreed under the Lomé Convention, and to customary local exports.

The original price band in the Agreement was US 11–21 cents per lb. Another significant difference was the introduction of the system of special stocks and the mechanism for maintaining and releasing these. Assuming an exchange rate of US$2.00 to the pound sterling, this would be equivalent to between £121 and £232 per tonne. When the agreement came into being the LDP was around £100 per tonne. The price continued to rise before eventually it moved into the price band where it stayed for only about ninety-five days before going beyond the upper target. The price collapse was equally dramatic and in June 1982 the price was again only around £100 per tonne.

Thus, International Sugar Agreements have been reached only when the world market has been in deep depression: in 1937, 1953, 1958, 1968 and 1977; and by and large they have had limited effect. The difficulties of managing such a volatile market have regularly been illustrated. However, the major problem confronting the 1977 ISA was the non-membership of the European Economic Community (EEC), the world's largest exporter with the current export potential of around 6.0 m. tonnes.

Table 1.11   *The original allocation of quotas under the Lomé Convention (in tonnes, white sugar value)*

Barbados	49,300	*Note*: Belize (39,400 tonnes), St Kitts (14,000 tonnes), Surinam (4,000 tonnes) and India (25,000 tonnes), though not ACP members, were accommodated under separate arrangements.
Congo	10,000	
Fiji	163,600	
Guyana	157,700	
Jamaica	118,300	
Kenya	5,000	There have been subsequent downwards adjustments to some of these quotas. Also, Zimbabwe will accede to the protocol with effect from 1 July 1982 with a quota of 25,000 tonnes.
Malagasy	10,000	
Malawi	20,000	
Mauritius	487,200	
Swaziland	116,400	
Tanzania	10,000	
Trinidad and Tobago	69,000	
Uganda	5,000	

The Lomé Convention

At Lomé, the capital of Togo, a unique convention between the EEC and forty-six countries in Africa, the Caribbean and the Pacific (ACP) was signed in February 1975. The convention covers a broad range of trade and aid issues and one of the protocols negotiated deals specifically with the guaranteed access for 1,304,000 tonnes WSE of preferential sugar. The EEC undertake, for a guaranteed period, to purchase and import each year up to 1.3047 m. tonnes of sugar from ACP countries at a guaranteed price (Table 1.11). The price, expressed in units of account, is negotiated annually and is within the range paid for beet and cane sugar within the EEC. Like the NPQ price of the now defunct CSA, and to some extent as its replacement, it is profitable to efficient producers.

Resuscitated industries, new industries and those in decline

Resuscitated industries

The great industries of the Philippines and Taiwan were resuscitated after the end of the Second World War, but there was only partial recovery in Indonesia. Production in the Philippines rose from 1.4 m. tonnes in 1960 to 3.0 m. tonnes in 1976 (2.3 m. tonnes 1980). The corresponding figures for local consumption were 300,000 tonnes and 850,000 tonnes (1.2 m tonnes 1980), and exports rose from 1 m. tonnes to 1.7 m. tonnes. In Indonesia recovery was less spectacular: from 600,000 tonnes in 1960 to 1.4 m. tonnes in 1976, but production by 1980 was still not sufficient to meet local demands. However, expansion is now taking place both in Java and Sumatra. In Taiwan annual production has been roughly 800,000 tonnes for the past twenty years, but increased local consumption has reduced exports to 400,000 tonnes.

New or developing industries

Even more important was the development of completely new industries, and also vast expansion elsewhere. To meet high local consumption, production in some countries has increased and still is increasing at a high rate: for example although 650,000 tonnes of centrifugal sugar (and much larger quantities of *panela*) were already being made each year in Colombia, the rate of growth in the 1970s was roughly 10 per cent *per annum*. In Australia output doubled from 1.5 m. tonnes in 1960 to 3 m. tonnes in 1974 and 3.5 m. tonnes in 1980. During this period annual exports increased from 800,000 tonnes to 2.5 m tonnes. The main markets for Australian sugar are now in the Pacific Ocean area: Japan, Malaysia and New Zealand. China remains a net importer in spite of quadrupling production from 1 m. tonnes in 1960 to 4 m. tonnes in each of the years from 1974 to 1977. In Bolivia the increase was from 5,000 tonnes in 1957 to 94,000 tonnes, sufficient to meet the demands of the home market, in 1964, and 280,000 tonnes in 1976, which was maintained up to 1980. Bolivia is now participating in the export market despite an increase between 1960 and 1970 of 30 per cent in population and 50 per cent in consumption per head. Elsewhere, in Belize, Malawi, Nigeria, Ivory Coast, Kenya, Sudan, Zambia, and far too many other countries to be mentioned individually, sugar industries were either started or vastly increased during the 1960s and 1970s.

Industries in decline

In several West Indian countries, however, the industry has contracted. Puerto Rico, for reasons of economy (its industry is so heavily subsidized), has withdrawn from the export market and in future will produce only sufficient sugar to meet the needs of local consumption. There has also been a decline in the fortunes of the sugar industries in the former British West Indian territories.

The liquidation of Empire took place during the period under review and has given rise to a paradox. Whereas one of the first requirements of a newly independent non-sugar-producing country is to become a sugar producer, many of those with flourishing established industries have sought to diversify their agriculture, but with as little success as that achieved in Cuba. The rapidly increasing availability of refined beet sugar on the world market has caused exports from the traditional centres of the refining industry to be reduced: for example those from the UK fell from 553,000 tonnes, raw value, in 1960 to 209,940 tonnes in 1970.

World production and consumption since 1840

An attempt to quantify world production of sugar, both beet and cane, is given in Table 1.12. To the many qualifications made in notes at the bottom of the table must be added yet another: the figures refer only to the production of centrifugal sugar; the vast quantities of low-grade non-centrifugal direct-consumption sugars made in Africa, Asia and South America are ignored.

The annual consumption of centrifugal sugar per head increased from 16.5 kg in 1960 to 20 kg in 1970, and although the upward trend continues, the excessively high prices of 1974 and 1975 caused a temporary decline in demand, especially in North America. Data for the years 1960–80 are given in Fig. 1.6 and Table 1.13.

Developments in factory technology

There have been several developments in factory technology since 1950. Turbines have gradually replaced steam-engines; automation is being introduced, especially to control the operation of boilers and centrifugals; and much investigation of the process of milling and diffusion has taken place. Nevertheless, in 1950 the techniques of manufacture already were highly developed and now juice extractions, by and large, leaves little to be desired; effective chemical process control is almost universal; and the handling of sugar in bulk rather than in bags is the rule and not the exception. It was in the field, rather than in the factory, that advanced technology was not everywhere accepted. This imbalance has now been corrected.

The large increase in yield achieved in the 1950s and 1960s by planting hybrid varieties has been difficult to maintain; the genetic combinations of material available to the cane breeders appears to have been well explored. More effective herbicides and insecticides are developed year by year, but it is in agricultural engineering that the most spectacular advances of the second half of the twentieth

Table 1.12   *World production of centrifugal sugar 1840–1977 (in metric tonnes, raw value)*

Year	Cane	Beet	Total	Cane % total
1840	788,000	48,198	830,198	93.0
1850	1,043,000	159,435	1,202,435	86.5
1860	1,376,000	351,602	1,727,602	79.7
1870	1,662,000	939,096	2,601,096	64.0
1880	1,883,000	1,857,210	3,740,210	50.2
1890	2,597,000	3,679,800	6,276,800	41.2
1900	5,252,987	6,005,868	11,258,855	46.6
1910	8,155,837	8,667,980	16,823,817	48.5
1920	11,924,813	4,906,266	16,831,079	70.8
1930	15,942,438	11,910,883	27,853,321	57.2
1940	18,620,162	11,530,549	30,150,711	61.7
1950	22,542,769	13,541,751	36,084,520	62.5
1960	29,461,000	22,838,000	52,299,000	56.3
1970	43,588,679	29,362,026	72,950,705	59.8
1980	51,164,744	33,227,687	84,392,431	60.6

Sources: 1840–1930, Deerr, *History of Sugar*, Vol. II.
1940 and 1950, C. Czarnikow Ltd, London.
1960, International Sugar Council, *Sugar Year Book*.
1970 and 1980, International Sugar Organization, *Sugar Year Book*.

Deerr made the following qualifications:
1. Some of the early statistics refer to exports only; local consumption is ignored.
2. Russian and Belgian production are not included in 1840.
3. Belgian production is not included in 1850.
4. Indian production is not included for the period 1840–90.
5. Production of South China and other Far Eastern countries is not included.
6. Central and South American production is included only in later years.
7. No attempt is made to distinguish between short, long and metric tons.
Deerr's qualifications do not apply to the data for later years. Figures for 1940 and 1950 have been converted to tonnes, and those for 1960, 1970 and 1980 are reproduced in their original form as tonnes, raw value.

century have been and are being made. Although more efficient cultivation, irrigation and drainage systems have been installed, their contributions to the economics of industry are small compared with those which flow from the widespread use of mechanical harvesters and modern means of transport. These are described in detail in Chapter 9.

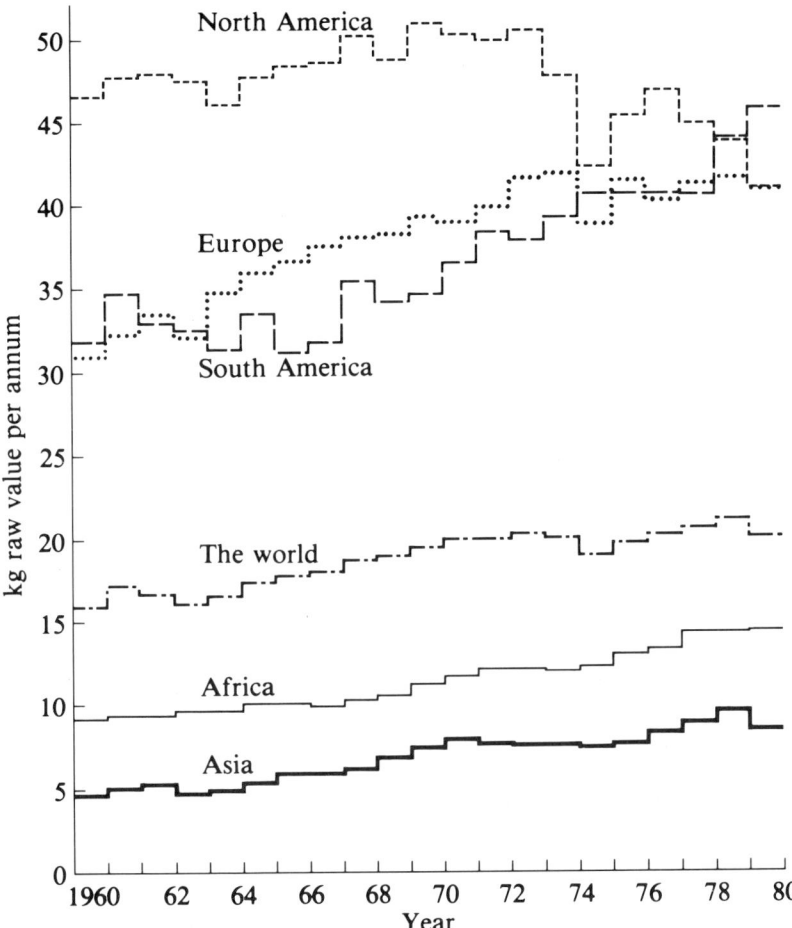

Fig. 1.6 Consumption per head of centrifugal sugar, 1960–80 (ISO Yearbooks)

Table 1.13 *World consumption per head of centrifugal sugar 1960–80 (in kg raw value)*

Year	Consumption	Year	Consumption	Year	Consumption
1960	16.4	1967	18.6	1974	20.0
1961	17.6	1968	19.2	1975	18.9
1962	17.6	1969	19.5	1976	19.7
1963	16.8	1970	19.9	1977	20.2
1964	17.2	1971	20.3	1978	20.7
1965	18.0	1972	20.4	1979	21.2
1966	18.4	1973	20.7	1980	20.0

Source: ISO Yearbooks.

Chapter 2

Botany

Sugar-cane is a tall perennial tropical grass which tillers at the base to produce unbranched stems, 3–4 m or more tall, and about 5 cm in diameter. It is cultivated for these thick stems or canes, from which sugar is extracted.

Barber in India and Jeswiet in Java pioneered the study of the morphology of sugar-cane. Artschwager carried on this work in the USA, first by recommending the standardization of taxonomic descriptions (1939), then by describing the vegetative characteristics of wild forms of *Saccharum* (1948) and finally, with Brandes (1958), by giving the origin, characteristics and descriptions of representative clones of *S. officinarum*. Meanwhile van Dillewijn (1952) had written a definitive book on the botany of sugar-cane.

The stem

The solid unbranched stem, roughly circular or oval in cross-section, is clearly differentiated into joints, each comprising a node and an internode. A node consists of a lateral bud in a leaf axil which leaves a scar if the leaf is removed, a band containing root primordia, and a growth ring (Fig. 2.1). The small buds are closely appressed to the stem on alternate sides of the stem and are not vulnerable to damage when planting. Several types of nodes and internodes are illustrated in Fig. 2.2.

Generally the nodes are spaced at intervals of 0.15–0.25 m, but are much closer at the top of the stem, where elongation is taking place, than at the bottom, where they form part of the rootstock and are essential to the formation of tillers. In commercial production, sugar-cane is propagated from stem cuttings (setts, or seed-pieces), each having two or more buds. The buds develop to give primary stems, the basal buds of which form secondary stems and so on (Fig. 2.3).

Sugar accumulates in the stems (canes), of which the internodes (joints) vary in length (5–25 cm), girth (1.5–6 cm in diameter),

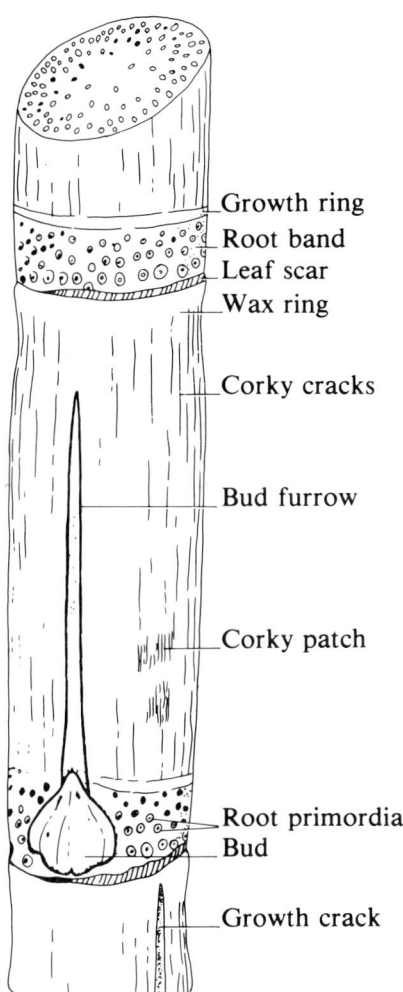

Fig. 2.1 Stem of sugar-cane (After Artschwager & Brandes 1958)

shape (cylindrical, conoidal, barrel or bobbin and circular or oval in cross-section), colour (yellow, green, red, purple, black, striped, variegated) and hardness according to variety and growing conditions. Each stem has a hard, wax-covered rind (epidermis) surrounding a mass of softer tissue (parenchyma) which is interspersed with fibres. The fibres are more abundant towards the periphery than in the centre, so the mechanical structure is fundamentally tube-like. The wax prevents loss of water by evaporation, the fibrous rind gives strength and rigidity to the stem and the thin-

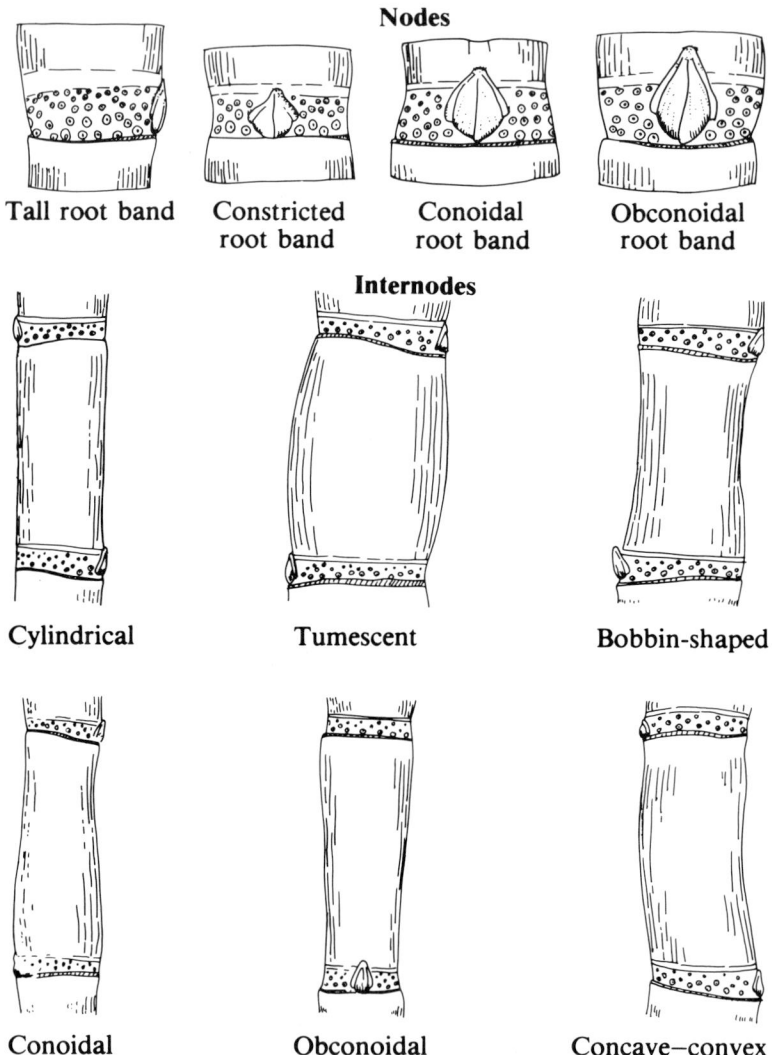

Fig. 2.2 Types of nodes and internodes (After Artschwager & Brandes 1958)

walled cells of the parenchyma store the juice. The construction of the stem and its differentiation into three distinct entities (dermax, comrind and comfith) is the basis of the separation process of manufacture described in Chapter 10.

Clones of *S. officinarum* (noble canes) normally have thick internodes with girths often more than 50 mm in diameter, whereas those of other species of *Saccharum* like Uba are much thinner and there-

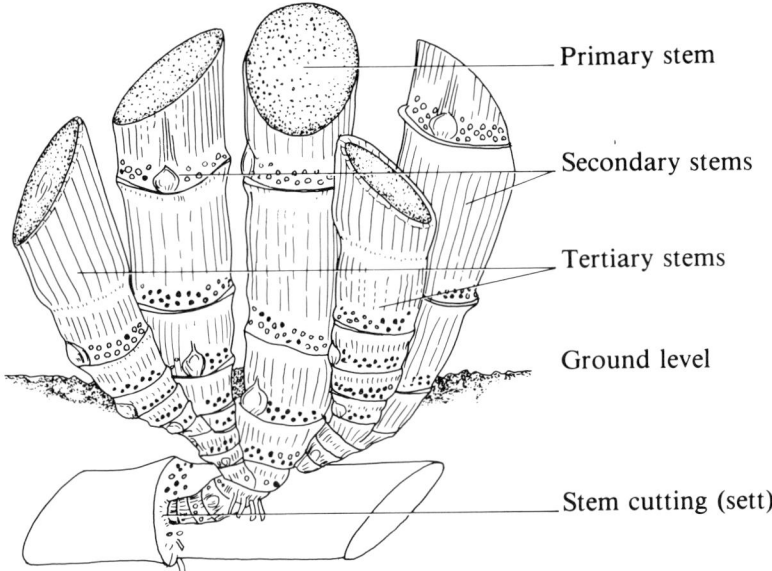

Primary stem

Secondary stems

Tertiary stems

Ground level

Stem cutting (sett)

Fig. 2.3 The formation of tillers

fore have higher ratios of rind tissue to parenchyma. The hardness of the rind affects the milling quality (millability) of the cane and, where harvesting by hand is still practised, thin canes are tiresome and expensive to reap. More importantly, the sugar content of thin canes is, in general, less than that of noble canes. Canes with a soft rind like 'Otaheite' are chosen for chewing.

The length of the internodes, and their girth, are also affected by other factors of which nutrition, temperature and water supply are the most important. For example, a stem might have short and thin internodes in its central portion, reflecting retarded growth during dry weather, with much longer and thicker ones (above and below) which were developed during wet periods.

There are large differences between varieties in the thickness of the wax coating, but it is more heavily deposited immediately below the node than elsewhere. The composition and extraction of cane wax are discussed in Chapter 11.

The colour of the stem depends upon many factors. Two basic pigments are involved: red and blue anthocyanins in epidermal cells and green chlorophyll in deeper tissue. When both are absent the stem is yellow. The colour is usually subdued if an internode is covered by its leaf sheath, but becomes distinct on exposure to sunlight. The immature top joints are pale yellow. Stevenson (1965) illustrated a range of colours in a photograph of the stems of twelve different varieties, which also showed that striped and variegated forms occur.

The leaf

Leaves are attached to the stem at the bases of the nodes, alternately in two rows on opposite sides of the stem. Each leaf consists of two parts: a sheath and a blade (Fig. 2.4).

The sheath, tubular in shape but broader at the base than at the top, fits closely to the stem and is separated from the long, tapering, pointed leaf blade by a ligule and two dewlaps. The sheath or leaf base is a thin structure closely overlapped at the base but opening out higher up. The free margins of the sheath stand on the opposite side from the bud, which it surrounds and protects; these margins dry out becoming brittle. The ligule, a membranous appendage to the sheath, is formed from elongated parenchyma cells without vascular bundles. Translucent and hyaline when young, it dries, changes colour and becomes torn with age. Nevertheless it is an

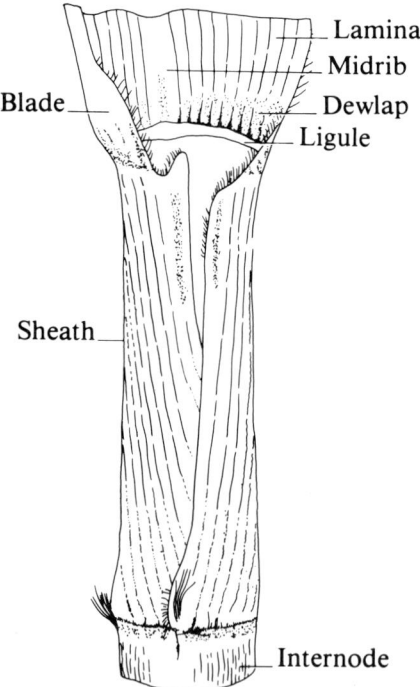

Fig. 2.4 The structure of a leaf

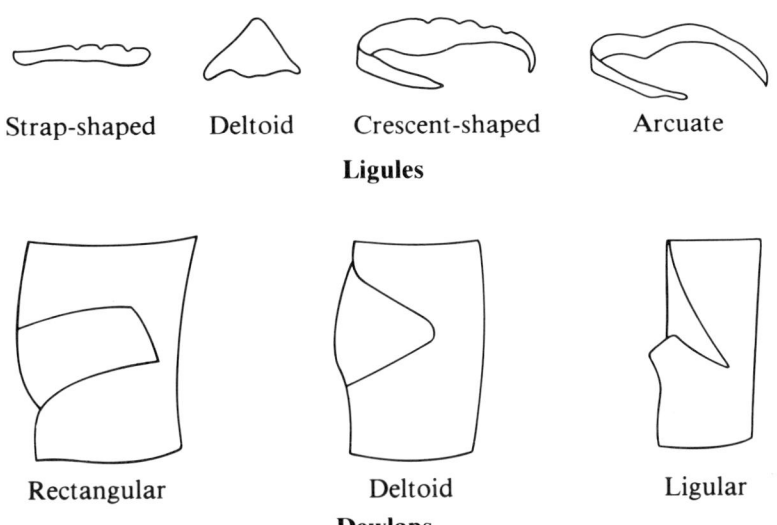

| Strap-shaped | Deltoid | Crescent-shaped | Arcuate |

Ligules

| Rectangular | Deltoid | Ligular |

Dewlaps

Fig. 2.5 Types of ligules and dewlaps

important diagnostic feature and four types are recognized: linear (strap-shaped), deltoid (triangular), crescent-shaped, and arcuate (bow-shaped) (Fig. 2.5). The dewlaps are wedge-shaped hinges made flexible by their collenchyma. They also are characteristic of each variety, and three main categories, with many intermediate forms, are recognized: rectangular, deltoid and ligular (Fig. 2.5).

The leaf has a strong midrib, white and concave on the upper surface, convex and green below. Along the midrib are motor cells which cause the leaf to inroll, up in some cultivars and down in others. Stomata occur on both sides of the leaf, but are about twice as numerous on the lower than on the upper surface. A fully expanded leaf has about 30 m stomata. The leaf anatomy is well described by Clements (1980). The leaf blade broadens from the ligule to as much as 10 cm in width, and then narrows towards its pointed tip. Although there is considerable variation, leaves may be as much as 1 m in length.

As the older leaves become moribund, the sheath with the blade attached may or may not adhere to the stem (free-trashing). This is a varietal characteristic of some practical importance because dead leaves (trash), if retained, impede reaping and may shelter pests. Another undesirable feature is the development of siliceous cells to form hairs or spines on the leaves, or to give sharp cutting edges to them. The more coarsely toothed cultivars are unsuitable for forage. According to Parris (1954) it was the sharp edges of the leaves of the cane seedlings which allowed Harper, in Barbados in 1858, to distinguish them from Guinea grass (*Paspalum maximum* Jacq.) and thereby to make the historic discovery that the sugar-cane was fertile.

The roots

Two types of roots develop shortly after a sett has been planted: those from the primordia of the cutting, which are thin and

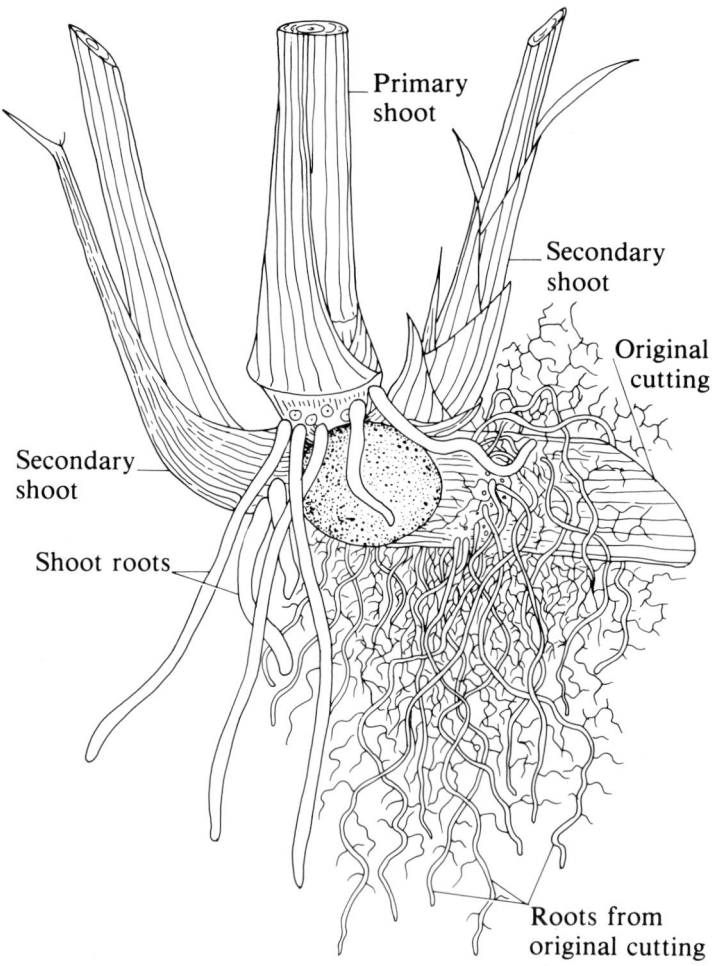

Fig. 2.6 Young cane plant showing two kinds of roots: sett roots from the root primordia of the cutting, and shoot roots originating from the root primordia of the shoots (After Martin 1938)

branched; and those from the primordia of the tillers, which are thick, fleshy and much less branched. At first the newly planted seed-piece depends entirely on its own roots for the uptake of water and nutrients. Later this function is taken over by roots formed by the tillers, and the sett roots die. Indeed, with age, all roots become brown, shrivel and die. Each shoot produces its own root system (Fig. 2.6).

When further growth takes place, the shape of the root system is determined by the condition of the soil in which the cane is planted. The roots proliferate wherever conditions of available water and soil aeration are favourable. Depth of cultivation, soil profile and moisture are of particular importance. Evans (1935, 1936) in Mauritius examined the roots of several varieties in detail and divided them into three categories: superficial roots, buttress roots and rope systems. The superficial roots absorb moisture and nutrients; the buttress roots, as their name implies, provide stability; and the rope systems penetrate to depths of 3–6 m, where the soil remains moist even in times of severe drought (Fig. 2.7). However, it should be emphasized that the precise pattern of development is

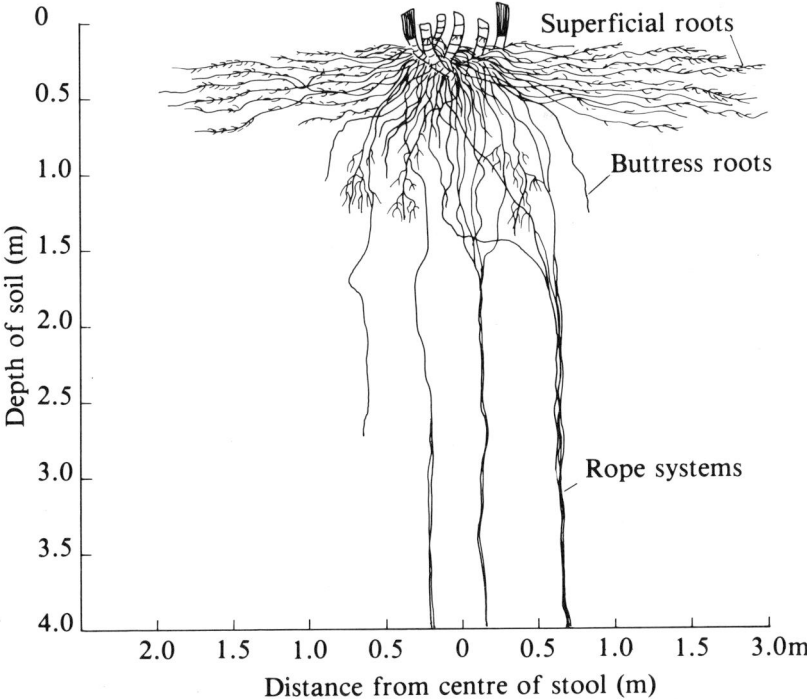

Fig. 2.7 The roots of an established cane stool (After Evans 1936)

peculiar to local soil conditions: what is characteristic of Mauritius may not be apparent elsewhere. In practice, it is often difficult to distinguish between superficial and buttress roots, and rope systems are extremely rare. Whatever the pattern, roughly 50 per cent by weight of the roots occur in the top 20 cm of soil, and 85 per cent in the top 60 cm.

Sugar-cane roots can penetrate downwards where the water potential of the soil is less than −15 to −20 bars, provided that the main root mass has adequate water. Similarly, a few main roots may transport water to the leaves through 2 or 3 m of very dry soil. Root growth is not only affected by soil moisture but also by soil temperature and the volume of soil available for the roots to spread. High soil temperatures reduce root growth.

The inflorescence

The factors which cause a stem to change from the vegetative to the reproductive phase are discussed in greater detail in Chapter 3. This transformation was described by Stevenson (1965). Most *Saccharum* varieties will not flower on daylengths longer than about twelve hours, nor if given light in the middle of the dark period. There are exceptions, some *S. spontaneum* varieties behave as long-day plants. Generally twelve and a half hours' daylength and 20–25 °C night temperature induces floral initiation if enough inductive cycles are given, probably at least ten (Bull & Glasziou 1975). Coleman (1968) suggested that a quantitative amount of stimulus is accumulated to start the differentiation of floral primordia. In the vegetative state the unexpanded sheaths near the apex are shorter than the expanded ones below. The first sign of flowering is that successive sheaths become longer and blades shorter. The terminal meristem which is surrounded by a leaf sheath ceases to form leaves and develops into an inflorescence primordia about three months before flowering. Then the distinctive last leaf (flag or boenting) appears. Its sheath, which encloses the young panicle, is 90 cm or more in length and its blade, shaped like a pennon, only about 15 cm long. Finally, the stalk elongates and pushes the panicle out (Fig. 2.8). The inflorescence, which is known as the arrow, emerges above the mass of foliage, presumably an adaptation to wind pollination. The loose terminal panicle is 25–50 cm long with a silky appearance due to rings of long hairs below each spikelet.

The morphology of the inflorescence was described by Artschwager *et al.* (1929).

The main axis of the panicle arises almost imperceptibly from the terminal internode, and gradually narrows until it merges into the terminal rachis of spikelets. The surface of the main axis is slightly

Fig. 2.8 Flowering sugar-cane (Photograph: A. L. Down)

furrowed and the lateral branches arise at the nodes, not always at the same level, some members of the whorl arising above and others below the main nodal region.

The bases of the panicle branches are swollen and thinly covered with short white hairs. At the base of the panicle the primary branches are about 15 cm long, but shorter above. The secondary branches tend to arise in two rows, alternately along the primary branches and may carry tertiary branches. The ultimate branches bear the pairs of spikelets, one of which is sessile and the other on a stiff pedicel (Fig. 2.9). At the base of each spikelet in the pair is

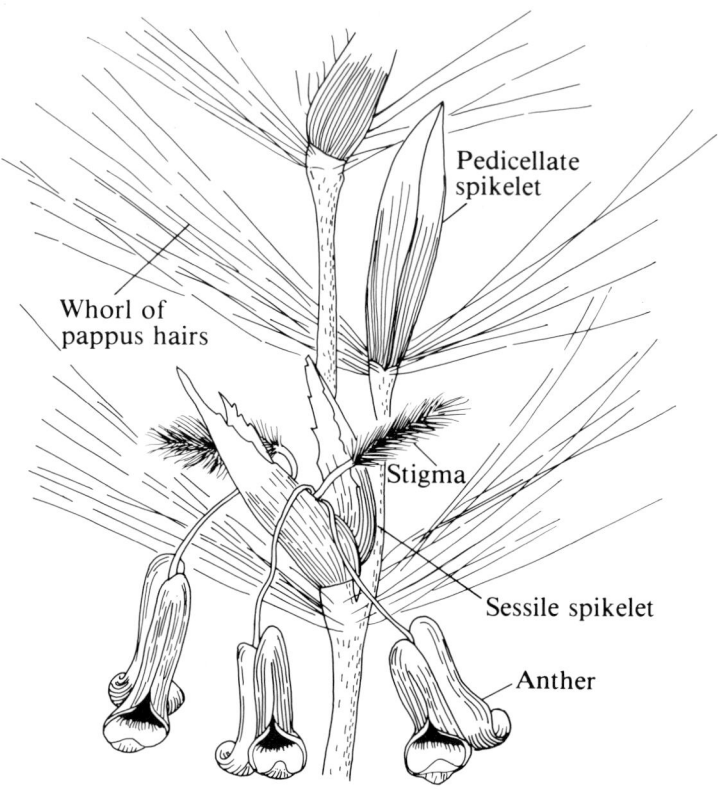

Fig. 2.9 Part of a lateral axis of an inflorescence of *S. robustum* (51 NG 71), showing the arrangement of spikelets (From Stevenson 1965)

a ring of silky white hairs which are longer than the spikelet. These hairs give the arrows their characteristic appearance.

Both spikelets have two florets, the lower one of which is sterile and represented by a delicate pointed lemma or third glume which is shorter than the glumes. The structure of both spikelets is similar, with a pair of hard boat-shaped glumes protecting the developing flowers. The upper floret of each spikelet is hermaphrodite, with no lemma except in *S. spontaneum* and some of its hybrids. When present the lemma is a narrow scale with fine hairs at the top.

At the base of the ovary opposite the palea are two short wedge-shaped lodicules. The three stamens with large bilobed anthers are in one whorl. Indehiscent anthers are usually yellow or pale orange, while dehiscent anthers are brown or purple. The ovary is round, flattened on the ventral surface with a single anatropous ovule. The pistil has two long terminal styles each with a large brush-like feathery stigma, plum red in colour. The spikelets open during the night or early morning, beginning at the top of the panicle and progressing downwards and inwards over a period of one or two weeks. The lodicules swell and push the glumes apart and the stigmas are extruded. The flowers are protogynous, the anthers appearing about three hours later when the filaments have lengthened so the anthers hang down well clear of the floret. High humidity delays anthesis. Natural pollination is by wind. The pollen grains remain viable only for a short time and the anthers fall from the filaments soon after anthesis, but the stigmas are persistent. The cultivated varieties exhibit considerable sterility of both pollen and ovules. The flowers rarely set seed.

After pollination it takes twenty-one to twenty-five days for the seed to fill and mature. Though the production of the inflorescence ends the production of leaves on that stem, vegetative branches may develop at lower nodes. These tillers have a lower sugar content than the main stem. As the production of the inflorescence also reduces the sucrose content of the stem, efforts are often made to delay flowering to maintain sucrose yield.

The seed

The seed is a dry one-seeded fruit or caryopsis formed from a single carpel, the ovary wall (pericarp) being united with the seed-coat (testa). The seeds are ovate, yellowish brown and very small, about 1 mm long. The withered stigma persists at the tip, and at the base are whorls of silky hairs for wind dispersal. The seeds are shed within the spikelet, individual florets breaking off at the nodes. Collectively they are known as 'fuzz'. The seeds soon lose their viability, but if freeze dried remain viable for a long time. For planting within two weeks the seed should be kept in a desiccator.

Except in Taiwan (Lee & Loo 1958), until recent years little effort seems to have been made to quantify fertility, even though it is important to cane breeding. Methods of seed storage, on the other hand, have been investigated since the 1930s. Rao (1980) discussed both these topics and Walker (1980) has given the following summary of present knowledge:

Complete inflorescences of commercial hybrids have been estimated to contain 25,000 florets, but the number of fertile florets is always much lower than this (e.g. 3 per cent in NCo 310 and 33 per cent in B 7264). As many as 700 seedlings per g of fuzz have germinated, but most varieties produce much less. Mature naked seeds weigh around 0.4–0.5 mg; hence at best only 35 per cent by weight of the fuzz is true filled seed.

Seed viability falls rapidly in ambient tropical conditions but, by suitable drying and cold storage, viability can be maintained at a high level for at least three years. Such seed storage is increasingly being used as part of breeding strategy, and a longer-term storage, as used in other cultivated cereals and grasses, is certainly feasible.

Germination of seed in soil takes from two to eight days at 35 °C. Young seedlings are delicate up to the four-leaf stage, but thereafter grow rapidly.

As the fuzz germinates better in light it is placed on the surface of sterilized compost in shallow trays and kept at a high humidity. The seedlings can be transplanted six weeks after germination.

Chapter 3

Sugar-cane and its environment

Commercially viable sugar-cane has been grown on a wide range of landform and soils. Successful ventures have been established on land with slopes varying from flat to steeply undulating and on soils derived from alluvium, and from virtually all rock types except the ultra-basic group with its inherent problem of heavy metal toxicities.

It is notable that many of the larger and long-established industries have used deep soils derived from fertile volcanic materials; examples include large sections of Hawaii, Cuba, Australia, Brazil (São Paulo) and Mauritius. Other long-established industries have traditionally selected fertile alluvial soils; many of the old Caribbean industries and Java are examples of this. No doubt, the fertility of these soils, before fertilizers were available, was the reason for their selection. Cane is, however, a highly versatile plant, and can be grown successfully under a wide range of conditions.

Landform

Ideally the land should have long, regular, smooth slopes of up to 1°–3°, the higher value being more acceptable with heavier soils. This permits the widest choice of field layout design and the most economical cultivation and harvesting techniques. Completely flat land is less suitable in view of the inherent difficulty of surface runoff and the necessity, on irrigated schemes, of establishing a slope for water-distribution purposes.

Slopes of up to 10 per cent remain suitable for development using contour ploughing and water-distribution systems, but restrictions on field design inevitably arise. Steeper and more irregular slopes can be, and are, used; but land preparation is more expensive and the choice of layout, husbandry and harvesting methods are restricted.

Soils

Sugar-cane is grown on soils which vary in texture from light sands to heavy clays, but in all they must have optimum access to the main elements of nutrition: nitrogen, phosphorus and potassium, and also to many trace elements. Sugar-cane is also tolerant to wide variations in acidity and alkalinity: most of Guyana's sugar is grown on highly acidic clays, of pH 4, whereas the Barbados industry was established on alkaline *rendzina* and *terra rossa* soils derived from coral limestone. Nevertheless, damage at both extremes of the acidity scale have been reported. More than fifty years ago McGeorge (1925) referred to the contribution made by aluminium, manganese and iron to the infertility of acidic soils in Hawaii; and more recently Evans (1960) described a similar condition (aluminium toxicity) in the *pegasse* soils of Guyana. Stevenson (1957), on the other hand, reported that iron deficiency under highly alkaline conditions caused severe chlorosis.

Cultivation techniques are determined largely by the physical characteristics of the soil. Where a deep and stable crumb structure exists, as in some of the soils of volcanic origin in Hawaii and St Kitts, preparation for planting is limited to little more than the formation of furrows at suitable intervals. Where the soils are more intractable, as in Florida and Guyana, more eleborate treatment is necessary. The methods used are described in Chapter 5.

Soil physical characteristics

The best rooting medium for sugar-cane is probably provided by more than 1 m of stable well-structured loam to clay loam soil. Sugar-cane has the potential for very deep rooting (over 5 m), and cane planted on such soils can show marked drought resistance. The soil would normally have a bulk density below 1.4 g/cc and a pore space of at least 50 per cent, which at field capacity would be half occupied by air and half by water. The desirable bulk density values depend on soil type; Trouse & Humbert (1961) reported severe suppression of rooting at bulk densities of 1.52 in alluvial soils, and of 1.05 in hydrol humic latosols.

The soil mass should have rapid surface infiltration rates and free internal drainage so that rain or irrigation water can be readily absorbed and any excess can drain away rapidly. Ideally, the ground-water table should be at not less than 1.5–2 m depth below the surface or, if higher than this in the natural state, the water table should be amenable to being lowered and maintained at below the rooting depth with a moderate investment in drainage facilities.

Ideally the available water capacity of the soil should be 15 cm/m or more to ensure an adequate reservoir of water available to the plant roots between rainfall or irrigation cycles. Irrigation technique must be adapted to the characteristics of the soil.

The further the soil characteristics depart from these optimum values, the more skilled the soil management needs to be in order to minimize consequent reduction in sugar-cane yields. However, in practice, sugar-cane is an adaptable crop and is grown in both very coarse and very fine soils and in soils with very high organic matter content. The main constraints that affect sugar-cane growth in these extreme soil environments are summarized below.

Coarse textured soils

1. Nematode infestation is a hazard where the topsoil is a loamy sand or coarser. The problem tends to be progressive as the eelworm population increases year by year in absence of treatment. In South Africa, yield increases approaching 100 per cent are not abnormal when nematodes are controlled with chemicals such as Temik®.
2. The available water capacity of sandy soils can be 6/cm/m or less, in comparison with values which are often 20 cm/m or more in medium and heavy textured soils. Low available water is an obvious disadvantage in areas reliant on rainfall; in irrigated schemes, it limits the range of equipment which can be used.
3. Rapid movement of water through the soil increases the loss by leaching of applied fertilizer. More frequent application of smaller amounts of fertilizer reduces this loss but increases husbandry costs.

Fine-textured soils

1. Where fine-textured soils are only slowly permeable, too much of the pore space in the root zone may be occupied by water and hence the cane may suffer from lack of aeration, or the penetration of water may be so slow that the plants may suffer from drought conditions. When surface infiltration rates are greater than internal permeabilities, anaerobic conditions will induce symptoms of drought.
2. Soil capping, which generally occurs in soils with a high silt content, causes the soils to be unstable at the surface when wet and very hard and impermeable when dry.
3. Where soils have a high bulk density root penetration and hence plant growth is restricted.

Organic soils

Oxidation of the organic matter after drainage progressively lowers the land surface which disrupts the drainage system and causes cultivation difficulties. Soil shrinkage can ultimately bring subsoil into the root zone. This is often raw clay with completely different management and drainage requirements to the overlying organic soil.

Soil chemical characteristics

The optimum soil pH for sugar-cane is about 6.5, i.e. very slightly acid. However, cane will grow over the pH range 4–8.5 with increasing yield reductions towards the extremes of the range. Neither the excess acidity nor alkalinity at the lower and upper ends of the pH range of themselves cause the reduction in growth, but rather the associated chemical environment.

At low pH values, aluminium, among other minerals, becomes increasingly soluble and is generally found to be toxic to sugar-cane when the amount held in the soil exchange complex exceeds 60 per cent. Manganese behaves in a comparable way. Low levels of silica in strongly acid soils exacerbate the toxicity effects of aluminium, iron and manganese in sugar-cane. A special case of low soil pH in sugar-cane occurs in Guyana where some of the soils used for cane have organic surface horizons overlying 'cat clay' subsoils. The latter have pH values when waterlogged of 4–5, but after drainage and oxidation of the sulphides, which are relics of the original marine deposition, soil pH values can be as low as 2. By leaching out the toxic iron, aluminium and sulphide/sulphate with sea-water, even some of this land has been brought into use for sugar-cane, with yields of 150 tonnes cane/ha/per year in the first cycle.

Cane is successfully grown in highly calcareous soils, but special precautions need to be taken to overcome deficiencies of phosphate, iron and zinc, and to avoid the volatilization of nitrogen fertilizers. In one extreme case, foliar fertilization prior to the cutting of seed-cane was necessary to ensure adequate germination, but more normal fertilizer applications were adequate for established cane. Iron chlorosis, corrected by foliar sprays, is a common but not serious occurrence in these soils.

Sugar-cane makes heavy demands on plant nutrients. The literature contains numerous estimates of the total nutrients removed by a crop; the estimates of de Gues (1967) approximate the medians of the values given by other workers. These are that a 100 tonne/ha crop requires an average of 120 kg N, 75 kg P_2O_5 and 150 kg K_2O. Preliminary soil analysis can give an indication of the likely response to different types of fertilizer, but cannot take the place of fertilizer crop trials for the determination of the optimum application rates required.

Different extraction systems are used according to the nature of the soils, and the interpretation of critical limits varies somewhat from place to place, dependent on the observed response of the plant in crop trials. The levels given below should be considered as a first approximation only. Cane grown on soils with values lower than those quoted are likely to show a positive response to application of the appropriate fertilizer, while on soils with equivalent or higher levels, a response is less likely.

Available phosphorus	(p.p.m.)	25 (Truog), 5 (Olsen)
Available potassium	(p.p.m.)	60–80 (light soils)
		250+ (heavy clays)
Available magnesium	(p.p.m.)	70 or more
Exchangeable	(meq/100 g	3.0–3.5
calcium	soil)	

However, once sugar-cane is established, foliar or tissue analysis is a far better diagnostic tool for estimating fertilizer requirements for both macro- and micronutrients. Foliar analysis is now used in the more sophisticated cane-growing areas and correlation between the amounts of nutrient elements in specified plant parts and fertilizer requirements for each main soil type and area have been established.

Cation exchange capacities of suitable soils for sugar-cane are generally greater than 15 meq/100 g soil. This property is an approximate measure of the retention of applied fertilizers against leaching losses. Where exchange capacities are below 15 meq/100 g soil, leaching losses are likely to be high with corresponding economic losses, together with less freedom of choice in the method of applying fertilizers.

The organic matter content of soils is important both in the retention and supply of plant nutrients and in the development and maintenance of favourable soil structure for plant growth. The desirable amount of organic matter in soils is difficult to quantify. Generally, values of between 2 and 4 per cent are considered adequate. The maintenance of such levels in soils supporting sugar-cane may be assisted by the incorporation of materials such as filter press cake from the factory into the soil at regular intervals.

Salinity

By and large cane is also more tolerant than most crops to saline conditions; but there are considerable differences between varieties in their ability to grow in soil which has a high salt content. For example, B 42231 proved highly tolerant, and in Jamaica was widely planted in saline areas being brought into cultivation for the first time.

The symptoms of salt damage appear on the leaves, which become pale green or yellow in colour, similar to those suffering from nitrogen deficiency, but also have scorched tips and margins. The

roots are deformed and, in general, the crop has an unthrifty·
appearance. It is widely accepted that the damage is caused by the
accumulation in the soil of chloride rather than sodium ions. Similar
damage in Venezuela has been attributed to high concentrations of
potassium sulphate in the ground-water.

The leaves of cane planted close to the sea on windward coasts
often have brown tips and margins caused by damage from salt-
water spray.

The reclamation of saline soils

A number of research workers have measured the effects of salinity
on sugar-cane growth by applying salt to plants grown in a variety
of rooting mediums in containers for limited periods of time. The
main advantage of this type of trial is that salt-free yields can be
accurately measured; however, these trials cannot measure the
indirect effects of salinity such as the consequent deterioration of
soil structure. Other workers have measured yields of cane of plots
in commercial fields which are variably affected by salinity. This
technique takes full account of the indirect factors, but introduces
difficulties in estimating yields under non-saline conditions, in alkalinity, bicarbonate accumulation, pH and nutrition. In general, it
might be expected that 'pot-trials' will indicate higher salt tolerances
than the field trials. These reservations need to be kept in mind
when the summary of published results given in Table 3.1 is examined. The consensus of these data suggest that yields are likely to
be virtually unaffected where EC_{SE} (electrical conductivity of satu-
rated extract) values of less than 2–3 are encountered, that 50 per
cent yield reduction is likely at EC_{SE} values of about 7, and that total
growth failure is likely at values of 11–12. Between the lower and
upper critical salinity levels, selection of varieties which are resistant
to salinity can significantly improve yields; differential varietal
tolerances are not of any practical importance, however, at the
lower and upper critical salinity levels (Nickell 1977).

No single value for electrical conductivity (EC) can be universally
applicable, as the influence of salts on yields will be affected by the
nature of those salts, the amount and pattern of rainfall, the irrigation system and the soil type (including its susceptibility to struc-
tural modification by Na). Thus, for example, Valdivia & Pinna
(1974) calculated that the critical lower EC levels would be 2.2 and
3.5 mS/m respectively, for two alluvial soils which had different pF
curves and different levels of aeration. These complications are of
special importance in defining the acceptable level of salinity in
irrigation water. Earlier reports, for example Bonnet (1968) and
SASA (Anon 1969–70) tended to agree that irrigation waters should
not exceed a value of about 1 mS/cm. Later, for typical Queensland

Table 3.1 *Effects of salinity level (in mS/cm EC) on cane yield (from published data)*

	Reduction in cane yield		
	10%	*50%*	*75–100%*
EC, pot tests or induced salinity			
Soil	2	9	12
Sand (100 days)	1.2–5*	7	12
Soil (100 days)	2.5–3*	7–10*	?
Gravel (max. EC 3.5)	2.0–3*	?	?
Soil (plant)	3.0–5*	7–9*	over 10
Soil (ratoon)	2	5–6*	about 7
Soil (max EC 8)	2	8 or higher*	?
EC, existing salinized soil			
	0.5	?	?
	1.5	5.0	7
	3.0	8.5	?
	2.0	2–8.0	8
	4.0	8.0	16
	2.0	6.5	10
Max. EC 6.0	2.5	?	?
Min. EC 3.0	?	?	11
ESP, existing sodicized soil			
	4	(20?)	(70?)
	5	20	40
	14	26	45
	10	15	?

* Range associated with varietal differences.

conditions, the acceptable level of salinity of irrigation water was calculated from the frequency of occurrence of undesirably high salt levels in the soil (Ridge & Kelly 1975). For free-draining soils, 'acceptable' water quality was calculated as 2,000 p.p.m. salt (3.7 mS); previously, a maximum of 1,600 p.p.m. (3 mS) had been advised. The above values are based on traditional furrow-type irrigation. However, assuming conditions of permanent trickle irrigation on a coarse, freely draining, aerated soil, the limit of water salinity before inducing any reduction in yield is about 1.5 bars (Robinson, 1963), or 4.0 mS. Similar irrigation systems will also reduce the leaching factor, reduce the dissolution of sulphates and bicarbonates and encourage greater water abstraction from the upper, less saline, soil strata (McNeal 1976).

The data on ESP* (exchangeable sodium percentage) values are confusing, probably because they were obtained from soils with widely differing Ca + Mg contents. Extrapolation of data referring to other crops in California (McNeal 1976) suggests that SAR*

* For a more detailed explanation of these terms see Russell (1973: 758).

(sodium adsorption ratio) of about 20 is likely to cause a 50 per cent yield reduction; if the composition of the soils was such that a 1 : 1 ratio of ESP : SAR is obtained then the data in Bonnet (1953) and Fogliata & Aso (1965a) appear to be correct. This is supported by one set of data from South Africa (Anon 1970–71) which may be summarized as follows after averaging soil analysis data over 0–90 cm:

	Cane healthy	Cane slightly affected	Cane severely affected	Cane dead
SAR	13	18	35	41
pH	8.2	8.3	9.1	9.6

With regard to cane quality the work of Janes (1966) is of fundamental importance. Irrespective of the cause of the increased osmotic potential of the rooting medium, plants respond by increasing their internal osmotic potentials. Under non-saline conditions, the mechanism is basically one of increasing soluble carbohydrates in the cell sap, which is the classical method of ripening through drying off. Where salinity is the causal agent, the salts are absorbed to increase cell osmotic potential; this will reduce purity and, thus, reduce the recovery of the sugar. In spite of the increase in non-sugar solids in the juice, total solids (Brix) are reduced by salinity due to a reduction in sucrose (Fogliata & Aso 1965b; Shukla & Prasad 1974). The salt contents of cane juice (as p.p.m.), according to Bonnet (1968) ranges from 900 to 1,900 on non-saline soils to 4,000–4,500 on saline soils; much greater ranges are quoted by Fogliata & Aso (1965b) (from 600 to 12,000 p.p.m.), but these data appear inconsistent; Bernstein *et al.* (1966) give analysis for Cl, Na, K, Ca and Mg, and the total of these values, over a normal range of salinities, approximates to the values given by Bonnet (1968), though much higher levels (of 9,000 p.p.m.) were obtained in a rooting medium with EC 10.0.

Data from Haft Tappeh (Mehrad 1968) show that recoverable sugar per cent cane is reduced by 0.23 unit for each increase of 1.0 mS in the rooting medium; approximate conversion of the meq/litre NaCl (1 : 5 extract) data from Argentina presented in Fogliata & Aso (1965b) to EC_{SE} gives a value of 0.3 unit recoverable sugar lost per 1.0 mS increase, which supports the Haft Tappeh regression.

The germination of cane setts may be tolerant of much higher levels of salinity than in subsequent growth. Levels of EC up to 12 did not affect germination in one trial (Leverington 1960) or only delayed germination in another trial (Shen & Tung 1964), though subsequent growth rates were markedly reduced, as indicated in Table 3.1. Root primordia tended to remain dormant at higher salt levels (Shen & Tung 1964). In contrast, however, Bernstein *et al.*

(1966) found that NCo 293 was severely inhibited in germination by salinities of ECs as low as 3.6; they concluded that soil salinities should not exceed 4.0 mS for uniform and rapid establishment.

Excesses of soluble salts occur in low-lying areas of poor natural drainage. Similar means are used to reclaim such soils (saline or alkaline) for the cultivation of most crops, and sugar-cane is not unique. The unwanted chemicals are leached from the soil, carried away in the drainage water and further accumulation is prevented.

For example, in Trinidad a plan was devised to recover roughly 500 ha of saline swamp for cane cultivation. The area was to some extent protected from being flooded by a number of regional drains, which nevertheless at high tide brought in sea-water. Therefore embankments were made around the areas to be reclaimed and freshwater drainage schemes within the polders were established. Pumps located at the lowest points were then installed to lift fresh water over the embankments into the area drainage system (Fig. 3.1). A salt-tolerant variety was planted and, although the first crops reaped were poor in yield, within five years the land was

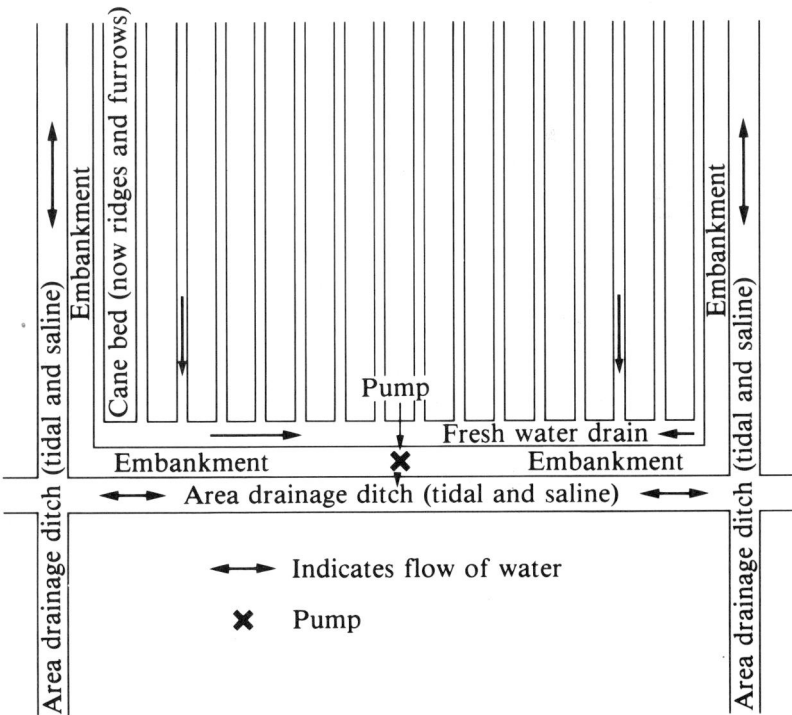

Fig. 3.1 The reclamation of a low-lying saline swamp

fertile. Pride in this achievement was somewhat diminished when, in the last stage of its development, a sluice gate (*koker*) was discovered on the banks of the Caroni River. No doubt an enterprising Dutchman, or someone with experience of Guyana, had brought this land into cultivation, perhaps a century ago, and had used the *koker* to discharge excess water at low tide.

Having been planted in soil of good tilth, the cane must have access to the main elements of nutrition, be free from the competition of weeds and, where necessary, be supplied with irrigation water.

Nutrition

Sugar is a carbohydrate, i.e. a compound of carbon, hydrogen and oxygen, the hydrogen and oxygen being in the same atomic proportion in which they occur in water. Therefore, provided that all the by-products of manufacture are returned to the soil, there should be no diminution of its mineral content. However, by-products are frequently used for other purposes (Ch. 11) and depletion of the soil reserves of essential nutrients must be made good. Moreover, soils differ markedly in their composition (for example many light soils are deficient in potassium) and, regardless of the fate of by-products, must be dressed with appropriate fertilizer if a satisfactory crop is to be grown.

In general only nitrogen, potassium and sometimes phosphorus and magnesium, are required. The minor (trace) elements are important in cane nutrition only where the crop is grown on special soils (for example the *pegasse* soils of Guyana and the muck soils of Florida).

Nitrogen

An adequate supply of nitrogen is required to stimulate vigorous vegetative growth. If the soil is deficient in this element, the plants are stunted and their leaves pale green or yellow–green in colour.

Choice of fertilizer

Some crops show marked differences in their response to various nitrogenous fertilizers. Grist (1975) described the reaction of rice and referred to three experiments in which, if the relative efficiency of ammonium sulphate was 100, that of seven other nitrogenous compounds varied from 40 (sodium nitrate) to 110 (urea). Sugar-cane does not show such marked differences; therefore the fertilizers most commonly used are those which have the cheapest cost per unit of nitrogen. They are ammonium sulphate (21 per cent nitrogen) and urea (42 per cent nitrogen). Because of its lower nitrogen content, it is more expensive to transport and apply ammonium sulphate than urea. Ammonium sulphate is marketed as large, free-flowing, transparent crystals; and urea as pellets.

Losses of nitrogen, from urea applied to the soil surface, of the order of 15 per cent, have been recorded compared with surface applications of sulphate of ammonia but, when the urea was covered with soil it was found to be equally efficient as sulphate of ammonia as a source of nitrogen.

Rate of application

When virgin land is cultivated, or green manures are grown between cane cycles, the decomposition of organic residues usually supplies sufficient nitrogen to sustain the first (plant) crop. Thereafter appropriate dressings are determined by carrying out fertilizer trials (often supplemented by leaf analysis).

Ideally this information should be derived from the statistical analysis of the results of standard factorial trials (including potassium and phosphorus as well as nitrogen), and, in most well-established cane-producing areas, fertilizer requirements are continuously under review, either as part of the research programme on individual properties or that of a central research organization.

When new cane areas are being developed there is often a lack of basic and precise information on the fertilizer requirements, particularly related to the nitrogen levels which should be applied on initial planting and to ratoon crops, as well as information on the phosphate needs of particular soil types. In new developments it is not often possible to cover the total range of conditions with a series of standard fertilizer trials and, where it is important to gain, in as short a time as possible, an overall assessment of, for instance, nitrogen or phosphate needs, simple skirmish plots have been found to be very useful. With the skirmish plot technique the objective is to lay down a large number of plots having rather restricted treatments, sacrificing statistical accuracy for intensive coverage of the soil conditions within the area.

Thompson (1981) has described the technique as developed in Jamaica. For example, to test the effect of four levels of nitrogen, sixteen sub-plots are arranged as shown in Fig. 3.2. Each plot consists of eight rows of cane, 10 m in length, readily accessible and clearly demarcated. For the plant crop, the sub-plots in row A receive the nitrogen treatments as indicated; in all other respects they receive the same treatment as the rest of the field, as do the sub-plots in rows B, C and D, including the field treatment for nitrogen. For the first ratoon crop, the sub-plots in row B are given the differential nitrogen applications, while the sub-plots in rows A, C and D receive the normal field dressing. In the second and third ratoon crops similar special treatment is given, respectively, to the sub-plots in rows C and D.

N_3	N_0	N_1	N_2	A
				B
				C
				D

Fig. 3.2 The design of a simple skirmish plot

The plots or sub-plots can be split to determine the effect of other treatments, such as time of application or a cross-dressing with, for instance, phosphate, but the most important consideration is to establish as many skirmish trials as possible to cover the range of soil conditions. Tiller counts, growth rates, Brix level and so forth can be recorded and leaf samples taken for analysis and, where the extra information would justify the work involved, the plots can be weighed at reaping. One of the advantages of the technique is that the treatments, in ratoons, are applied to plots which have had the same treatment as the rest of the field.

Evans (1960) wrote: 'It is probably correct to state that more investigations have been carried out on the nitrogen, potassium and phosphorus nutrition of sugar cane than of all other tropical crops', and few would doubt him. However, the results of some of the experiments should be treated with caution. For example in Trinidad it was reported that, although under normal circumstances cane did not respond to light dressings of phosphorus fertilizers, large increases in yield were achieved in heavy applications of both nitrogen and phosphorus were made. This finding has yet to be confirmed by field experience.

In commercial practice the rate of application of nitrogen varies according to local circumstances, such as the rainfall received and whether or not irrigation is available but, where annual cropping is practised, applications are usually in the range 100–140 kg N/ha (90–155 lb/acre). In general, responses to nitrogen are significantly greater in ratoons than in plant cane, and cane growing in inherently fertile soil responds well to the higher levels of application.

Where there is complete dependence upon rainfall, the rate of application must be the optimum for average conditions: in a wet year it will be less than could profitably be assimilated by the plant; and in a dry year it will be in excess of the requirement. In Barbados the standard dressing for ratoons in low rainfall areas is 105 kg N/ha (94 lb nitrogen/acre), but in high rainfall areas it is 131 kg N/ha (117 lb nitrogen/acre).

Timing and placement are important. Nitrogen promotes vegetative growth and inhibits the storage of sucrose in the parenchyma of the stems. Consequently nitrogenous fertilizers are applied to ratoon shoots as soon as possible after the previous crop has been reaped, and all nitrogen should be applied five months before the cane will be harvested. If the dressing is too heavy, or its application unduly delayed, the sucrose content of the juice will be depressed.

Nitrogen in usually applied as a band along the cane row, but equally good responses have been obtained when, as sulphate of ammonia, it has been broadcast or even applied on top of the cane trash. Where irrigation is used the application should be confined, where possible, to the wetted areas.

The application may be made by machine or by hand. Where cane is grown on ridges and machines are used it is convenient and effective to discharge the fertilizer from a hopper placed immediately in front of the discs, which cover it as they gather soil to restore the shape of the ridges.

In many countries the application to plant cane is split. At the time of planting, half of the fertilizer requirement is placed in bands parallel with the rows of setts and roughly 5 cm from them. The second dressing is made shortly after the shoots have appeared.

Different methods must be used where, as in Hawaii, the crop is grown for two years or more before being reaped. At the end of the first year the mass of vegetation cannot easily be penetrated by people or by machines. Therefore fertilizers (including nitrogen) are applied either in solution with the irrigation water or in pellet form from the air.

The yield responses to applications of even quite small quantities of filter mud, particularly on problem soil areas, are often very much greater than could be expected from the actual nutrient content of the muds, and much greater use should be made of this valuable material, in spite of the high costs of distribution.

Phosphorus

It has been estimated that 100 tonnes of cane remove 50 kg (110 lb) of phosphorus (as P_2O_5) from the soil. At the factory the phosphorus is expressed in the raw juice, precipitated in the clarifiers (Ch. 10) and removed in the filter mud. The concentration of phosphorus (as P_2O_5) is 0.40 per cent of the total soluble solids in the raw juice, but only 0.08 per cent in clarified juice. The filter mud, and therefore the phosphorus, is not returned evenly to the land from which the cane was reaped. Its disposal is costly and fields close to a factory usually receive more than their fair share of the mud.

Phosphorus is indispensable to the formation of a healthy root system. Consequently the growth of sugar-cane on phosphate-deficient soils is inhibited and the number of tillers (roots which arise from the buds of the rootstocks) is limited. The deep purple discoloration of the leaf tips, typical of phosphate deficiency in many crops, is also shown by sugar-cane, especially by indicator varieties such as B 3337.

Choice of fertilizer, rate of application and placement

The main source of phosphorus fertilizer, phosphate rock, is but slightly soluble in water. Therefore, before being placed on the market it is usually treated with either sulphuric acid to make the more soluble superphosphate, or with phosphoric acid to form triple superphosphate.

The value of a phosphatic fertilizer is usually judged by its content of water-soluble P_2O_5, though in some countries the criterion is solubility in a 2 per cent solution of ammonium citrate. There are phosphorus fertilizers other than superphosphate and triple superphosphate (TSP), for example bone meal and ammonium phosphate, but superphosphate and TSP are the forms most widely used in the sugar-cane industry. Phosphate is usually applied at time of planting, and the range of quantities applied varies rather widely, depending on factors such as 'fixation' with a range from about 50 kg P_2O_5/ha to as much as 250 kg (20–100 kg/acre). Dressings to ratoons are rarely given, except where compound fertilizers are used, or to soils particularly deficient and responsive to phosphate.

In some soils the soluble phosphate of the fertilizer quickly forms insoluble compounds, especially in combination with iron and aluminium, and becomes unavailable to the plant. It has become 'fixed'. The problem of phosphate fixation is well known to agronomists and is not confined to the growing of sugar-cane. Where it occurs, it is overcome:

1. By placement of the fertilizer in the furrows (rooting zones of the cane) at the time of planting.
2. By using pellets rather than finely divided forms of fertilizer. By these means the phosphorus is gradually released into the soil close to the developing root system which it is intended to nourish.

Potassium

Potassium, like nitrogen and phosphorus, is essential to the healthy growth of sugar-cane. It is also intimately connected with the formation by photosynthesis of carbohydrates in the leaves and the subsequent translocation of sucrose to the parenchyma of the stems.

Deficiency symptoms

Slight potassium deficiency can only be diagnosed by carrying out carefully designed fertilizer trials in combination with soil and juice (and/or leaf) analyses. Gross deficiency, on the other hand, is shown by typical and unmistakable symptoms which appear on the older leaves. Their margins and tips become brown, but the pattern of necrosis is quite different from that caused by salt damage. The dead tissue is not evenly distributed around the edges of the leaves, but extends downwards on each side of the midribs. The cause of the premature death of leaf tissue is that potassium, being in short supply and mobile within the plant, is withdrawn from the periphery and concentrated at the growing points.

There is another feature of serious potassium deficiency which is easily recognizable: a striking deep red discoloration, in blotches, on the upper surfaces of the midribs. This must not be confused with a similar discoloration, caused by injury, which is diffuse through the midribs and not confined to their surfaces.

The leaf symptoms of potassium deficiency, unlike those of phosphorus deficiency, appear on most varieties; therefore it is unnecessary to use an indicator variety for diagnostic purposes. The symptoms were described by Honert, on POJ 2878 in Java, as long ago as 1932; by Martin (1934) on H 109, the main variety grown in Hawaii at that time; and by Blackburn on B 3337, B 34104, B 37161, and B 37172, the varieties cultivated in Trinidad, in 1947. In Martin's work there are photographs in colour of the symptoms shown by leaves in response to the absence of each of the elements of nutrition required by sugar-cane.

Choice of fertilizer and rate of application

When sugar-cane is processed, most of the potassium passes through the various stages of manufacture in the juice and syrup and eventually leaves the factory in the molasses. Irvine, in Meade & Chen (1977), quoted analyses of potassium (as K_2O) as 1.31 per cent of total solids in raw juice and 6.55 per cent in molasses. Therefore, each crop of 75 tonnes cane/ha (30 tons/acre) removed 92 kgs/ha (84 lb/acre) of potassium (as K_2O) from the soil. Little of the molasses or the potassium-bearing lees (dunder) from the distilleries is returned to the land (Ch. 11). Therefore the loss must be made good by the application of fertilizers.

Potassic fertilizers occur naturally in mineral deposits, usually as the chloride or sulphate; and also, combined with nitrogen and phosphorus, in the organic manure, guano (2–3 per cent K_2O). Like nitrogen, their effect on the growth of sugar-cane is not influenced by their molecular structure and, in practice, their value is determined by their content of potassium, expressed in terms of its oxide, K_2O, and called potash. Potassium chloride (muriate of potash) is the cheapest form of the fertilizer and therefore the most widely used in the sugar-cane industry.

Application may be by hand or machine, and alone or in combination with other fertilizers. There is little to be gained by special placement, therefore the cheapest method of application is the best.

Where annual cropping is practised, the standard dressing is 1.5 cwt/ acre of muriate of potash (60 per cent K_2O), which is roughly 115 kg K_2O/ha (100 lb/acre). Proportionately larger applications are given, in split doses, to heavier crops which have longer periods of growth.

Although potassium is readily available to the plant, quantities in excess of the requirement for growth are retained in the soil. In Barbados, pen manure and guano were replaced in the late 1920s by synthetic fertilizers plus a mulch of sour grass (*Andropogon pertusus* (L.) Willd.). It was thought by some that this drastic change in policy would impoverish the soil. However, some twenty years later (using the same soil-sampling technique in the same fields, and subjecting the samples to the same method of analysis), it was found that there had been an increase in the exchangeable potash content of the soil. It may be that the rate of application, mentioned above and used in many countries, is too high. Nevertheless, two considerations should be borne in mind:

1. On most soils there is an interaction between nitrogen and potassium on the growth of sugar-cane. Heavy dressings of nitrogen, with consequent prolific vegetative growth, can be given without causing deterioration in juice quality provided that sufficient potassium is present. Therefore, under no circumstances should the soil be deficient in potassium.
2. Having established a reserve of potassium in the soil, it is possible to draw on it, without affecting the growth of the crop, when financial difficulties arise.

In addition to the main elements of nutrition (nitrogen, phosphorus and potassium) several others may be necessary, in special circumstances, for the successful growth of sugar-cane. Evans (1960) reviewed the literature on this subject and described his experience in Guyana.

Calcium

The influence of calcium on the nutrition of sugar-cane has been widely investigated, especially in relation to acidity and alkalinity; to toxicity caused by aluminium, iron and manganese; and to the availability of phosphorus and molybdenum. All appear to be interrelated.

Irvine, in Meade & Chen (1977), quoted the concentration of calcium (as CaO) as 0.29 per cent in raw juice, 0.30 per cent in clarified juice and 0.35 per cent in syrup. Yet considerable quantities of calcium (as milk of lime) are added to the raw juice in the course of manufacture to cause defecation (Ch. 10). The flocs which are formed are precipitated in various forms of chemical combination as filter mud, and then are incorporated with the soil. Therefore, if the filter mud was evenly distributed over the arable area, calcium deficiency should never arise. In practice this does not happen. The high cost of transport causes fields close to the factories to receive preferential treatment, and often the more distant land is neglected.

Photographs of the leaf symptoms of calcium and trace element deficiencies in sugar-cane were illustrated in black and white by Evans (1960) and in colour by Martin (1934). When calcium deficiency arises, chlorotic spots first appear. Later their centres become dark red–brown as the tissue dies. The necrotic areas frequently coalesce to give a rusty look to the leaves. Such symptoms are rarely seen in the field and positive responses to calcium (usually applied as ground limestone) in fertilizer trials are uncommon.

When ammonium sulphate was the most widely used nitrogenous fertilizer (it is now being replaced by urea), there was considerable apprehension concerning the long-term effect which it might have in increasing the acidity of heavy clay soils. Consequently, to counteract this possible trend, for many years dressings of ground limestone were broadcast on the surfaces of fields immediately before they were ploughed. In Trinidad this operation was discontinued thirty years ago, with no deleterious effect on cane growth or increase in the acidity of the soil.

Evans (1960) discussed the work carried out in Hawaii and Guyana and concluded that, where only calcium nutrition is concerned, the critical level in the soil is 100 p.p.m of exchangeable calcium. However, where the effect of calcium is to increase the availability of phosphorus, to lower the toxic levels of iron or aluminium, or to reduce acidity, positive responses by sugar-cane to dressings of limestone are likely even where the exchangeable calcium is as high as 500 p.p.m.

Sodium clays

When land is inundated with salt water, the base exchange capacity of the soil is limited largely to sodium ions. Sodium clays are difficult to work; they are extremely sticky when wet and set into concrete-like masses when dry. The reclamation of such land can be achieved by replacing the sodium ions with calcium ions by incorporating a suitable calcium compound with the soil. It was stated by Barnes (1974) that the sodium soil of an old lake bed in Kenya was made productive by broadcasting calcium sulphate (gypsum) at the rate of 4.8 tonnes/ha (2 tons/acre), followed by the application of 1.2 tonnes/ha (0.5 ton/acre) in the furrows at the time of planting.

Magnesium

The leaf symptoms of magnesium deficiency, as shown by culture in nutrient solutions from which the element has been excluded, are similar to those caused by calcium deficiency. Chlorotic spots turn brown as the tissue dies, coalesce and give a rusty appearance to the leaves. Magnesium is mobile within the plant; therefore the deficiency symptoms are most severe on the older leaves.

There are very few reports of increases in yield caused by the application of magnesium; indeed, in Hawaii there was no response even when the exchangeable magnesium in the soil was as low as 30 and 50 p.p.m.

Minor (trace) elements
Iron

The leaf symptoms of iron deficiency are similar in most members of the Gramineae. In the young leaves pale longitudinal stripes, almost devoid of chlorophyll, appear between parallel dark green veins. As the deficiency increases in severity, the interveinal areas become white and the larger veins are sharply demarcated. In the most advanced cases the whole of the young leaves, including the main veins, are white. Iron has low mobility within the plant; therefore older leaves may remain deep green in colour while the young ones are chlorotic. Iron deficiency also restricts the development of an adequate root system.

Damage by iron deficiency is largely confined to sugar-cane grown on limestone soils, but only because the element has been made unavailable to the plant by its conversion to insoluble ferrous carbonate. Similar damage may occur if the iron : manganese ratio in leaf tissue is less than 1: 1. It is moot point whether such a condition should be described as iron deficiency or manganese toxicity. Thus, three causes of iron deficiency may be distinguished:
  1. Soil deficiency.
  2. Lack of mobility within the tissue.
  3. A low ratio of iron : manganese (and possibly other elements).

The remedy, in all cases, is to spray the plants with a dilute (0.5–1.0 per cent) aqueous solution of ferrous sulphate or a 0.1–0.5 solution of an iron chelate.

Sugar-cane is susceptible to iron toxicity as well as to deficiency. In Java the condition is known as Kalimati disease (the name of the most important plantation on which it occurs). The symptoms are similar to those of potassium deficiency, and in Java they disappear if potassium is applied (Van Dillewijn 1952). It is now generally agreed that potassium deficiency interferes with the mobility of iron within the plant; but Evans (1960) reported iron toxicity in the presence of adequate, or even high levels of available potassium. The problem is neither widespread nor serious.

Manganese

Manganese deficiency, on the other hand, occurs much more frequently: it was the cause of Pahala blight in Hawaii and, in association with deficiencies of copper and zinc, was responsible for the poor growth of sugar-cane on the organic soils of Florida.

The leaf symptoms of manganese deficiency are similar to those of iron deficiency, except that the chlorotic longitudinal stripes rarely extend to the full length of the leaf. In acute cases the chlorotic tissue may die, turn brown and split along the lines of necrosis.

The deficiency may be corrected by the application to the soil of 34–56 kg/ha (30–50 lb/acre) of manganese sulphate. Pahala blight in Hawaii is controlled by incorporating manganese in the fertilizer. In Florida it has been found necessary also to correct alkalinity when manganese is applied.

Excessive concentration of manganese in the soil can cause toxicity, the symptom of which is mild interveinal chlorosis of the leaves. However, such conditions are rare and usually associated with acid conditions. One example was found by Evans in Fiji and was corrected by the application of potassium.

Copper

The status of copper as one of the most important micronutrients necessary for the satisfactory production of sugar-cane on certain special soils was first established by Allison and his co-workers in Florida (Allison *et al.* 1927). They found that, on the raw peat soils of the Everglades, no growth took place in the absence of copper. In more recent years, copper deficiency has also been recognized as the cause of crop failure on sandy soils in the Mossman and Innisfail districts of Queensland and on granitic soil in the south of Natal.

The symptoms of copper deficiency are poor growth with few tillers, and chlorotic spindles from which the leaves are reluctant to unroll. In more mature cane the leaves hang downwards and this, together with the unrolled spindles, gives a characteristic appearance to the cane which is called 'droopy top' in Queensland (Fig. 3.3). Leaf symptoms are sometimes not unlike those caused by mosaic disease: blotches of green (often rectangular in shape) surrounded by chlorotic tissue. On other varieties they may appear as pronounced stripes.

In Queensland large increases in yield were reported when copper sulphate was applied to the soil in a deficient area at the rate of 60 kg/ha (0.5 cwt/acre). In general, the rate of application is 23–57 kg/ha (20–50 lb/acre). Some workers have used chelates with success to remedy copper deficiency, but ultimately the choice of the most suitable treatment must be based on economic factors.

Zinc

Zinc nutrition is a serious problem only in the Florida Everglades and on some of the lighter soils in Queensland.

In nutrient cultures it has been shown that mild zinc deficiency

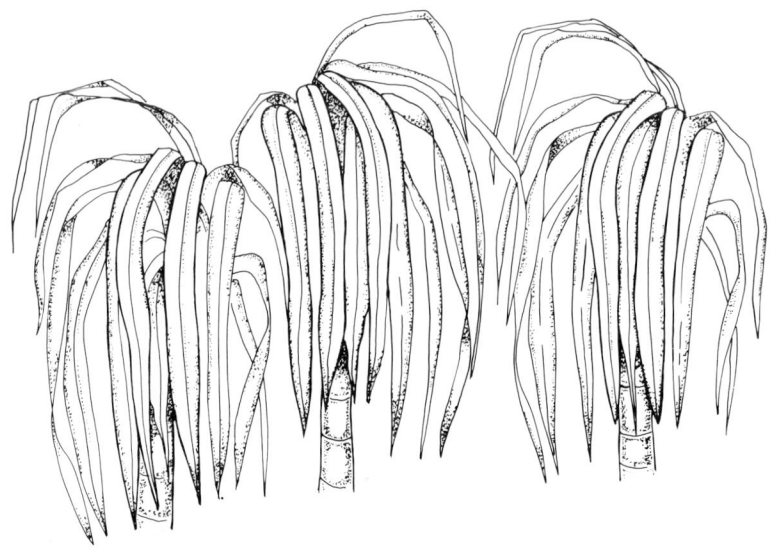

Fig. 3.3 Droopy top caused by copper deficiency

might be indicated by the development of anthocyanin pigments in the leaves; but this is not an infallible diagnosis. The most conspicuous and definite symptom of zinc deficiency is chlorosis of the veins (in contrast with iron and manganese deficiency, in which the striped chlorosis is interveinal). Veinal necrosis takes place as the deficiency becomes more acute.

The most widely used remedy for this condition in Florida, and also for manganese and copper deficiency, is to apply a dressing in the furrow at the time of planting of 84 kg/ha (75 lb/acre) of manganese sulphate with 34 kg/ha (30 lb/acre) of both copper sulphate and zinc sulphate.

Boron and molybdenum

Martin in Hawaii and Evans in Guyana described similar deficiency symptoms on the leaves of sugar-cane grown in nutrient cultures from which boron has been excluded. The growing points were also affected, and so distorted that they resembled those injured by the fungal disease, pokkah boeng (Ch. 7).

Evans also described yellow streaks on the leaves of cane grown on silty clay in Guyana and suggested that they might be symptomatic of molybdenum deficiency.

However, no response to the application of either boron or molybdenum has been reported and these elements are unimportant in commercial sugar-cane production.

Leaf analysis and the crop log (see Clements 1980)

From 1940, when Clements in Hawaii first devised the crop log technique for sugar-cane production, there has been increasing interest in the analysis of leaf tissue as a guide to the nutrient status of the plant; and also, in irrigated areas, to determine its water requirement. As yet there is no consensus concerning the method of sampling, or of the tissue to be analysed. In Guyana Evans's technique was to analyse the top visible dewlap (TVD); but Clements's method is to take the tops from five stems in five separate locations in fields as large as 120 ha (316 acres). The laminae of leaves three, four and five are then dried, ground in a Wiley mill and analysed after their water content has been determined.

Foliar diagnosis and the crop log are most widely practised in Hawaii and the methods used were described by Clements (1972, 1980).

Interaction of soils on mechanized systems

Traditional systems of cane cultivation, harvest and transport relied on human and animal power; the introduction of mechanized systems must take soil type and topography into consideration.

For example, cultivation of cane on slopes as steep as 30° as in some Caribbean islands cannot be continued once mechanization becomes essential. However, medium-steep slopes, up to 10°, can remain in cane. The combination of systems such as bench terracing and grassed waterways with the protection afforded by cane, can control erosion. Contour ploughing and planting, and the use of self-loading trailers, allow a high degree of mechanization.

The 'Reynoso' system of Java, and the 'cambered beds' of Guyana, both developed to provide adequate drainage on completely flat lands, cannot be mechanized. Alternative systems involving the provision of some gradient, and planting on high ridges, are being introduced in such areas.

The texture, in particular, the drainage characteristics, of the soil has a significant impact on types of cultivation, harvest and transport systems that can be used. The cost of modern sugar plant is such that the length of the processing season must be maximized. Thus it is often necessary to select field equipment which will cause the minimum of damage when operated under marginal conditions.

Climate

Sugar-cane is frequently considered to be a tropical crop. In fact, many of the largest and most successful industries are located in

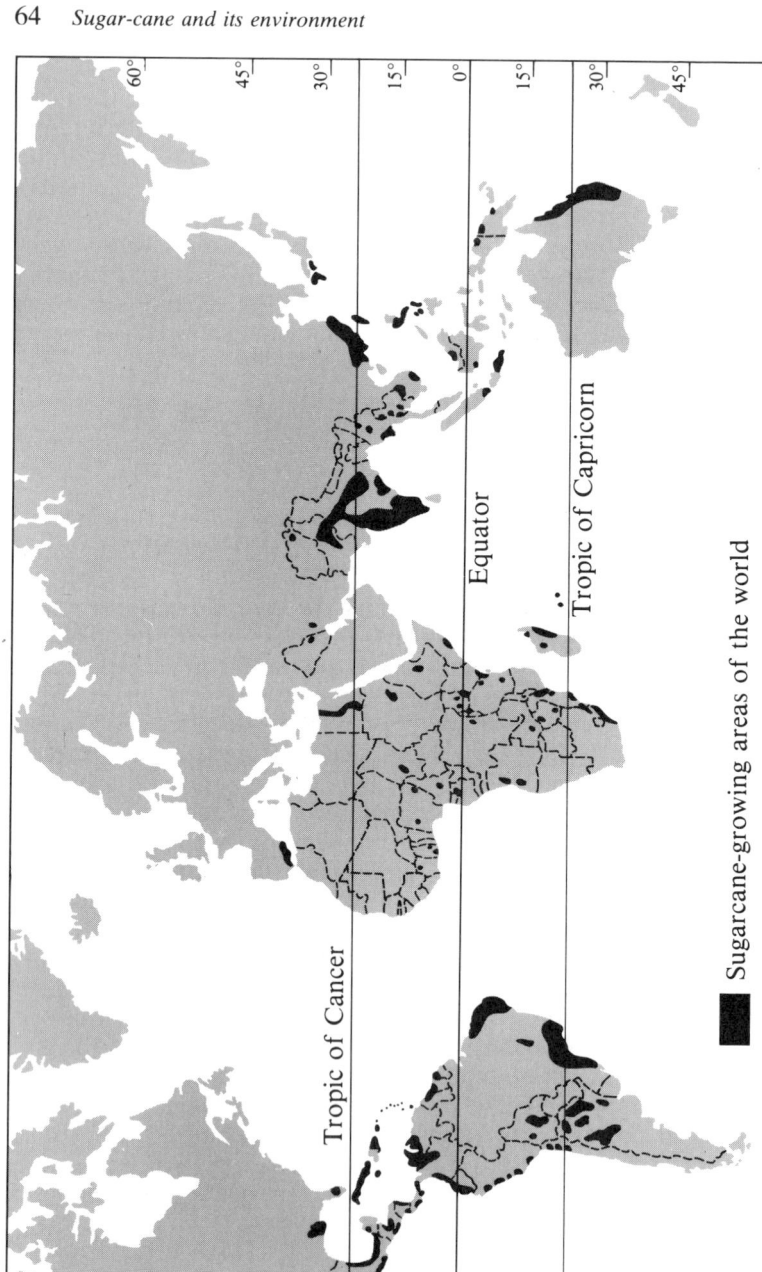

Fig. 3.4 Sugar-cane-growing areas of the world

subtropical areas, between, say 15° and 30° latitude (Fig. 3.4).
Examples include:

	Million tonnes sugar/annum
South/Central Brazil	5.2
Cuba	5.6
Mexico	2.6
South Africa	2.1
India	3.7
China	2.4
Australia	2.9
Hawaii	1.0

The ideal climate for sugar-cane may be defined as follows: the
growing season should be warm with mean day temperatures around
30 °C and with fully adequate moisture and high incident solar
radiation. The ripening and harvesting season should be cool, mean
day temperatures between 10 °C and 20 °C, but frost free, dry and,
again, with high incident radiation. Unfortunately, vigorous growth
of ratoon shoots requires adequate moisture, preferably provided by
controlled irrigation, during the harvest season; this controlled irrigation can also prevent excessive drying of cane prior to harvest.

Very few areas have such ideal conditions, and management
methods are adapted to suit them.

Temperature

The most important of the uncontrolled factors is temperature.

The minimum temperature for active growth is approximately
20 °C, but varietal and cultural factors modify this slightly. Thus, for
example, Bacchi *et al.* (1977) found that the critical temperatures for
the varieties grown in South Brazil were 19–20 °C (unirrigated) and
18–19 °C (irrigated). The difference with irrigation is probably due
to the importance of soil temperature: work in Hawaii (Mongelard
& Mimura 1971) showed that root temperatures are more important
than air temperatures (Fig. 3.5). This explains why growth rates are
best correlated with antecedent mean day temperatures. The effect
of day and night temperatures tend to be additive.

However, final cane yield values are not necessarily reduced
progressively by mean temperatures down to about 20 °C, as is
shown by the commercial data from two fully irrigated estates in
Ethiopia:

	Metahara	*Wonji-Shoa*
Altitude	950 m	1,500 m
Mean annual temperature (°C)	24.8	20.5
Mean sunshine (hours/day)	8.4	8.2

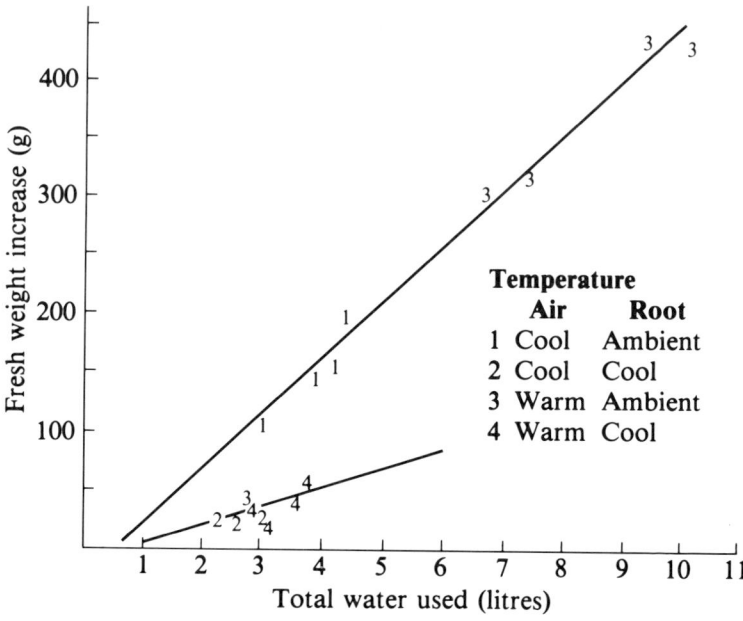

Fig. 3.5 Relationship between total water used and fresh weight increase. The position of each number represents the mean of four replicates (After Mongelard & Mimura 1971)

Mean stalk number/ha*	77,000	108,000
Mean cane yield (TCH/annum)	118	123
Mean cane quality (pol % cane)	14.3	15.1

These cane yield data appear to conflict with reports of linear responses in growth to temperatures (for example, White 1970); the explanation is the well-documented effect of lower temperatures increasing the degree of tillering, as is illustrated above.

The length of the season with temperatures significantly below 20 °C influences both the growing season and the ripening. Low temperatures are the most effective means of ripening cane, counteracting adverse factors such as excessive moisture or nitrogen; for example, Brickaville on the eastern coast of Madagascar used to average a value of about 12 per cent commercial sugar per cent cane in spite of its very wet environment (with 2,900 mm of rain) and only two months averaging less than 100 mm rain; mean day temperatures are about 20 °C for four months of the year. A report from

*Ratoons of CO 419 and B 41227

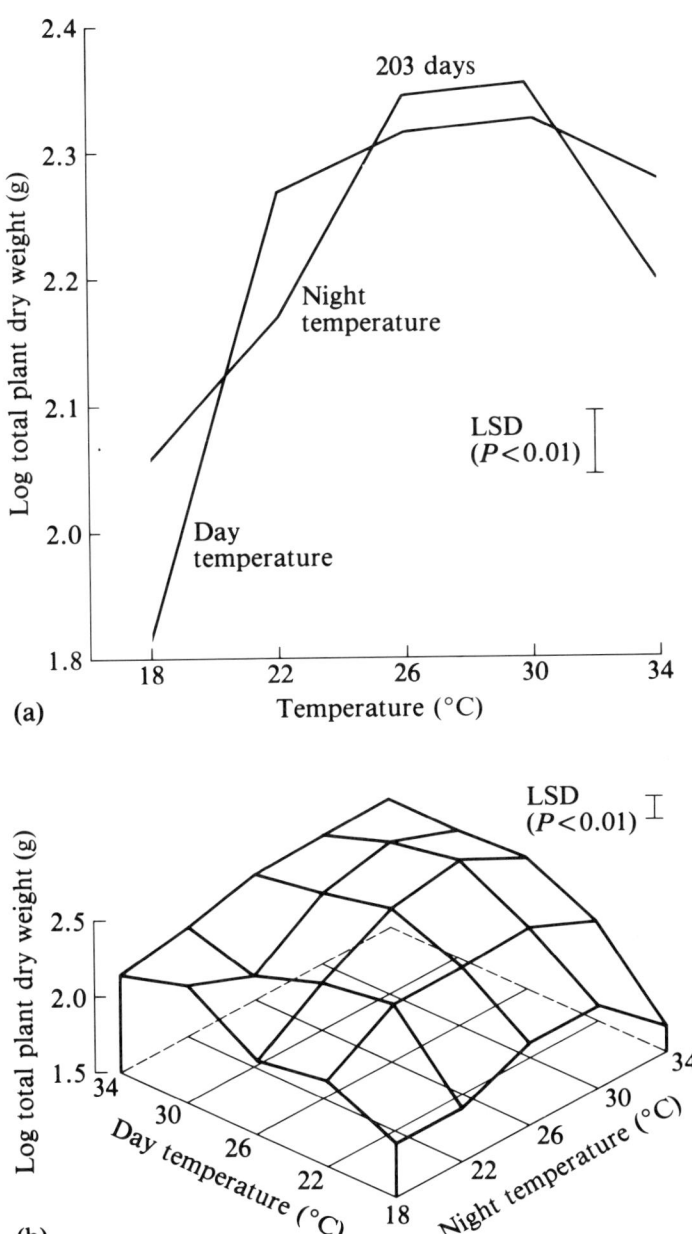

Fig. 3.6 Effects of day and night temperature on dry weight production. (a) Main effects; (b) interaction of day and night temperatures. Duration 209 days (After Glasziou *et al.* 1965).

Iran (Gowing *et al.* 1977) indicates that the ripening effect becomes less when temperatures fall below 10 °C; freezing conditions, of course, rupture cells and caused irrevocable deterioration.

The effects of high temperatures on cane are not well documented, but temperatures above 38 °C reduce photosynthetic rate (Kortschak 1972) while photorespiration increases with temperature (Chu & Kong 1971). In practice, cane appears wilted irrespective of water supply when temperatures approach 35 °C, so that growth during such periods is curtailed.

Sunshine

The extensive work on photosynthesis of, and light intensity on, sugar-cane can be summarized as follows: sugar-cane under field conditions, that is, where shading of some leaves occurs, can continue to increase its photosynthetic rate up to full natural light intensity, so that the greater the incident radiation, the higher the yields that may be expected. Plant breeders are now beginning to examine leaf geometry but, in the short term, significant yield increases can be obtained by proper weed control or, in areas with short growing seasons, by closer planting; typically, in areas with long growing seasons, there is a heavy mortality of shoots after the formation of a full canopy, so that the heavier density of shoots at the earlier stages cannot increase final yield.

Daylength is another important factor which is difficult to evaluate as it is confounded with winter-summer temperature changes at high latitudes. Thus, for example, cane yields in South Queensland (with a growing season, defined by temperature, of only seven months) are very similar to those in North Queensland, where temperature does not limit growth.

The flowering of sugar-cane is photoperiodically controlled (night period of 11 hr 30 min to 11 hr 15 min). There is considerable disagreement about the effect of flowering on yield, though it is safe to summarize various reports by saying that flowering has no adverse effect for some time (possibly one or two months), but has progressive deleterious effects after that. The release of non-flowering varieties is an objective of all breeding stations.

Frost

The limits of cane cultivation, by and large, are 30°N. and 30°S.; at higher latitudes the growing season is unduly restricted by the length of the cold season. In several countries, for example Argentina, Egypt, Iran, Pakistan, Zimbabwe, South Africa and continental

USA (Florida and Louisiana), the crop is often damaged by frost. In Louisiana, because of the short growing season, the stems are erect at harvest and therefore are ideal for reaping by mechanical means. The effect of frost can be mitigated to some extent by planting resistant varieties, or by irrigation, but all varieties are killed if the frost is severe. Browne (1939) reported that frost damage occurs in Louisiana and Florida roughly once in every five years, while Irvine (1969) stated that the freeze of 3 November 1966 caused losses of US $12 m. in Louisiana. Irvine (1967) also tested several varieties for cold tolerance and found that, although differences were slight, resistance to frost was most noticeable in the Dhaulu group of *Saccharum sinense*. It is generally agreed that, while temperatures of -1 °C to -2 °C will kill the leaves and even the meristems, the juice will not freeze and its quality will remain good for several months, provided ambient temperatures remain low. If, however, the temperature falls further (to -7 °C or -8 °C) the juice will freeze and destroy the cells, and even at such low temperatures sucrose will be hydrolysed into glucose and fructose.

Moisture

An adequate supply of water is essential for cane growth, and a dry season (or the withholding of water in irrigated areas) for ripening (the storage of sucrose in the stems) and for reaping.

In the equatorial zone the precipitation of conventional rain reaches maxima shortly after the passage of the zenithal sun (April and November at the equator). The region with no dry season extends only to 2°N. and 2°S. of the equator. From 2° to roughly 15° (N. and S.) there are two wet and two dry seasons; but beyond 15° the short dry season disappears, leaving one wet and one dry season. Consequently, there are two harvest periods in Guyana (6°N.) and other equatorial countries but only one elsewhere. In Trinidad (10°N.) the short autumnal dry season (the *petit careme*) is so uncertain that reaping is confined to the more reliably dry months of January to May. These climatic patterns are illustrated in Fig. 3.7. In the main cane-producing areas, rainfall varies from virtually nil in the Chicama, Lambayeque and Santa Catalina valleys of Peru, where there is complete dependence on irrigation, to more than 350 cm/year on the Hilo (Hamakua) coast of Hawaii Island. Wide variations occur between, and even within, some of the long-established sugar-producing countries (Table 3.2). The length of the dry (harvest) period(s) is important. If it is short, high capital investment per tonne sugar produced is required in factory, reaping and transport machinery; if it is long there is a corresponding reduction in this requirement, but expensive irrigation might be necessary.

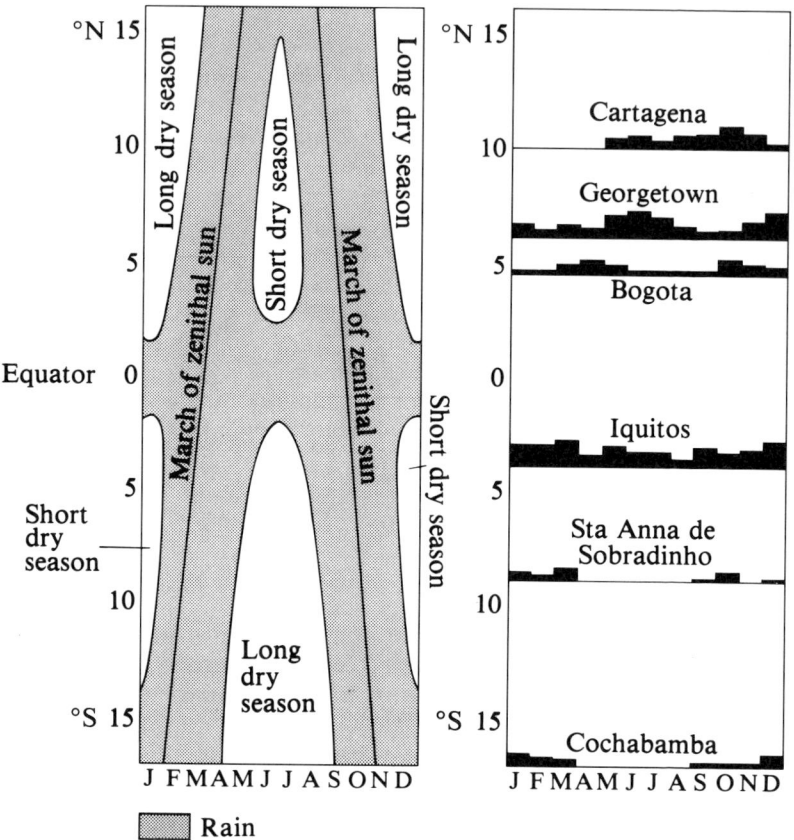

Fig. 3.7 Rainfall patterns in equatorial and tropical regions (After Miller 1931)

Rainfall

With respect to moisture, sugar-cane is more responsive than many other crops in that the vegetative portion is harvested: that is, maximum yields are obtained only when vegetative growth continues without check. Under reasonable soil conditions, and with reasonable management, 1 tonne/ha per 10 mm of evapotranspiration can be obtained.

At the same time, however, it shows remarkable resistance to drought, especially where the soils are of such a nature to allow the deep rooting of which the plant is capable. This point is of sufficient importance to quote three examples of long-established unirrigated commercial ventures which have remained viable under adverse conditions:

Table 3.2 *The average annual rainfall in some sugar-producing countries*

Country	Annual rainfall cm (in)	Country	Annual rainfall cm (in)
Antigua	109 (43)	Jamaica	
		Clarendon	76 (30)
		Duckinfield	229 (90)
St Kitts	137 (54)	Mauritius	
		leeward coast	89 (35)
Trinidad	152 (60)	windward coast	190 (75)
		central plateau	305–406 (120–160)
Barbados			
wet districts	190 (75)		
dry districts	127 (50)		
Guyana	254 (100)		

Note: Low rainfall contributed in part to the abandonment of the sugar industry of Antigua in the 1960s

Australia, Bundaberg region: average annual rainfall about 1,200 mm of which 60 per cent falls in four months (December–March); cane grown on a variety of soils including poor sandy soils, and has survived the 1968/69 year when only 571 mm of rainfall were recorded.

Tanzania, Mtibwa Estate: average rainfall about 1,200 mm, 78 per cent falling in six months (December–May); cane grown on a variety of soil types including compact clays and sandy soils, and has survived numerous seasons where only about 800 mm of rainfall were recorded.

Dominican Republic, Porvenir: average rainfall about 1,200 mm, 85 per cent occurring in six months (May–November); cane grown on a variety of soils including some extremely shallow soils over coralline rock, and has survived numerous (one in seven) years when annual rainfall is less than two-thirds of the long-term average.

The coincidence of 1,200 mm rainfall/year in these three examples is notable. A rare example where mainly rainfed cane is grown with less rainfall is at Cantarrnas in Honduras which has an average annual rainfall of about 850 mm, but the production history is too short to define the extreme level of drought tolerance (the first crop was processed in 1977). It would appear that generations of experience across the world's sugar-cane industry has shown that the crop can survive the normal variations in rainfall around a mean of 1,200 mm/year, but that lower rainfall zones are not suitable.

Other climatic effects

Many cane-growing areas are in cyclone- or hurricane-belts. The mechanical damage caused can be severe. Wind of more normal velocity is unimportant except where it affects irrigation systems.

High humidities encourage numerous fungal diseases of the leaf, sheath and root; the only practical control is the selection of resistant varieties. High humidity and high temperature vastly accelerate the rate of deterioration of cut cane; this can only be countered by efficient logistics. Rain and floodwaters assist in the transmission of numerous fungal, bacterial and virus diseases. The most striking example of disease transmission, however, is the transatlantic movement from Africa to the Caribbean of smut (*Ustilago scitaminea*).

Cyclones

Hailstones and wind (causing wind burn) occasionally damage cane in Queensland and in South Africa, but much more important are cyclones (hurricanes or typhoons) which occur there, in the Caribbean, the Indian Ocean and the China Sea.

Schömburgk (1848) described, in chronological order, 127 hurricanes and severe storms suffered in the Caribbean during the period 1494–1846. They were disastrous when the factories were damaged: '1667, September 1. A tremendous hurricane desolated the island of St Christopher's; not a house or a sugar-work remained standing . . .' However, provided that buildings do not suffer serious damage, the industry in the West Indies is little affected. Stems are blown over, become lodged and roots form at the nodes, but this is compensated to some extent by the accompanying rainfall. Even the most severe hurricane to devastate Barbados, that of 1831, caused production in 1832 to fall by less than 20 per cent. The most recent occurrence, that of 1955, had a similar effect (Table 3.3). When hurricanes reach Central America they are fully developed (Fig. 3.8), but nevertheless their effect on sugar-cane appears to be much the same as in the West Indies. Hurricanes Carmen and Fifi of 1974 caused production in Belize to be reduced by only 7 per cent (from 89,000 tonnes in 1974 to 83,000 tonnes in 1975).

In Mauritius and Réunion cyclones are interspersed between years of severe drought, and much greater damage is suffered there

Table 3.3   *The effect of hurricanes on sugar production in Barbados*

Year	Production (tonnes)	Year	Production (tonnes)
1831	16,397	1955	172,730
1832	13,437	1956	153,590
1833	19,659	1957	207,797

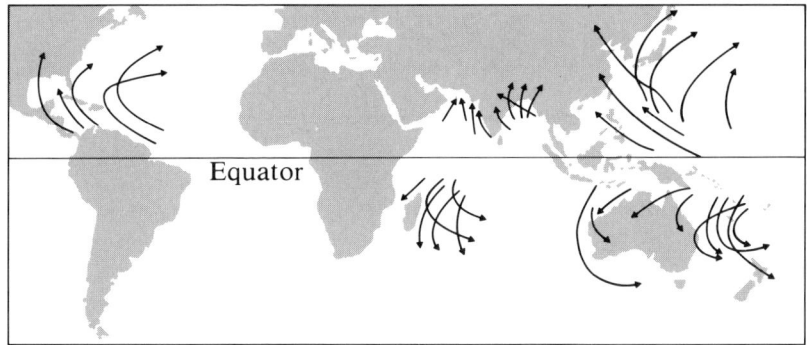

Fig. 3.8 Distribution of tropical cyclones.

than in the Caribbean. The frequency of occurrence is one cyclone in three to four years, and one drought in ten years. The devastation which they cause is illustrated by their effect on the Mauritius crops of 1945 (reduced by 54 per cent) and 1960 (reduced by 60 per cent). A Cyclone and Drought Insurance Fund was created in Mauritius in 1946 by an annual levy of 4.5 per cent of the average value of the last three normal crops. Details of the establishment and operation of the fund were given by O'Connor (1951). By 1950 Rs 23 mm. Had accrued and no payments had been made. Contributions were later increased: producers were required each year to pay into the fund 6.2 per cent of the insured value of their sugar. This proved to be a sound policy: by 1962 Rs 200 m. had been distributed in compensation for damage. Paturau (1963) stated that, except for the neighbouring island Réunion, Mauritius is more seriously affected by cyclones than any other cane-producing country in the world.

Cyclones of equal severity occur in the sugar-growing areas of Australia, but with much less frequency than in Mauritius. Jones (1951) recommended the establishment of group-microseisms at three or more centres to detect, well in advance, the development of cyclone centres.

Liu (1969) reported that 218 typhoons were recorded in Taiwan during the period 1897–1956, mostly in the months of July–September, and often caused damage to crops and buildings. These are the critical months throughout the Northern hemisphere and the subject of an old West Indian jingle:

> June, too soon.
> July, stand by!
> August, come it must.
> September, remember!
> October, all over.

Table 3.4   *The average annual frequency of tropical cyclones (as percentage of total)*

	J	F	M	A	M	J	J	A	S	O	N	D
Northern hemisphere	1	1	1	1	2	4	16	25	25	15	6	3
Southern hemisphere	26	22	23	9	4	0	0	0	0	1	4	11

Source: From Miller (1931).

In the Southern hemisphere, of course, cyclones occur most frequently during the months of December to April (Table 3.4).

Whirlwinds

When a whirlwind travels across a field of cane, the damage caused is severe, but is limited to the width of the 'twister'. The stems are broken off at ground level and the field looks as though a harvester had been driven through it. Fortunately such occurrences are rare.

Lightning

Damage by lightning has been reported from South Africa, Mauritius, Jamaica and many other areas, but in each case it was not regarded as serious.

The effect of climate on ripening

The ripening of cane in relation to reaping strategy is considered in Chapter 8. Ripening is influenced by rainfall, humidity, the incidence of sunshine, night length, altitude and temperature. In equatorial regions high temperature and rainfall, with heavy cloud cover and little difference in the lengths of the nights, cause vigorous vegetative growth and militate against ripening.

On the other hand, in subtropical areas cool and long nights immediately before and during harvest favour the accumulation of sucrose in the stems. In dry areas irrigation is generally discontinued about six weeks before reaping is due to take place in order to encourage ripening.

In general, therefore, other factors being equal, at sea-level cane quality is a function of latitude. When sucrose content and latitude are plotted against each other they give a bi-modal curve with its nadir at the equator and peaks at 18°N. and 18°S. (Fig. 3.9).

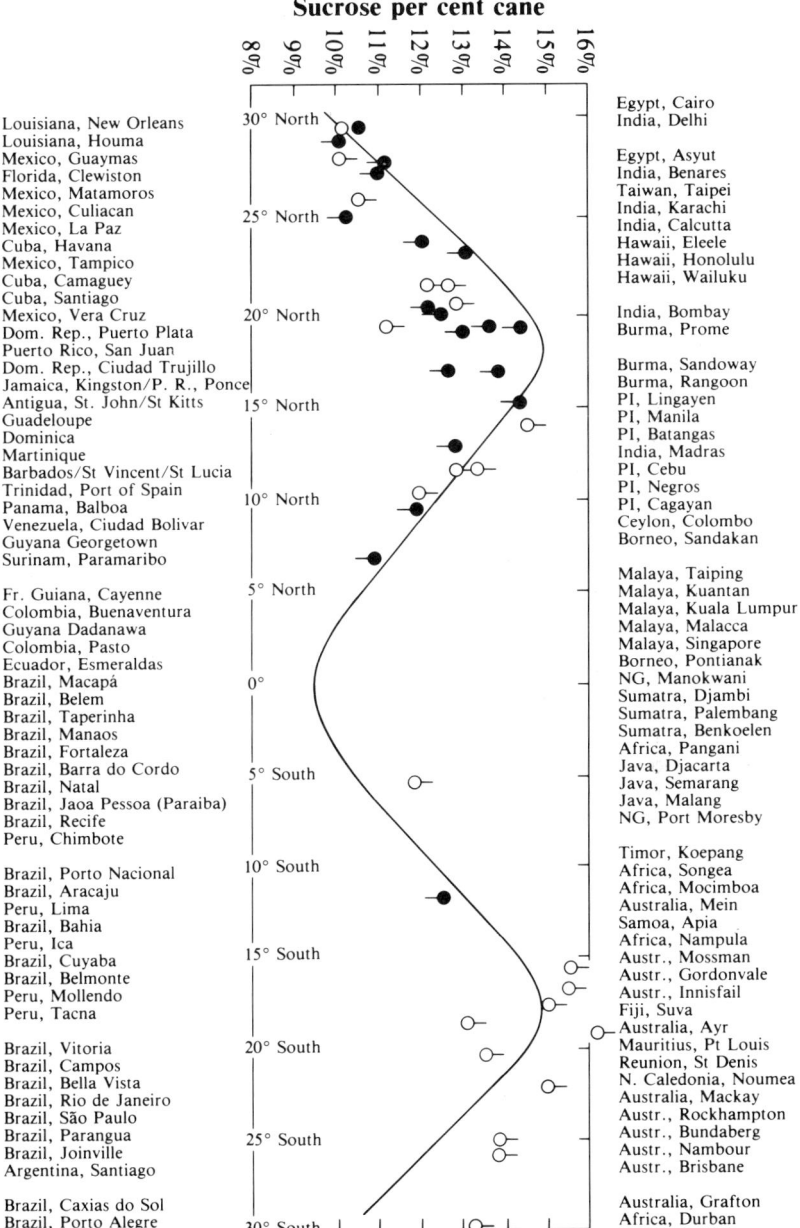

Fig. 3.9 The influence of latitude on sucrose content of cane at harvest (From Shaw 1954)

Factors influencing cane flowering
H. A. Thompson

Flowering is of considerable practical importance in sugar-cane production, since yields of sugar can be substantially reduced as a result of flowering. On the other hand, control of flowering is essential for the breeding and selection of new varieties of sugar-cane. A knowledge of the factors which regulate flowering is necessary not only to the plant breeder, who must be able to control the timing of flowering with precision, but also to the cane grower, who may suffer financial loss as a result of heavy flowering.

In some major cane-growing areas flowering is extremely rare, or does not occur, while in others almost 100 per cent of mature stalks will flower. In some instances flowering can be very localized, to the extent that in one portion of a field the cane may flower very regularly. Furthermore, a variety which flowers heavily in one environment may not do so in another, even when grown at almost the same latitude and altitude.

It is quite clear that flowering is inhibited or induced by a number of interacting factors, among which the most important established factors are photoperiod, temperature, variety, 'ripeness to flower', soil conditions, associated moisture states of the soil and cane and nutrition.

Photoperiod is a dominant factor in regulating flowering of cane and so has received much greater attention than any other factor, particularly by plant breeders, in their efforts to induce flowering in naturally shy flowering varieties. Sugar-cane is generally regarded as being a short-day plant, although certain varieties will only flower when the photoperiod occurs within a very narrow range; characteristic of intermediate or middle-day plants. The photoperiod which has been established as optimal, for moderate to heavy flowering varieties, is twelve and a half hours of daylight, followed by eleven and a half hours of darkness, with the greater stimulus being produced when daylengths are shortening, although this may not be the case in all circumstances. Burr *et al.* (1957) consider that the explanation for the lack of flowering in the spring, with increasing daylengths, is due to low temperatures, and seems more plausible than that shortening days are required.

The way clear-cut reactions in respect of flowering which are obtained in areas relatively far from the equator where minimum temperatures vary appreciably and where the optimal photoperiod is of very limited duration, are seldom obtained in areas closer to the equator, where the main differences between seasons are mainly rainfall and soil moisture conditions, rather than temperature. Temperature can, however, have a dominant role in inhibiting flowering and Coleman (1963) reports that night temperatures of 18.3 °C

(65 °F) can prevent flower initiation provided, under Louisiana conditions, six nights or more, between 1 and 25 September, have temperatures 18.3 °C (65 °F) or lower.

In some situations where marked variation in the extent of flowering occurs regularly, even within a single field, it has been shown that night temperatures are the prime cause. There are, however, very marked differences in the intensity of flowering experienced in areas where the night-time temperatures are consistently well above 18.3 °C (65 °F) and, in these circumstances, the moisture status of the soil is an important factor.

It has been shown, both in India and Hawaii (Burr *et al.* 1957), that the time of appearance of flowers bears little relation to time of initiation. So-called late and early varieties may actually initiate primordia on the same date and differ by months in time of flowering, which depends upon the entirely separate process of elongation of the flower stalk. Within a single variety (H 37–1933) the time of emergence of flowers extended from November to April, although randomly selected control all initiated in September.

An interaction of light, temperature and moisture can produce the effect which Coleman (1963) has called 'climate'. He found that climatic differences between two Hawaiian islands whose daylengths were the same, during September, had a marked influence on flowering. On the island whose weather favoured flowering, average minimum temperatures were higher, rainfall was higher but the average period of sunlight was less than in the island on which flowering was depressed.

Sugar-cane displays a 'ripeness-to-flower', in common with other plants, in that stalks which are too young cannot be induced to flower and there are varietal differences in the length of the juvenile phase (Mangelsdorf 1946, 1953). The variety POJ 2725 flowers if there are only two or three mature internodes present during the photo-inductive period, whereas, with BH 10/12 a greater number of mature internodes must be present.

In a young stool with stalks of many sizes only one or two of the largest may flower, so the term 'percentage of flowering stalks' is an expression of the number of stalks which had reached the 'ripeness-to-flower' stage, at the critical date for flower initiation.

Some practical effects of cane flowering and its influence on the yield of sugar

There are marked varietal differences in the physiological reaction of cane to flowering, just as there are differences in susceptibility to flowering. A well-grown annual cane, Co 421, can develop an open 'piped' pithiness, almost immediately at flowering, for about one-third of its length, making it unsuitable for use in some areas, in spite of its very good growth performance. On the other hand, NCo 310, growing in exactly the same conditions, does not develop equivalent pithiness, even after standing in a flowered condition for as long as five months.

The variety NCo 310 can also maintain, and improve, its juice quality over a long period with little loss in weight, as evidenced by its performance in Zambia, where it is grown at an elevation of about 1000 m. In November, having flowered in April, it can produce crushed juice Brix levels of 22–23° and purities in excess of 92 per cent. Minimum temperatures, of the order of 11–14 °C are generally low for much of the crop but, from September onwards, rise to about 18 °C. In the Caribbean, by contrast, at low altitudes and minimum temperatures of the order of 20–25 °C, the juice quality of flowered cane tends to deteriorate, in many varieties, after about six weeks.

Where conditions are favourable for the maintenance of juice quality and low night temperatures inhibit growth, the overall loss of sugar yield by heavy flowering can be low if, for instance, a shy-flowering variety is available for reaping during the latter part of the crop period. Where, however, a variety susceptible to flowering is reaped towards the end of crop and then flowers almost 100 per cent in the following year, at less than five months of age, cane yields can be reduced by 40 per cent or so, in a heavy-flowering year for the early flowered cane, compared· with the equivalent yields when flowering is slight.

In Hawaii, with a two-year crop cycle, yield loss of the order of 20 per cent can result from heavy flowering in the first year whereas, if the flowering occurs in the second year of growth sugar yield loss may be nil, or be increased as a result of flowering.

The variety Co 421, grown at a latitude of 4° S., in Tanzania, and at an altitude of 700 m does not flower whereas, in Zimbabwe, at a latitude of 16° S. and an altitude of 400 m it often flowered to almost 100 per cent. In Nigeria, at 9° N., Po 980 plant cane flowered completely, when many other varieties, regarded as being relatively susceptible to flowering and planted at the same time, did not flower. In contrast the variety B 37172 has not been known to flower, even in conditions where the flowering stimulus has been very strong.

Because of the need for cane breeders to achieve precise control of cane flowering, to be able to make specific crosses using shy flower-ing varieties as parents, by far the greatest proportion of the work on cane flowering has been on its induction rather than on its inhi-bition. In commercial production of sugar-cane the availability of heavy flowering can be beneficial, in that advantage can be taken of the tendency for a heavily flowering cane to have better juice quality early in crop. Similarly, where complete control of reaping schedules can be assured, the inclusion of shy flowering varieties to reap towards the end of crop can help in achieving optimum sugar output.

In areas where cane flowering causes consistent loss in sugar yield, the ultimate solution would be to eliminate flowering through variety selection. With cane reaped annually it is usually possible to devise a planting programme which includes a proportion of shy-flowering cane, which can be harvested during the latter part of the crop, when losses from early flowering can be at their greatest. Under irrigated conditions, where there is dry weather over the flower initiation period, a marked reduction in flowering can be achieved by withholding irrigation over a specified period. In Hawaii, if the critical round of irrigation is completed by 4 August, and irrigation withheld until early in September, flowering can be reduced quite significantly. In such circumstances factors such as an extension of the irrigation cycle times, because of excessive drying out of the soil, have to be taken into account when assessing the overall value of the flowering control achieved.

Other means of flowering control have been considered in areas where heavy flowering occurs, such as cutting back young plant or ratoon shoots. For instance, in Nigeria, it has been found that when certain varieties, such as B 47419, reaped in February/March, are cut back in early May flowering is suppressed, giving the possibility of harvesting the following ratoon as an annual cane or at seventeen months. The loss of about one month of growth, during the intensely dry crop season, could be more than made up during the good growth conditions of the wet season with the suppression of flowering and the facility of growing a proportion of varieties which were otherwise suitable, except for their flowering characteristics. With some varieties the critical date for cutting back was as late as mid-June.

Partial to complete inhibition of flowering can be achieved through the use of chemicals and in some circumstances treatment is worth while. When, however, flowering would not have occurred, without the treatment, there can be an actual loss of yield because of the effects of treatment as well as an added cost of treatment.

Sugar-cane as an irrigated crop
R. A. Yates, Booker Agriculture International Ltd

No factual data on the proportion of the total cane area, world-wide, which is irrigated are available. It is probable that most cane is grown without irrigation, but vast areas of cane are heavily reliant on irrigation in, for example, the Indian subcontinent, Egypt and Peru; many other areas use 'supplementary' irrigation.

The word 'irrigation' may imply total reliance on applied water (for example in the Indus Valley, the Nile Valley, and Peru); it may mean the application of water during the dry seasons (as in parts of Australia, in numerous countries in Africa and in some Caribbean islands); or it may be used for little other than the establishment of the plant crop (an extreme example being the famous industry in Java, where it is common to see more water being used for the unique system of land preparation than is used for the rest of the growing period).

Similarly, 'irrigation' may apply to the totally uncontrolled flooding of fields or 'bays', often with no attempt being made to remove surplus water (a system which is regrettably common in Pakistan, Sri Lanka and elsewhere); it most commonly applies to furrow or overhead sprinkler irrigation; and it may apply to a special form of drip irrigation which has been expanded very rapidly in Hawaii to the extent that, in 1980, it had been installed in 28,000 ha (or 50 per cent of the irrigated cane land).

The degree of sophistication in the control of irrigation varies as much as the techniques used. The literature on irrigation of cane is extensive but, unfortunately, much of it is of purely local interest only, as it refers to the frequency of wetting, using strictly local techniques under strictly local conditions. The bulk of the work which provides transferable data (such as soil moisture tensions, evaporation rates and growth rates as well as yields) has been done in subtropical zones such as Hawaii, Southern Africa and Australia; it is possible that, in particular, the Southern African data on the relationship between evapotranspiration and Class A pan evaporation has been accepted too readily by countries which do not have marked seasonal temperature differences.

Physiological aspects

Growth season

The harvested portion of sugar-cane is the stem which grows continuously (weather conditions permitting) for a period of at least twelve months, which is the approximate age of harvest in most countries. This vegetative growth is directly proportional to the water transpired, so that it is desirable to have adequate soil moisture throughout the growing season. Numerous workers have reported on the linear relationship between cane yield and evapotranspiration: they include Isobe (1969), Mongelard & Mimura (1971), Humbert (1971), Scott (1971) and Thompson and Boyce (1971). The production rates reported were all 1 tonne ± 0.2 tonne cane/ha per cm of evapotranspiration. It is notable that the correlation with sugar production is much poorer. This relationship holds for plant and first ratoon crops, but reduces by an average of about 10 per cent in subsequent ratoon crops (Thompson 1976).

Cane ripening

The ultimate commercial product is sugar, and the yield of sugar per unit area depends on the yield of cane and on the sugar content of that cane. Sugar contents increase when growth rates (and, thus, the respiratory losses of sugars) are reduced without any equivalent reduction in photosynthesis and translocation. Cool weather is the most effective ripening agent but, failing this, a gradual increase of moisture tension is very effective. At least four weeks of such stimuli are required to induce significant ripening (Anon 1959).

The work of Janes (1966) illustrates the mechanism involved: irrespective of the cause of the increased osmotic potential in the rooting medium, plants respond by increasing their internal osmotic potentials. Under non-saline conditions, they increase the soluble carbohydrates in the cell sap: that is, ripening by 'drying off'; under saline conditions, they respond by absorbing salts which interfere with the recovery of sugar.

It is of interest to note that many investigators have failed to distinguish between the real and apparent effects of 'drying off' on ripening: the partial desiccation of the stems will result in apparent ripening.

'Drying off' is easily accomplished on deep soils of good moisture-holding capacity: irrigation is simply discontinued for some weeks (typically six weeks) prior to harvest. On shallow, or otherwise unsuitable, soils adequate 'drying off' to induce ripening introduces serious problems, as the leaf canopy may be excessively damaged or, in extreme cases, the whole plant may be killed.

'Drying off' is also necessary for the practical purpose of allowing the mobility of in-field harvesting equipment.

Influence of temperature

Temperature affects irrigation requirements in at least three ways:
1. It influences the evaporative capacity of the atmosphere; for example Thompson (1976) reports evapotranspiration rates varying from 85 mm in midwinter to 200 mm in midsummer, per month.

2. Root temperatures have a greater effect than air temperatures on transpiration and growth (Mongelard & Mimura 1971); the difference between critical temperatures for growth between irrigated and non-irrigated cane reported by Bacchi *et al.* (1977) can be most readily explained by the effect of irrigation on soil temperatures.

3. As growth ceases when mean day temperatures fall below 19–20 °C, no response can be expected to irrigation in cold weather (Yates 1967); this need not imply that all irrigation be stopped as other factors (such as the maintenance of a reserve of moisture in the soil to reduce later peak requirements for irrigation) must be considered.

Type of root system

In most countries, sugar-cane is grown as a perennial crop. The root system is, however, annual (or biennial, etc. depending on the frequency of harvest). This is because each new stem produces its own root system from its basal nodes.

However, prior to the establishment of the new root system, the new stem draws on the total exploitive capacity of the old root system. At this stage, rate of growth of the new stem is limited because of the limited vascular connections between the germinated bud and the old root system; indeed an experimental technique for producing dwarf cane makes use of this feature. This transition phase allows the cane to show a remarkable degree of drought resistance; the new root system cannot be formed until the soil around the base of the new shoot is moistened and, consequently, the growth of that shoot and of its leaf canopy is curtailed. It is dangerous to allow any superficial wetting which can encourage superficial root growth and an increase in canopy which cannot be sustained by the total available soil moisture.

Determination of water requirements of sugar-cane

Direct measurement of soil moisture potentials with resistance blocks and, occasionally, ceramic tensiometers was used in Hawaii for many years (Humbert 1968); these techniques are still used for experimental work.

Various techniques have been developed for determining moisture status through measuring the moisture content of selected tissues; Clements (1948) selected certain leaf-sheaths; Baver (1960) selected immature internodes. Neither has found commercial application in the control of irrigation, but they are of value for diagnostic purposes: abnormal tissue moisture levels may indicate limiting factors such as poor drainage and inadequate aeration in the root zone; they are also influenced by levels of mineral nutrition (for example, normal tissue moisture levels are not attained if nitrogen is deficient). Evans (1965) and Singh (1976) used the exposed, rolled spindle as this proved to be a reliable indicator of moisture status which was much less influenced by nutritional factors. Again, this has not been used commercially, but has obvious attractions for experimental work both to check the efficiency of irrigation (maximum growth at 81 per cent moisture) and to control 'drying off' (as growth stops at 68–69 per cent moisture).

Many workers have installed cane lysimeters since the early 1950s, but much of the detailed knowledge of water use by cane has been derived from the large lysimeters operated in South Africa by G. D. Thompson and his colleagues, and in Hawaii by R. B. Campbell and his colleagues. These workers found a close correlation between the evapotranspiration of cane and the evaporation from the standard US Class A pan; naturally, any satisfactory estimation of open-water evaporation of such an instrument (e.g. through using the Penman formula) can also be used. Appropriate factors for the Penman formula are: roughness 12.6 cm, and surface resistance 0.75 sec/cm (Thompson 1976).

The $E_t : E_0$ ratios obtained in Hawaii rapidly increased to a value of 0.9 as the crop canopy increased to approximately full cover at five months; the ratio then increased to about 1.1 when the cane was between ten and fourteen months of age: after this, the ratio fell to around 1.0 when the cane was fifteen to twenty months of age.

More recently, Thompson (1976) has summarized much greater volumes of data obtained in South Africa. His results are very similar to those obtained in Hawaii:

1. Evaporation from bare soil depends on the frequency of wetting (reducing to 1 mm/day when a dry soil mulch is obtained); a mulch of dry cane leaves significantly reduces the loss from moist soil.
2. During the period prior to the attainment of full canopy, the $E_T : E_0$ ratio increases linearly up to 1.0. The period of time required to reach 1.0 varies from 70 to 190 days, depending on climatic factors (especially temperature).
3. At full canopy, in an erect crop, the ratio may increase to 1.2, presumably due to varying roughness factors; lodging will reduce E_T by as much as 25 per cent for three months, after which it recovers to normal levels (presumably because of the production of a secondary canopy). On average, an E_T/E_0 ratio of 1.0 can be taken.

The actual correlation reported between E_T (cane) and E_0 (Thompson & Boyce, 1971) was excellent:

$$E_T = 0.29 + 0.93E_0$$
with $r = 0.92$ for $n = 208$

Under certain circumstances, the E_T/E_0 ratio may reach much higher levels. Yates (1967) reported ratios reaching 2.0 for short periods: this was attributed to advective energy. More recently, Mongelard (1973) has presented indirect evidence of similar ratios from a trial where drip irrigation was used to keep soil moisture tensions below -0.2 bar; some of the results are illustrated in Fig. 3.10, though it should be noted that the small plots (of 20 ft

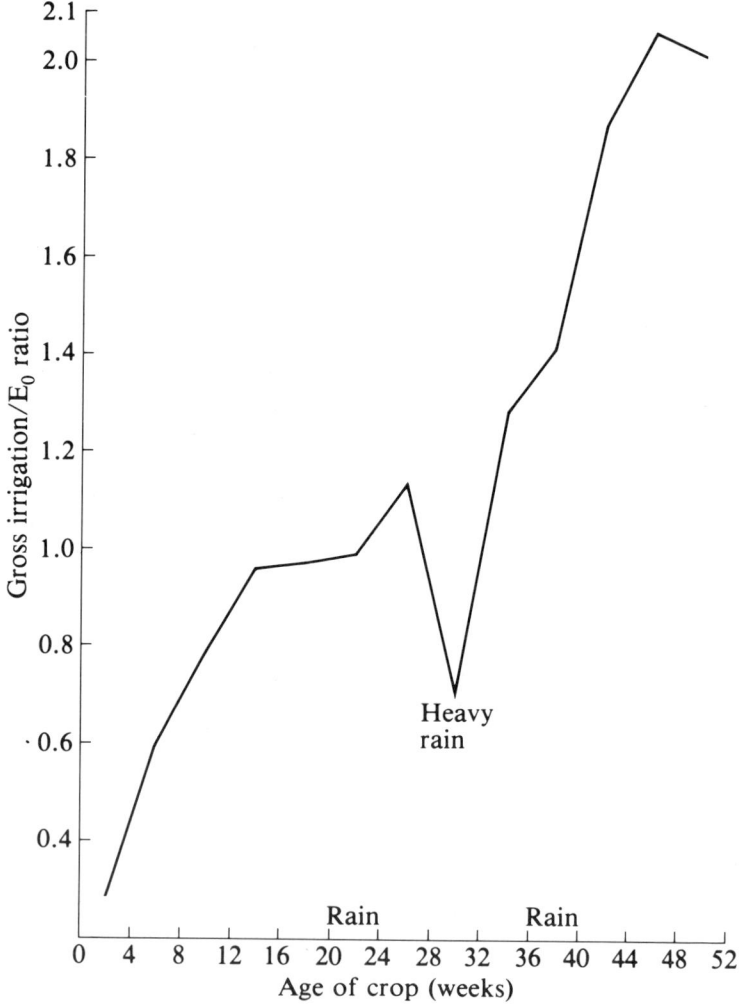

Fig. 3.10 Irrigation/pan evaporation ratio required to maintain high water potential in the root zone at different crop ages (From Mongelard 1973)

× 30 ft) separated by 10 ft pathways are likely to have influenced these data and contributed to the 170 tons cane/acre per twelve months recorded in this trial.

Total annual crop water use varies from around 1,250 mm in South Natal, through around 1,500 mm in Pongola (South Africa), Mauritius and Ayr (Australia) to over 2,000 mm in Hawaii; the last-mentioned figure is not directly comparable with the others as, because of a two-year interval between harvests, it would be less affected by the period of incomplete canopy.

Daily evapotranspiration values of 8–10 mm are frequently reported; occasional records of 10–12 mm, and one record of 15.7 mm/day (from Ayr, Australia, as an average over three to four weeks) have also been reported. (These data are summarized in the paper of Thompson 1976.)

Soil factors

Soil moisture and evapotranspiration

The evapotranspiration rates given above depend on water being freely available to the root system at all times.

Thompson (1976) defines total available moisture (TAM) as that released between 0.1 and 15 bars; freely available moisture (FAM) is released between 0.1 and 1.0 bar. Over a wide range of soil types in South Africa pressure-plate techniques indicated that FAM varied from 37 to 69 per cent (average 55 per cent) of TAM; actual values ranged from 41 to 72 mm FAM, or 89 to 127 mm TAM/m depth of soil: in all measures, the lowest values were on heavy clays. The E_t measurements indicated that the pressure-plate technique was reasonably accurate: that is, FAM averaged 60 per cent (not 55 per cent) of TAM. Measurements in Hawaii indicated that FAM was 72 per cent of TAM. On the heavy montmorillonitic clays of Guyana, Evans (1965) obtained large responses to irrigation if 66 per cent of TAM was exhausted between irrigations.

The highest rate of capillarity conductivity recorded in South Africa was 0.4 mm/day in moist soil (much reduced in dry soil). Capillary movement of water is, thus, unlikely to do anything more than prevent total desiccation of a crop.

Pattern of rooting

The potential rate of elongation of cane roots, in following a receding water table, is probably of much greater importance than the capillary movement of water. In the South African Root Laboratory, the average rate of root elongation in light soils was 40 mm/day (the maximum recorded rate being 75 mm/day); values of about 70 per cent of this were recorded in heavy soils (see

Thompson 1976). However, water tables can recede so rapidly that even sugar-cane can be killed. An example seen in Java illustrates this point: a field of cane on highly permeable soil was surrounded by ricefields; the high water table induced by the ricefields strictly limited the depth of rooting, and the cane could not adapt to the sudden drop of the water table at the time of the rice harvest. Perched water tables, where they prevent deep rooting, have the same effect. It is now accepted that the condition 'yellow wilt' is caused by such phenomena.

Where suitable soil conditions exist (permeable, friable soils with reasonable moisture contents, free of any serious chemical limitations), cane roots can penetrate to a depth of several metres. The degree of proliferation in various horizons can be controlled by a number of factors. Thus, for example, Baran *et al.* (1974) showed that cane given frequent (weekly) irrigation had about 70 per cent of its roots in the top 25 mm, while infrequently (three-weekly) irrigated cane had only 50 per cent of its roots to this depth (see Fig. 3.11). Similarly, improvement of the chemical status of the soil can result in a marked increase in root proliferation; as shown by Evans (1947, 1948), for example, when he added basaltic dust to leached volcanic soils in Mauritius. Again, Evans (1965) illustrated

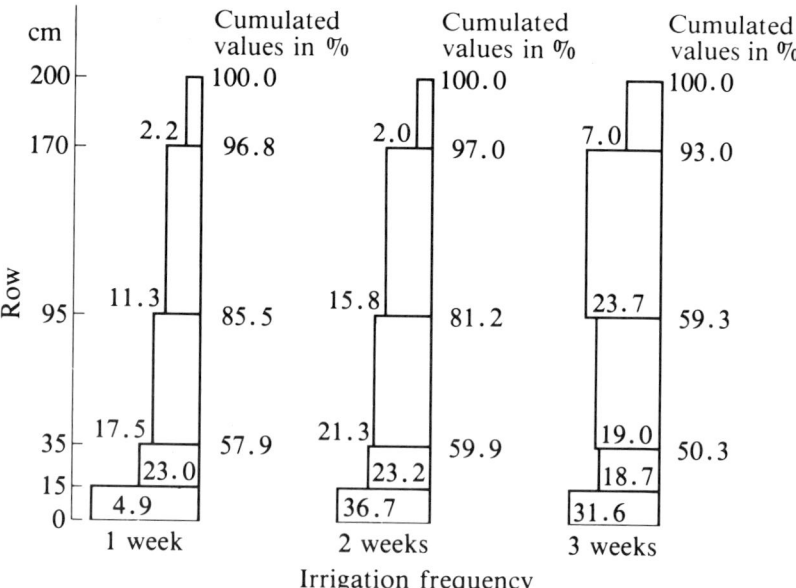

Fig. 3.11 Effect of irrigation frequency on root distribution of sugar cane (From Baran *et al.* 1974)

the influence of improved physical structure, caused by special soil-preparation techniques on the degree of proliferation of roots in heavy montmorillonitic soils; and Ricaud (1977) greatly improved the depth of rooting by penetrating a compact intermediate horizon with a 'vertical mulch'-type subsoil in the line of the cane rows.

Thompson (1976) concluded that the effective available moisture of any soil may be calculated by multiplying FAM by the total rooting depth, or by multiplying TAM by 50 per cent of the total rooting depth. This is a useful rule of thumb, but it cannot be universally applicable, as must be assumed from the complications noted above in this section, and as FAM as a percentage of TAM can vary considerably from one soil type to another.

Practical systems of irrigation in cane

Control of irrigation

It was noted above that the Class A evaporation pan is now accepted in most areas as the most convenient method of scheduling irrigation. A simple arithmetic budgeting system is used, with irrigation cycles adapted to match the average soil moisture parameters of the defined area; the inherent inflexibility of most irrigation systems implies that a significant degree of approximation is unavoidable.

In recent years, evidence has been accumulating (especially in South Africa) that maximum efficiency of water utilization is achieved when irrigation is scheduled to satisfy only $0.8 \times E_0$. A paper by Gosnell & Lonsdale (1977) illustrates this clearly; Fig. 3.12 summarizes five years of experiments.

In the above experiments, a four-month 'drying off' period was used for all treatments. Recommendations derived from these trials with respect to 'drying off' were that irrigation should be reduced to only $0.6 \times E_0$ for the last two months prior to harvest during the hotter months; during the cold winter months, irrigation could be reduced to $0.5 \times E_0$ for four months.

Irrigation systems

Virtually all known systems are used in sugar-cane: bay, furrow, portable overhead, solid-set overhead, centre pivot, rain-guns and drip. The criteria of slope, infiltration rates, drainage requirements, etc. are the same as for other crops. Special mention need be made only of drip irrigation where, because of the row-crop nature of sugar-cane, perforated tubes are used rather than discreet emitters. So far, it has not proved possible to prevent the destruction of the tubes at each harvest, so that their use has been adopted commercially only in Hawaii where cane is cut every two years; experiments with 'pineapple row' spacing are being conducted in an attempt to

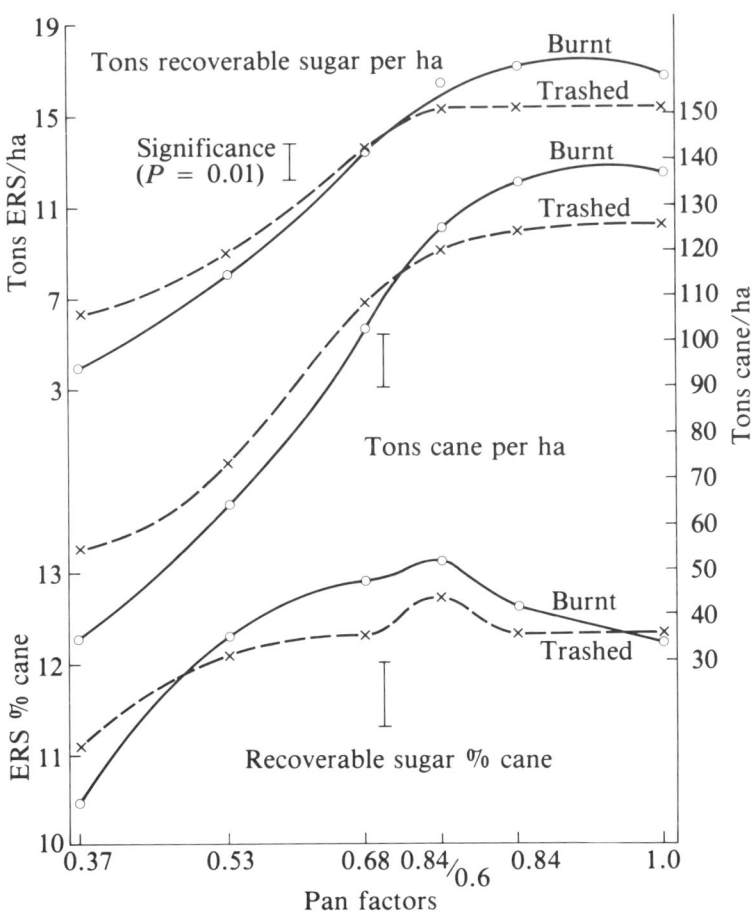

ERS = Estimated Recoverable Sugar

Fig. 3.12 Response of sugar-cane to irrigation (From Gosnell & Lonsdale 1977)

reduce the total length of tube per unit area. Gibson (1978) recently described the success of drip irrigation in Hawaii.

Economics of irrigation of sugar-cane

Because of the wide range of conditions, responses and techniques, it is impossible to generalize about the economics of irrigating sugar-cane. Each area has its own peculiarities which must be carefully evaluated. Factors that must be considered include:

• The source of water: surface, underground, and the pumping head required.

- The quantity of water: whether existing river flows (or storage structures) are adequate, or whether additional storage structures are required.
- The system of irrigation: its overall efficiency, and whether it can be operated night and day, or during daylight hours only.
- The quantity of water, associated with the nature of the soil, the natural rainfall and the system of irrigation, thus affecting leaching requirements.
- The length of the irrigation season (and the reliability of the rainy season).
- The response to be expected in terms of cane yield to irrigation:
- The value of each unit of cane produced, which is effectively the value of the equivalent amount of sugar less the costs of harvest, transport and processing.
- The cost of applying the irrigation water, which depend on the source of water (which may involve the construction of major dams), the cost of transmission, the cost of power (which may be subsidized, or local cheap hydroelectric power, or expensive diesel-generated power), the cost of labour and the system adopted.

Most of the items listed above are common to all irrigation enterprises. Two, however, justify some further explanation.

First, the planner must not assume that all irrigation will give of the order of 1 tonne cane/ha per cm of effective irrigation. To take the most obvious example, irrigation may not be justified on deep soils of good moisture-holding capacity where the dry season is sufficiently short. Thus, for example, a detailed examination of rainfall and soil physical conditions in Guyana indicated that irrigation would be unnecessary if an active rooting depth of 1.5 m could be encouraged (as there would be only two weeks of severe moisture stress, on average); a rooting depth of 0.5 m would result in an average of twelve weeks of severe moisture stress. Similarly, Yates (1967) reported responses of only 0.15 ton/acre per in on heavy black clays with saline subsoils compared to about 1 ton/acre per in on deep red volcanic soils.

Secondly, the reliability of the rainy season is of considerable importance in semi-arid areas; but it may not be of importance in wetter areas. Thus, for example, in parts of Tanzania where annual rainfall averages around 1,200 mm, a reduction of 30 per cent (to 800 mm may cause complete crop loss. If, in contrast, the annual rainfall exceeds 2,000 mm and is well distributed (as, for example, on the west coast of Madagascar), a 30 per cent reduction in rainfall is unlikely to induce any significant moisture stress.

Chapter 4

Varieties of sugar-cane

Classification and cytology

All varieties of sugar-cane are species or hybrids of the genus *Saccharum* (Gramineae: Andropogoneae). Stevenson's (1965) key to the classification of the species is based on the work of Jeswiet (1925) but also includes *S. robustum* and *S. edule*, discovered later and assigned in accordance with the principles laid down by Artschwager & Brandes (1958). Nevertheless, classification is still the subject of much discussion and debate (Grassl 1969). Stevenson's key is as follows:

A. Main axis of inflorescence and cluster axes with long hairs. Glumes always four. Lodicules either ciliate or not. If the spikelets of the same pair do not flower simultaneously, the pedicellate one always blooms first. Culms green, greyish green, greenish bronze, ivory or white.
    1. Lodicules ciliate. Long subterranean runners present. Growing wild.
                                              S. spontaneum L.
    2. Lodicules not ciliate. Subterranean runners short. Sugar-producing cultivated plants.
        a. Leaves broad (up to 50 mm). Tall cane species. Nodes all bobbin-shaped, greenish bronze (among others, Uba cane).
                    S. sinense Roxb. emend. Jeswiet
        b. Leaves narrow. Short inconspicuous cane species. Nodes usually cylindrical, greyish green, white or ivory. Rather limited to India and Pakistan (among others Chunnee cane).
                                S. barberi Jeswiet
B. Main axis of the inflorescence never having long hairs, often glabrate; rachis nodes glabrate or with very few hairs. Glumes generally three, sometimes four. Lodicules not ciliate. If the spikelets of the same pair bloom at different times, the sessile one is always the first. Culms pale or dark green to dark yellow, dark red or violet.
    1. Wild plants up to 10 m tall. Culms with relatively high fibre and low sucrose content: generally river-bank perennials.
             S. robustum Brandes *et* Jeswiet *ex* Grassl

2. Cultivated plants
   a. Inflorescences aborted, leaves more or less pubescent.

   S. edule Hassk.

   b. Inflorescences normal. Culms with relatively low fibre and high sucrose content.

   S. officinarum L.

Saccharum officinarum L.

As described in Chapter 1, clones of *S. officinarum*, the 'noble' canes were for centuries the source of most of the cane sugar of international trade. Now they have been replaced in commercial production by hybrid varieties and are grown only on a garden scale in some of the islands of the Pacific (and elsewhere) to be chewed or to provide material for the manufacture of illicit rum.

The geographical origin of *S. officinarum*, New Guinea, and the routes by which it spread throughout the tropics and subtropics (see Fig. 1.1) are generally agreed. However, there is still some doubt and speculation concerning its botanical lineage, though it is accepted that *S. robustum* and *Erianthus maximus* Brogn. are involved in its ancestry. Several expeditions to New Guinea have been made to collect clones of *S. officinarum* for use in cane breeding and Badila (NG 15), found there by Tryon in 1895, was still being grown in parts of Queensland in the early 1960s. Accounts of three of these expeditions were given by Brandes & Sartoris (1936), Buzacott & Hughes (1951) and Warner & Grassl (1960) (Fig. 4.1). Before the end of the nineteenth century, when organized cane breeding started, only noble canes from the Pacific islands, and the Creole, were available to the planter (except in Asia, where *S. sinense* and *S. barberi* were grown). The most important of these were the Creole, the Otaheite or Bourbon, the Cheribon or Transparent, and the Tanna or Caledonia.

The Otaheite or Bourbon

The Otaheite or Bourbon was the first variety to be grown on a world-wide scale. It was the standard variety in Mauritius from the inception of the industry there, *ca.* 1793, until the onset of gumming disease *ca.* 1840. In 1789 it was transported to Martinique and Guadeloupe in the French West Indies, where it was called Bourbon (after Réunion). In 1793 it was introduced from Tahiti into St Vincent and Jamaica by Captain William Bligh. In the same cargo were cuttings of the breadfruit *Artocarpus altilis* (Park.) Fosberg (syn *A. communis* Forst.), which were to be grown and propagated to provide a staple food for the slaves on the plantations. In a description of his more notorious visits to Tahiti on HMS *Bounty*

Fig. 4.1 *Saccharum officinarum* being examined by Buzacott (Buzacott & Hughes, 1951) during an expedition to the highlands of New Guinea (From Stevenson 1965)

in 1788–9, when sugar-cane was not one of the plants to be collected, Captain Bligh wrote of being given stem cuttings which were 6 in. (15 cm) in circumference. He added that the local chief, Tinah, was astonished to find that this was the raw material used in the manufacture of the highly prized loaf sugar, and was 'desirous to discover the means' by which it was done.

The Otaheite or Bourbon became the most widely grown cane in the Caribbean area, Mexico, Brazil and Hawaii, where it was named Lahaina (its port of entry on the island of Maui). It was also widely planted in India and Java. In most countries, as in Mauritius, it was replaced because of susceptibility to disease.

Except in Java, the Otaheite has been little used in cane-breeding programmes, but it is a remote ancestor of POJ 2878.

The Cheribon or Transparent

The Cheribon or Transparent cane was described by Stevenson as probably the most important in the world. It replaced the Otaheite when the latter was overcome by disease, and is a progenitor of most of the varieties which are now cultivated. It has given rise, by genetic mutation, to several colour forms with synonyms as follows:

Light Cheribon:	White Transparent, Cristalina, Light Preanger, Caledonian Queen, Rose Bamboo, Diard Rose, Rappoe, Naga, Namuri, Hope, Java, Burke and Meligeli.
Striped Cheribon:	Red Ribbon, Louisiana Ribbon, Rayada, Striped Preanger, Diard Rayee, Bamboo Rayee, Guinghan, Striped Singapore and Seete.
Black Cheribon:	Louisiana Purple, Morada, Black Java, Beloguet, Diard, Purple Transparent, Purple Mauritius, Queensland Creole and Meera.

The Tanna or Caledonia

One of the first references to the magnificent canes growing on the island of Tanna in the New Hebrides was made in 1774 in an account of Captain Cook's voyages of discovery.

The Tanna cane, like the Cheribon series, exists in a colour cycle of striped, light and dark forms. The striped form was introduced into Mauritius in 1869 and produced White Tanna as a mutant in 1892. Later a black form arose, also a mutant, and was followed by a succession of interrelated Tannas. White Tanna was a great success. It is hardy, resistant to gumming disease and adapts well to cooler and wetter climates. Consequently, by 1925, it occupied 63 per cent of the land planted with cane in Mauritius. The Tanna was widely grown also in Hawaii and Fiji. Under the name Yellow Caledonia it was taken to Hawaii in 1881 and was the standard variety in unirrigated areas until the 1920s. It was imported into Fiji under yet another name, Malabar, and was cultivated there with equal success.

The Tanna has not been used much in cane breeding because it is almost sterile.

Seedling varieties of S. officinarum

For many years sugar-cane was thought to be sterile. Seedlings were first recognized by Iran Aeus Harper, a man of African ancestry, at Highlands Estate in Barbados in 1858. Harper drew these to the attention of the owner, James Parris, who reported the observation in a letter published in the *Barbados Liberal* of 12 February 1859 (Parris 1954). This knowledge was not used for the production of better varieties until 1888, when cane breeding was started simutaneously in Barbados and Java. The programme carried out in Barbados led ultimately to the selection of BH 10(12), Ba 11569 and B 2935, all of which were of commercial importance. The policy was based on line-breeding (and consequently on some degree of inbreeding and homozygosity), members of each line showing common characteristics. Therefore, it is not surprising that B 2935, typical of the last group of noble seedlings to be widely planted, was bred from a cross between varieties from different genealogical lines.

Eventually it became clear that new varieties of wide adaptability, and especially of improved resistance to disease, would not be developed within the genetic range of *S. officinarum*. Nevertheless, many noble canes of high quality had been bred and were excellent parent material for the production of hybrid varieties by crossing with other forms of *Saccharum* and then back-crossing the progeny with noble canes. The first and most important work on the production of such new varieties, called nobilized seedlings, was undertaken in Java. The legendary 'wonder cane', POJ 2878, was bred in the early 1920s and survives in some countries as a standard commercial variety. Similar work was undertaken at other breeding stations and the era of the noble cane had ended.

Characteristics of S. officinarum

In general, noble canes are rich in sugar and have thick stems with soft rinds. They grow well only under favourable conditions.

Saccharum spontaneum L.

Saccharum spontaneum is an extremely variable species widely distributed in Africa, through India to China and the islands of the Pacific; and even in Turkmenistan, USSR. A large collection, kept at the Sugarcane Breeding Institute, Coimbatore, India, includes 400 clones of which 300 are from India. They are numbered according to an SES system (*Spontaneum* Expedition Scheme) and the most promising clones are sent to overseas breeding stations for inclusion in nobilization programmes. Panje (1933) recognized two subspecies: *indicum* and *aegypticum*.

Characteristics

Saccharum spontaneum grows vigorously to produce thin, fibrous stems which have juice of low sugar content. It tillers profusely (Fig. 4.2)

Fig. 4.2 Various forms of *S. spontaneum*. (a) Bushy habit of clone collected in northern India; (b) medium-erect type from peninsular India; (c) tall, erect type, with thicker stems, from eastern India, particularly Assam. (From Stevenson 1965)

Fig. 4.2(b)

Saccharum barberi Jeswiet

From the earliest times, the canes grown in northern India and what is now Bangladesh were indigenous subtropical forms. Barber (1918) divided them into five groups, based on vegetative characteristics: Mungo, Nargori, Saretha, Sunnabile and Pansahi. Clones of Pansahi

Fig. 4.2(c)

were also found in Indochina, China and Taiwan, and Jeswiet (1925) included them in *S. sinense*, leaving the four other groups to form *S. barberi*. One argument that has been put forward is that all the groups of *S. barberi* should be transferred to *S. sinense*; another is that *S. barberi* should be restricted to the Saretha group. However, with the reservation that all the groups are of hybrid origin, Stevenson favoured the retention of Jeswiet's classification for practical purposes because of long usage.

Whatever their origin and correct classification, and these will no doubt be discussed for many years, to cane breeders clones of the

(a)

Fig. 4.3 Clones of *S. barberi*. (a) Chunnee; (b) Kansar (From Stevenson 1965)

Saretha group are by far the most important and have been used by them in Barbados and Java as well as in India. One of the parents of Co 213, cultivated for many years in several countries, is the Saretha variety Kansar; and the other, POJ 213, was bred by crossing a noble cane with another Saretha variety.

(b)

Characteristics

Saccharum barberi, suited to cultivation in the temperate and subtropical zones, tillers profusely to form stems intermediate in thickness between those of *S. officinarum* and *S. spontaneum*. In general the sucrose content of its juice is lower, and the fibre content of its stems much higher, than those of *S. officinarum* (Fig. 4.3).

Saccharum sinense Roxb. emend. Jeswiet

Roxburgh (1819), collecting material in 1796 for the *Flora of India*, received canes from an officer of the East India Company stationed in Canton, noticed peculiarities in one of them and created from it a new species, *S. sinense*. This new variety was propagated and grown in India under the name Chinea. Later it was found that the same variety had already been grown for more than a century in Taiwan, where it was known as Tekcha. Barber included Chinea in his Pansahi group of *S. barberi* which, as already discussed, is better referred to as *S. sinense*.

Uba, the best known clone of *S. sinense*, was for many years grown on a world-wide scale. It was planted extensively in Jamaica, Louisiana and Puerto Rico when, *ca.* 1920, mosaic disease caused severe damage to the noble canes previously grown in those territories. Nevertheless, the poor quality of its juice and undesirable milling qualities caused it to be replaced within a few years by nobilized seedlings from Java. In Trinidad, Uba was grown in the 1930s and 1940s in infertile areas where cane was also subject to heavy froghopper infestation. However, Uba was established on the largest scale in Natal, South Africa, even before the appearance there of mosaic disease.

Characteristics

Saccharum sinense is vigorous and widely adaptable, but has thin stems and narrow leaves. The stems have a high fibre content and yield poor-quality juice. *Saccharum sinense* is immune to gummosis, mosaic and sereh but susceptible to red rot, rust and streak diseases. Resistance to frost, though slight in all *Saccharum* spp., is most noticeable in the Dhaulu group of *S. sinense* (Ch. 2).

Saccharum robustum Brandes *et* Jeswiet *ex* Grassl

Brandes and Jeswiet discovered a new species of *Saccharum* on a visit to New Guinea in 1928. Ten clones were collected and were named *S. robustum* by Brandes (1929). Nineteen years later, Grassl (1946) gave a botanical description of the species and divided it into *S. robustum* proper, typified by 28 NG 251; and *S. robustum* f. *sanguineum*, typified by 28 NG 219. The latter has blood-red tissue

Fig. 4.4 *Saccharum robustum* growing on the banks of the Laloki River in New Guinea. This is the source of the type 28 NG 251 (From Stevenson 1965)

in its stems, especially near the rind, and deep-red buds, dewlaps and growth rings. Both forms are indigenous to New Guinea and the neighbouring islands of New Britain and the New Hebrides (Fig. 4.4). *Saccharum robustum* has been used without success in cane-breeding programmes in Barbados and Florida, and with limited success in Australia. It is, however, in the ancestry of the Hawaiian 37–1933 which was for many years widely planted in irrigated areas.

Characteristics

The growth of *S. robustum* is vigorous and luxuriant. Many tillers are formed and may grow to a height of 10 m. The stems are of medium girth, but have a high fibre content and low-quality juice. *Saccharum robustum* is susceptible to Fiji disease, downy mildew and root rot. Its reaction to gummosis is not consistent: some clones are susceptible and others are resistant.

Saccharum edule Hassk.

For some time *S. edule* was thought to have arisen as a mutation of *S. robustum*, to which it is very similar. Its vegetative characteristics were described by Grassl (1946). When Warner & Grassl (1960) went to New Guinea in 1957 they found that *S. edule* was in general cultivation as a vegetable crop, the edible part of the plant being its swollen and aborted inflorescences. Following later studies, Grassl (1967) stated that *S. edule* is a hybrid of *S. robustum* and *Miscanthus floridulus* (Labill.) Warburg *ex* K. Schum. and Lauterb., and recommended its removal from the genus *Saccharum*. Grassl (1969) followed this by giving an interpretation, in detail, of the origin and names of all species of *Saccharum*. Purseglove (1972) took note of Grassl's findings but nevertheless considered that *S. edule* is probably a sterile form of *S. robustum*. He also pointed out that the genus *Erianthus* (excluded by Grassl) is now sometimes absorbed in the genus *Saccharum*.

Characteristics

The charactertistics of *S. edule* are similar to those of *S. robustum*, except that *S. edule* rarely produces normal inflorescences. Consequently, it is not used in cane breeding.

Sugar-cane cytology

The definitive work on the genetics and breeding of sugar-cane was written by Stevenson (1965). His work has been used freely in the construction of this chapter.

The chromosomes of sugar-cane are numerous and small, and therefore give much discouragement to geneticists who try to study them. Somatic chromosomes are best examined in preparations made from root-tips or very young leaves. Although the methods used must be modified to suit local circumstances. Stevenson found that the leaf-squash technique developed by Price (1956) and his own root section squash technique (Stevenson 1957) gave the best results. He described both methods in detail.

Chromosomes in *Saccharum* species
Saccharum officinarum

Bremer (1929) stated that most varieties of *S. officinarum* have 40 chromosomes in the haploid phase and 80 in the somatic cells of the root-tips ($2n = 80$). Since the basic chromosome number of the Andropogoneae is often 10, noble canes are presumably octoploid. Price (1960) confirmed this by reporting that of 144 noble clones, 129 had 80 chromosomes and the rest were probably of hybrid origin. In strictly botanical terms, therefore, the name *S. officinarum* is now confined to forms with 80 chromosomes, though for many years it has been synonymous with all 'noble' canes and still is used to describe them.

Saccharum spontaneum

Janaki Ammal (1936) examined six forms of *S. spontaneum* and

Table 4.1 *The chromosome numbers of forms of S. spontaneum*

Type	Chromosome number (2n)	Type	Chromosome number (2n)
Lahore	48	Relagaddi	64
Dehra Dun	56	Bihar	64
Coimbatore	64	Dacca	80

Source: From *Janaki Annual* 1926.

concluded that this species is a polypoid with a basic number of eight chromosomes (Table 4.1). Conflicting reports were largely resolved by the authoritative paper of Panje & Babu (1960). From clones grown in Africa, the Mediterranean, India, South-East Asia and the islands of the Pacific Ocean, they found well-marked modal frequencies for 2n numbers of 54, 60, 64, 80, 104, 112, 120 and 128 chromosomes. Multiples of eight were more frequent than multiples of six and ten. Aneuploid numbers (not multiples of six, eight or ten) occurred more frequently in the central area, from which it was inferred that, in regard to evolution, India was the most significant region.

Saccharum barberi and S. sinense

A cytological study of the four groups of *S. barberi* (the fifth, Pansahi, having been transferred to *S. sinense*) was made by Bremer (1932). The results (Table 4.2) confirmed the earlier classification (by morphological features) of Barber who regarded Rakhara, Rakri and Dhaulu, the 82 chromosome members of the Sunnabile group, as atypical.

In the Pansahi group of *S. sinense* Bremer found 118 chromosomes in Uba and Kavangire.

Saccharum robustum

There was much confusion concerning the cytology of *S. robustum* until Price (1957) distinguished two types, based on geographical distribution: those with 2n = 60 chromosomes from Borneo, Celebes and Australian New Guinea; and those with 2n = 80 from New Britain, the New Hebrides and both Australian and Dutch New Guinea. Bremer (1961) suggested a further simplification by transferring the 60-chromosome types from *S. robustum* to *S. spontaneum*. Examination of more collections from Borneo, Celebes and New Guinea will be needed to determine whether their continued inclusion in *S. robustum* is justified.

Interspecific and intergeneric crosses

Stevenson also described the chromosome patterns of interspecific

Table 4.2　*The chromosome numbers of forms of S. barberi*

Mungo group		Sunnabile group	
Type	Chromosome number (2n)	Type	Chromosome number (2n)
Mango	82	Naanal	116
Hemja	82	Khadya	116
Burli	82	Bansa	116
Katara	82	Sunnabile	116
Sarauti	82	Rakhra	82
Buxaria	82	Rakri	82
Matna	82	Dhaulu	82
Rheora	82		
Reora	82		

Saretha group		Nargori group	
Type	Chromosome number (2n)	Type	Chromosome number (2n)
Chunnee	about 91	Katai	124
Katha	about 90	Sararo	124
Kansar	92	Hathooni	124
Saretha	92	Baruk	124
Burra Chunnee	90 or 91	Manga	107
Nagori Behar	89–91	Kewali	107
Maneria	91		

Source: After Bremer (1932).

and intergeneric crosses. The intergeneric hybrids are of *Saccharum* with *Narenga*, *Sorghum*, *Sclerostachya*, *Miscanthus*, *Miscanthidium*, *Imperata* and *Zea*.

Improvement by selection and breeding

The Bourbon variety was devastated by gumming disease in Mauritius *ca*. 1840 and in Brazil in 1869; and by a complex of diseases in Barbados in 1890. Stevenson suggested that the Barbados epidemic was caused inadvertently by infected cuttings sent from Mauritius to Jamaica in 1882, of which twelve were received in Barbados in 1884. Because both the Bourbon and the Creole (which it replaced in the West Indies) are male-sterile, it was wrongly assumed that all varieties were sterile. Therefore a replacement for the Bourbon was sought not by breeding but by the discovery of naturally occurring resistant varieties. Expeditions were made from Mauritius to several islands in the Pacific and a comprehensive collection of noble canes was established there in the Botanic

Gardens. Eventually the Bourbon was succeeded in most countries by noble canes of the Cheribon series.

Although Parris's (or Harper's) discovery (1859) of the fertility of sugar-cane was soon confirmed in several other countries, in general the seedlings produced were regarded as curiosities rather than as potential new varieties. An exception was Parris, who multiplied seven seedlings, but later abandoned their cultivation because they showed undesirable characteristics. Thirty years later, however, the significance of seed-fertility and the possibility of its deliberate use to breed better varieties was recognized by Soltwedel in Java (Kobus 1893), and independently, though almost simultaneously by Harrison & Bovell (1888) in Barbados (following an observation made by Pilgrim).

Organized breeding to produce varieties with higher yields of sugar was soon established in both islands. By the turn of the century, throughout the world there were six independent breeding stations and at least fifteen by 1930. They met with immediate success. In Java, the Otaheite and Cheribon gave BK seedlings and SW 111; in Queensland Badila produced Q 813 and HQ 409; in Hawaii Lahaina, probably crossed with Rose Bamboo, gave H 109; and in British Guiana (now Guyana) D 109, D 145 and D 625 were raised from forms of Cheribon. All were of considerable commercial importance. However, the largest programme was undertaken in Barbados during the period 1887–1928. There, propagation of seedlings from open-pollinated arrows favoured selfing and five genealogical lines were recognized. Line 4 produced Ba 11569, itself widely planted and a parent of B 2935, B 3013 and a number of later clones. Ba 11569 was badly affected by gumming disease in the 1930s and was replaced by resistant varieties. The variety BH 10(12), in line 5 of the Barbados programme, was also grown on a large scale in many countries.

The modern era of nobilized hybrid varieties

Breeding programmes based purely on noble canes gave rise to varieties with many good qualities; but they were not hardy, did not thrive in unfavourable ecological conditions and, in particular, were susceptible to many diseases. Since improved varieties could not be expected within the genetic variability of *S. officinarum*, other species of *Saccharum* were brought into the programmes. As indicated earlier this new approach, called nobilization, was first adopted in Java and, by 1925, nobilized varieties were in large-scale cultivation; indeed, by 1929, one of them, POJ 2878, occupied 90 per cent of the area planted with cane in Java. Similar changes took place elsewhere, sometimes (as in Louisiana) speeded by outbreaks of mosaic disease. In Hawaii, POJ nobilizations from Java were

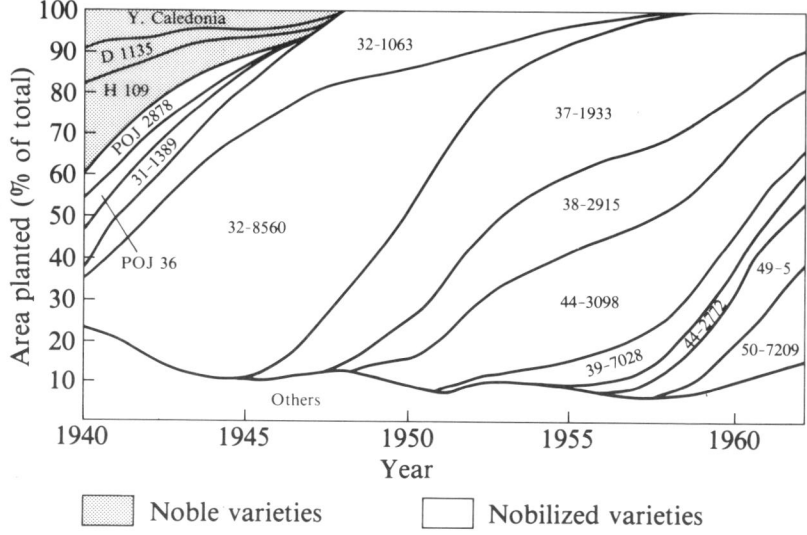

Fig. 4.5 The expansion of the area planted with H 32-8560 and other nobilized varieties in Hawaii

brought into cultivation *ca.* 1930, and these in turn were succeeded (*ca.* 1938) by locally produced nobilized varieties, especially 32-8560 (Fig 4.5). By 1941, 20,000 ha had been planted with POJ 2878 in British Guiana, but noble canes continued to be planted for several years more in Barbados and Jamaica. However, by 1948 the inevitable change had taken place and only nobilized hybrids were being grown. The expansion of B 37161 in Barbados was spectacular: from 1.5 per cent of the area under cane in 1942 to 63 per cent in 1945 and 92 per cent in 1949 (Fig. 4.6). Similar changes took place in Trinidad where not only noble canes but also Uba (*S. sinense*) were replaced by nobilized varieties.

The cane breeding stations
Java

Cane breeding started in Java immediately after Soltwedel's appreciation of the importance of the fertility of sugar-canes (Ch. 4). The main purpose was to find a variety resistant to sereh disease, which at that time was causing great damage; but although some seedlings gave higher yields of sugar, all were susceptible to the disease. Then Krüger found a wild cane, called Kassoer, and was impressed by its vigorous and luxuriant growth. Wilbrink crossed Kassoer with noble canes and found that the progeny were of no commercial value. However, Jeswiet back-crossed them with noble canes, a process which he aptly named 'nobilization', with great success. Seedlings

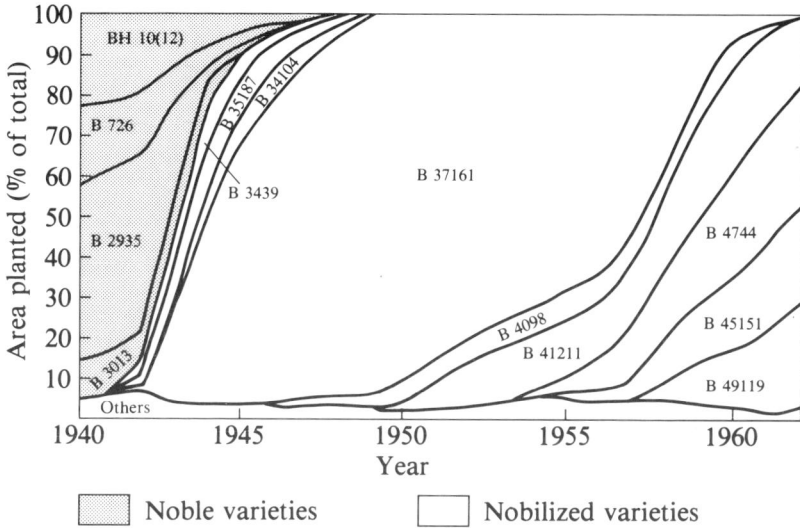

Fig. 4.6 The expansion of the area planted with B 37161 and other nobilized varieties in Barbados

which combined the hardiness and disease resistance of the wild cane with the more desirable characteristics of high-quality juice, low fibre content and good habit of the noble cane were selected for propagation. The concept of nobilization was a significant advance in cane breeding, and its practice was adopted at many other breeding stations. Jeswiet, at the Proefstation Oost Java, produced several excellent nobilized varieties, the most important being POJ 2878 (Fig. 4.7). Mangelsdorf (1960) described in detail the pedigree of POJ 2878 and stated, *inter alia*, that all the commercial varieties of Hawaii, as well as those of many other regions, were descended from it.

Cane breeding in Java was discontinued during the Japanese occupation of the island during the Second World war, but later was resumed under the auspices of the Republic of Indonesia at Balai Penjelidikan Perusahaan Gula, which succeeded the Proefstation Oost Java.

Barbados

In the beginning, cane breeding in Barbados was confined to the selection of seedlings grown from the fuzz of open-pollinated arrows. Bovell, who was in charge of this work, recognized the disadvantages of the system and, in 1902, began controlled breeding, using three different techniques:
1. Varieties which shed little or no pollen, used as female parents,

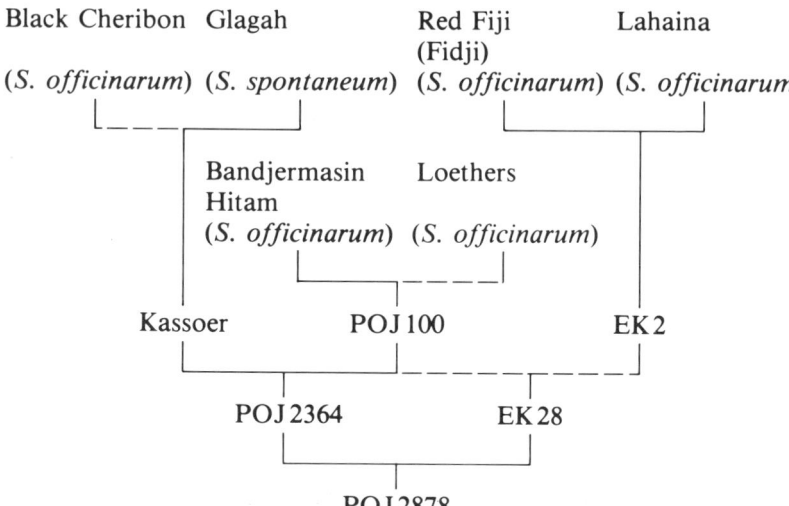

The broken lines indicate lack of complete authenticity

Fig. 4.7 The genealogy of POJ 2878

were planted alternately with male parents which shed pollen profusely. The progeny were given the prefix BNH (Barbados Natural Hybrid), but none was a commercial success.

2. Emasculated female parents were crossed with pollen from selected male parents. This was done with great care and delicacy under a dissection microscope on a platform 2.5 m high. The seedlings bore the symbol BH (Barbados Hybrid) and one of them, BH 10(12), was grown extensively in many parts of the world. Bovell ceased to use this technique in 1917.

3. The third approach was selfing. The significance of achieving homozygosity by selfing, and the commercial success of the hybrid progeny of inbred lines, was not yet known. The purpose of this work was to evaluate Barbados seedlings for specific characteristics.

Bovell abandoned controlled breeding in 1919; thereafter, until his retirement in 1925, he reverted to the original method of raising seedlings only from open-pollinated arrows.

In 1932 the breeding work which had been carried out since 1919 by the Barbados Department of Science and Agriculture was transferred to the newly created British West Indies Central Sugar Cane Breeding Station (WICSCBS), also located in Barbados. The new organization was required to provide varieties for all the British colonies in the Caribbean area. McIntosh, the geneticist-in-charge, reviewing past performance noted that only one important variety,

B 726, had been produced between 1913 and 1925. He concluded that the genetic variation of the material available in Barbados had been well explored and that a degree of homozygosity had been achieved by selfing in open-pollinated arrows. Five lines had been developed, each well suited to local ecological conditions, of good quality and resistant to gumming disease. He therefore introduced nobilized varieties from Java and India, crossed them with representatives of the best of the established Barbados lines and soon bred a highly successful series of Barbados nobilized varieties. The policy and methods were described by McIntosh (1944) and Inniss (1954).

The first important nobilized Barbados variety planted on a large scale, B 34104, was a disappointment. It succumbed to mosaic disease in Jamaica, to leaf scald in British Guiana and had little ability to recover from froghooper attack in Trinidad. Then came B 37161, highly successful in Barbados (Fig. 4.6), Antigua, St Kitts and in the drier areas of Jamaica; B 41227, equally successful in Guyana and Trinidad, where it is still the most widely grown variety; the early ripening B 4098 and B 4362; and the all-purpose variety (but with a hard rind) B 49119. Stevenson who succeeded McIntosh, investigated the genetics and cytology of selfing with a view to its commercial application (Stevenson 1960).

In the 1960s it was decided that, because of different ecological conditions, selection of varieties to be grown in some of the countries served by the WICSCBS should be made locally. Therefore, in addition to the bud-chips and stem cuttings of promising varieties chosen in Barbados, fuzz was sent to Guyana, Jamaica, Trinidad and the Dominican Republic (Central Romana) for germination and selection from the seedling population. The chief advantage of this scheme is that more land is available for the evaluation of each cross. Although some local selections have shown promise, none had yet been planted on a large scale. Indeed doubt was cast on the validity of the basic premise by Blackburn (1967), who described the remarkable similarity of performance of the main commercial varieties in all WISA territories. The same had been previously noted by Mangelsdorf in regard to varieties grown in the several islands of Hawaii.

The international character of cane breeding is illustrated by the diversity of the countries which support the WICSCBS. Apart from the members of WISA, financial contributions were received from Belize, Panama, El Salvador, Nigeria and Zambia in 1973–74. There is also ready exchange of varieties between stations, but this requires the provision of adequate quarantine regulations and facilities. These are provided for the WICSCBS in the island of St Vincent, where the sugar industry had been abandoned for several years. Difficulties may arise from the decision made in 1979 to resuscitate the industry and to relocate the Forres Park factory from Trinidad at Mount Bentinck in St Vincent.

Guyana

Harrison, who was associated with Bovell in the early days in Barbados, when transferred to British Guiana in 1889, took with him twelve seedlings and started the D (for Demerara) series: D 109, D 145 and D 625 were grown locally; and D 74, D 95 and D 1135 both locally and abroad. All are noble canes. Diamond 10 bred at (Diamond Estate), also a noble cane, was grown in British Guiana during the decade 1935–45. All were replaced by hybrids: first by POJ 2878 and then by Barbados varieties. For many years Guyana has been closely connected with WICSCBS and although cane breeding in that country ceased for some years, now also makes some crosses of its own.

Australia

In Australia there are two cane-breeding organizations. One, controlled by the Government of Queensland, is part of the Bureau of Sugar Experiment Stations and produces varieties which have the prefix Q; the other, that of the Colonial Sugar Refining Co. Ltd, has stations at Macknade, Queensland, and Lautoka, Fiji; it gives such names as Trojan, Janus and Pindar to its varieties.

As elsewhere, noble canes have been replaced by nobilized varieties whose main characteristics are resistance to many of the major diseases (most of which are present in Australia), high sugar content and erect habit. Two factors have influenced breeding policy in Australia: the restriction of the growing season because of the probability of cold weather and frost; and the Queensland law that cane from only 85 per cent of the assigned area may be reaped each year. The effect of the second stipulation is that the number of ratoon crops taken is limited to two or three.

Descriptions of the methods used and results achieved were given by Gard (1954) and Buzacott (1960). Suffice it to say that they have been highly successful: varieties have been bred which are free from many diseases and give the remarkable average yield of roughly 10 tonnes sugar/ha after a growing period of only nine months.

Mauritius

Although cane breeding in Mauritius started in 1891, controlled crossing began with establishment of the Sugarcane Research Station of the Department of Agriculture in 1930. Nobilization was first used in 1931 and in 1932, produced M 134/32 from a cross between POJ 2878 and D 109. Eventually, M 134/32 was grown almost to the exclusion of all other varieties. Uba Marot, a natural hybrid of *S. spontaneum* and *S. officinarum* found locally by

Stevenson, is in the genealogy of M 147/44 and other varieties which, to a large extent, have replaced M 134/32.

Hawaii

Cane breeding in Hawaii began in 1904 but, shortly afterwards, the leafhopper *Perkinsiella saccharicida* Kirk. was accidentally introduced in a consignment of varieties from Australia or Fiji. Consequently, further importations were prohibited and the ban was not lifted until 1923. Meanwhile, H 109, a seedling of Lahaina, had been selected and was widely grown in the four sugar-producing islands of the group. A large collection of breeding material was imported immediately after the restraint had been removed and work began on nobilization and on crosses between nobilization lines. This was highly successful and soon produced several commercial varieties, of which 32-8560, selected from a cross between Co 213 and POJ 2878, was the most important. The variety 32-8560 is also a parent, or in the genealogy, of most of the Hawaiian varieties grown in recent years.

The qualities sought in Hawaiian varieties are unique. Because cane is grown for two years or more before harvest is taken, great importance attaches to resistance to rotting, rooting at the nodes, rind cracks, the growth pattern of recumbent stems, amenability to mechanical reaping and propensity to form arrows. This has led to the development of a distinct 'Hawaiian type' variety, bred for the special requirements of the local industry but rarely successful in other countries.

Verret and Mangelsdorf made a significant contribution to cane breeding when they discovered that arrows could be cut and preserved, for long enough to bear seed, if they were placed in a dilute solution containing 150 p.p.m. of sulphurous acid (as sulphur dioxide) and 85 p.p.m. of phosphoric acid. Warner (1954a) described this technique and also the specialized methods by which seedlings are selected in Hawaii.

India

In India sugar-cane is grown in vast subtropical areas of relatively infertile soil where it must also withstand the rigours of a continental type of climate: large diurnal and seasonal differences in temperature. Such ecological conditions are unsuited to the cultivation of *S. officinarum* but are good enough for the growth of hardy, indigenous canes of relatively low sucrose content.

A breeding station was established at Coimbatore, Madras, in 1912. Seed produced there is distributed to the main centres of cultivation for propagation and selection. Barber, who was the first to be in charge of this work, crossed *S. spontaneum* with noble canes and produced a commercial variety, Co 205, in the first generation.

Although fairly rich in sugar, Co 205 is susceptible to mosaic disease. Later, nobilizations of *S. spontaneum* were crossed with nobilizations of *S. barberi* (especially Chunnee, Saretha and Kansar) to give highly successful tri-species hybrids. Two of the most important, Co 281 and Co 290, were widely planted throughout the subtropics: in India, Argentina, Australia, Brazil, Louisiana and South Africa. Diagrams, with coloured illustrations, describing the nomenclature of the vegetative characteristics and the genealogy of these, and five other outstanding Coimbatore varieties, were produced by Venkatraman & Thomas (1931) and Venkatraman & Rao (1928).

Varieties to be grown in tropical India are also bred at Coimbatore. Varieties Co 419 and Co 421, which have thick stems, were grown on a limited scale in British Guiana in the 1940s and were used in breeding work elsewhere.

Dutt & Rao (1950) described the morphology and agricultural characteristics of all Coimbatore canes in cultivation at that time.

Pakistan

Partition of India in 1947 deprived the three experiment stations in Pakistan of their source of fuzz, namely Coimbatore. However, re-examination of the seed remaining at Lyallpur led to the selection of two commercial varieties, CoL 44 and CoL 54. A substation has now been established in the Murree Hills, where the environment encourages the formation of arrows.

South Africa

Under normal conditions few varieties flower satisfactorily in Natal, the main area of South African sugar production. The reason for this is that low night temperatures cause pollen sterility and sometimes also ovule sterility (Brett 1954a). Therefore, when the Sugar Experiment Station was established at Mount Edgecombe in 1929, seed was imported from Coimbatore (and some from Mauritius) in order that seedlings might be selected locally. This scheme produced NCo 310, one of the most famous of all varieties, from a cross between Co 421 and Co 312. In South Africa, NCo 310 quickly replaced the Coimbatore varieties which had succeeded Uba (which, in their time, had been planted when the noble canes succumbed to mosaic disease). It was also widely grown in Australia, the Philippines, Louisiana and Taiwan. The characteristics of NCo 310 were described by Barnes (1974).

Brett (1954b) studied the factors which affect sterility in South Africa and eventually developed a technique which allows crosses to be made locally. Marcotted canes to be used in crossing are kept in a heated greenhouse. They are planted in compost and treated with the standard sulphurous acid: phosphoric acid solution, first used in Hawaii to preserve arrows, to which 4.5 p.p.m. of methoxyethylmercuric chloride is added to promote the formation of roots. Fertile flowers are produced and crosses can then be made. Several seedlings from local crosses have shown promise, but so far none has been planted on a large scale.

Puerto Rico

There have been three main breeding stations in Puerto Rico. The first, established at Fajardo Central in 1913, was financed by the Sugar Producers' Association. The second, opened in 1917 under the sponsorship of the Government of Puerto Rico, was located at Mayaguez (the Insular Station). The third, the Federal Station at Rio Piedras, is controlled by the US Department of Agriculture. Work has now ceased at Fajardo and Mayaguez and all cane breeding is done at Rio Piedras. The best varieties produced by the now defunct stations were probably FC 916, M 28 and M 42, all of which were planted on a limited scale.

When (*ca.* 1920) mosaic disease threatened the existence of the industry, Uba was planted extensively, as in similar circumstances in South Africa, in spite of its well-known disadvantages. Nobilizations of *S. spontaneum* were then introduced from Java and soon proved their worth. Later they were crossed with tri-species hybrids from Coimbatore, such as Co 281, to give the Rio Piedras (prefix PR) varieties which are now grown throughout Puerto Rico.

Reference was made in Chapter 1 to the decline of sugar production in Puerto Rico: from 952,768 tonnes in 1960 to 407,000 tonnes in 1970 and 240,826 tonnes in 1977.

Florida

A cane-breeding station was opened by the US Department of Agriculture at Canal Point, Florida, in 1918, in order to provide new varieties for Louisiana, where the crop was being devastated by mosaic disease. Several strains of the virus were present and the production of resistant varieties, satisfactory in all other qualities, was by no means easy. Edgerton (1955) gave a general description of the outbreak and the methods by which it was controlled. Several successful seedlings were produced at the new station, but the imported varieties Co 281 and Co 290 were of greater commercial value. Apart from being resistant to mosaic disease, canes grown in Louisiana must also be resistant to red rot and ratoon stunting, and tolerant of cold weather. The variety CP 44–101 met these conditions, was cultivated on a large scale for many years, but now has been replaced by even better CP varieties.

Canal Point is the home of the World Reference Collection (WRC) of varieties from which material is made available by the US authorities, without restriction, for breeding and research work.

Belcher (1969) gave a list of the varieties of sugar-cane and related grasses in the WRC in 1967. It comprised 655 varieties of *S. officinarum*, 65 of *S. robustum*, 101 of *S. sinense* (including *S. barberi*), 150 of *S. spontaneum*, 27 of *S. edule*, 51 crosses between *Saccharum* and *Miscanthus*, 254 hybrids of commercial or historical importance, 76 of *Erianthus* spp., 5 of *Miscanthidium* spp., 15 of *Miscanthus* spp., 4 of *Narenga porphyrocoma* (Hance) Bor., and 3 of *Sclerostachya fusca* A. Camus. A duplicate collection is kept at Coimbatore.

Two other breeding stations in Florida provide varieties for the local industry. They are the University of Florida Everglades Experiment Station at Belle Glade, whose varieties bear the prefix F; and the Research Division of the US Sugar Corporation at Clewiston, where the Cl series is bred. Although the climate is not as adverse as that of Louisiana, the peat or 'muck' soils on which much of the cane is grown present problems. Bourne (1936, 1961) was associated with both stations, first at Belle Glade and later at Clewiston. The most important varieties produced to suit these unusual soils are F 31–962, F 36–819 and Cl 41–233.

Cuba

In general, research in Cuba has concentrated on factory rather than field technology. Consequently, despite being one of the world's largest producers of sugar, Cuban cane breeding has been trivial. Fors (1960) gave two main reasons for this: inadequate financial support from Government authorities; and, perhaps more important, lack of interest and co-operation by factory owners and cane growers. Two locally produced varieties, ML 3–18 and Pepe Cuca, have been planted to some extent, but cannot be regarded as major varieties.

This description of several cane-breeding stations is by no means exhaustive. Similar work has taken place in many other countries, especially in Brazil, and consequently a multiplicity of symbols is used to indicate the origin of varieties. Several of them are listed in Table 4.3

Cane crossing methods

The first sugar-cane seedlings recognized as such were derived from open-pollinated arrows. Selection from the products of open pollination was adopted by Bovell in Barbados and by Harrison in British Guiana, where organized breeding began at the turn of the century. Several excellent commercial varieties were produced, a tribute to the flair of those making the selections. There is no doubt that much selfing occurred, with consequent inbreeding and the additional disadvantage that, because of doubtful parentage,

Table 4.3   *Symbols used to identify the varieties produced by cane breeding stations*

A	Antigua
B	Barbados
Ba	Barbados
BNH	Barbados Natural Hybrid
BSF	Barbados Self-Fertilized
BH	Barbados Hybrid (noble seedlings of known parentage)
BJ	Varieties grown and selected in Jamaica, Central Romana
BR	(Dominican Republic) Trinidad and Guyana respectively, from fuzz
BT	imported from Barbados.
B	After a number, which is usually preceded by J., seedlings raised by Bouricius in Java
BO	Province of Bihar and Orissa, India
C	Cuba
CB	Estacão Experimentalde Campos, Brazil
CH	Cuban Hybrid
Cl	US Sugar Corporation, Clewiston, Florida, USA
Co	Coimbatore, India
CoL	Varieties bred at Coimbatore, India, and selected at Lyallpur, Pakistan, from seedlings introduced before partition
CP	Canal Point, Florida, USA
D	Demerara (Guyana)
DI	Demak Idjo, Java
Diamond	Diamond Plantation, Demerara (Guyana)
Ebène	Mauritius
EK	Seedlings raised by E. Karthaus, Java
EPC	Estación Experimental, Palmira, Colombia
F	University of Florida Everglades Experiment Station, Belle Glade, Florida, USA
F	Formosa (Taiwan)
Fabri	Fabri Mill, Java
FC	Fajardo Central, Puerto Rico
G	Guadeloupe
GC	Guernica Central, Puerto Rico
H	Hawaii
HJ	Varieties raised in Jamaica from seed imported from Hawaii
HM	Hebbal, Mysore, India
HQ	Hambledon Sugar Co., Queensland, Australia
IAC	Institute Agronomico Campinas, Brazil
J	Jamaica
J	Java (formerly used instead of POJ)
L	Louisiana, USA
LC	La Carlotta, Philippines
M	Madras, India
M	Mauritius
M	Federal Experiment Station, Mayaguez, Puerto Rico
MC	Marguerita Central Peru
ML	Media Luna, Cuba
MP	Seedlings raised by Perromat, Mauritius
MQ	(between two numbers) Macknade, Queensland, Australia
N	Natal, South Africa
N : Co	Varieties raised in Natal from seed imported from Coimbatore, India
NG	New Guinea
P	Peru

Table 4.3   *(continued)*

PB	Pernambuco, Brazil
POJ	Proefstation Oost Java (East Java Experiment Station)
PR	Insular Experiment Station, Rio Piedras, Puerto Rico
PSA	Philippine Sugar Association, Philippines
PT	Pingtung, Taiwan
Q	Queensland. Seedlings from Bureau of Sugar Experiment Stations, Queensland, Australia
R	Réunion
R	(between two numbers). Rarawai, Fiji
RP	Seedlings raised by a planter in Demerara
Sa	These letters are the first two in the names of varieties produced at the South African Sugar Experiment Station, Natal
SC	Saint Croix (Santa Cruz), Virgin Islands
SJ	South Johnstone, Queensland, Australia
SN	Varieties raised at Broadwater, New South Wales, from crosses made at Macknade, Queensland, by the Colonial Sugar Refining Co. Ltd
SW	Sempel Wadak, Java
Tjep	Tjepering, Java
TUC	Tucuman, Argentina
UD	Hawaii (from crosses between Uba and D 1135)
US	US experiment Station, Canal Point, Florida (now replaced by CP)

planned programmes could not be pursued. Methods of controlled crossing, therefore, had to be devised. Unfortunately, they were either tedious or unproductive (the chess-board planting of male-sterile varieties with those which shed pollen profusely). Consequently, for some years there was a reversion to open pollination.

The 'lantern' method

A major advance towards controlled breeding was the recognition that a male-sterile variety, as female, could be fertilized with pollen from the desired male parent to give progeny of known parentage. Crossing took place in the field: the female arrow was covered by a 'lantern' to exclude wind-blown pollen and, at the right moment, pollen from the intended male parent was introduced. Lanterns of different sizes and shapes were used in Java, Barbados (Fig. 4.9), India (Fig. 4.8), and several other countries. They differed only in detail. The Barbados lantern (it has now been superseded) was an open box, roughly 1 yd (0.91 m) × 1 yd × 1 yd, glazed on three sides and on the top. It was supported on adjustable stilts, roughly 2 m high, and from it was suspended a cloth skirt, 1–2 m in length, which was draped around the cane stool. The fourth side of the box was covered with cloth. The weave of the material used was sufficiently close to prevent the intrusion of unwanted pollen. Male arrows were collected and, at appropriate times, their pollen was dusted over the female parents. The seed was then allowed to mature and, after roughly three weeks, the ripe arrows were collected and the seed extracted and sown. Descriptions of this technique and its development were given by McIntosh (1935) and Stevenson (1965). The original 'lantern', developed by D'Albequerque and Skeete, consisted of a wooden framework 2 ft (0.6 m) × 2 ft at the top, widening to 4 ft (1.2 m) × 4 ft at the bottom, and glazed on three sides; hence the name 'lantern'.

Marcotting

Another development in controlled cane-breeding was the use of marcotting. The root primordia, which occur at each node, can be

Fig. 4.8 A light plastic field 'lantern' in use in India (From Stevenson 1965)

Fig. 4.9 The last type of field 'lantern' used in Barbados (From Stevenson 1965)

induced to grow if they are placed in a suitable environment. This is done in the field by surrounding the selected nodes with an aluminium or plastic container filled with an appropriate medium such as moist soil, peat or coconut fibre. A rooting stimulant may then be injected into the stem between the third and fourth nodes below the marcotted area. Shortly before the arrow is due to emerge, the stem is cut immediately below the container. The plastic or aluminium cover is removed and the marcotted upper part of the stem, with well-formed roots, is transferred to a pot or tub filled with soil (India), sulphurous acid preserving solution (Mauritius and South Africa) or aerated tap water (Puerto Rico); or it may be set in a trough through which water is circulated (Canal Point, Florida). Whichever method is used, the inflorescence emerges and controlled crosses can then be made. Indeed, in South Africa, where low temperature affects fertility, it is only by transferring the stems to a heated greenhouse that breeding is possible. Marcotted stems, before and after separation, are shown in Figs. 4.10 and 4.11.

The Hawaiian solution method

Reference has been made to the development by Verret and Mangelsdorf in Hawaii of a successful, labour-saving method of making crosses. It had long been known that if nearly mature hermaphrodite arrows (plus the upper joints of the stems) were cut and placed in water, their flowers would open normally and shed

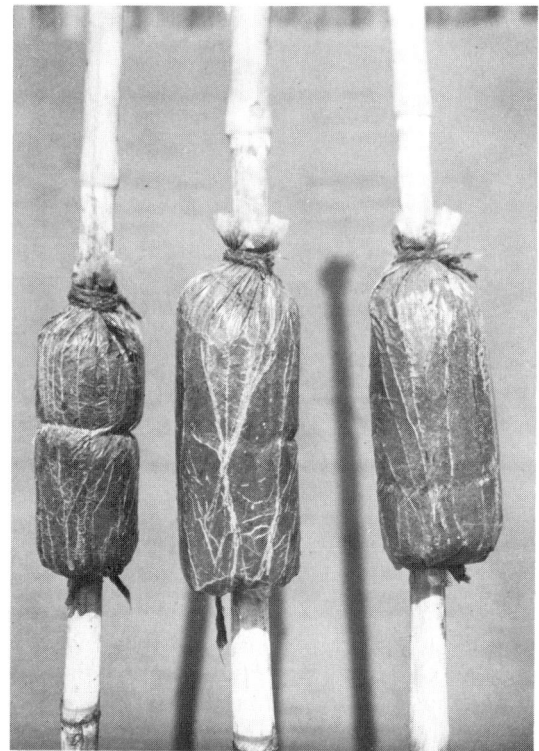

Fig. 4.10 Marcotted stems in India (From Stevenson 1965)

Fig. 4.11 Marcotted canes in a greenhouse in Mauritius (From Stevenson 1965)

pollen. However, despite daily recutting of the stems, they soon died and it was necessary to change the arrows every few days. Verret and his co-workers assumed that the premature death of the arrows was due to blockage of the xylem by bacterial action. They therefore assessed the effect of several bactericidal solutions on the longevity of cut arrows, both male and female. They found that, if the cut stems were placed in an aqueous solution containing 150 p.p.m. of sulphurous acid (as sulphur dioxide) and 85 p.p.m. of phosphoric acid, the arrows usually remained fresh for very much longer. This was of immediate practical importance and is the basis of the Hawaiian solution method of breeding.

Crossing is carried out by placing cut arrows in easily handled pails of 19 litres (US 5 gallons) capacity, the male flowers being secured in positions slightly higher than those of the female parents. After flowering is complete, the male arrows are discarded and the female ones are taken away to ripen. During the whole of this period, the preserving solution is changed at intervals of two or three days. When the seed is ripe, the arrows are 'papered up' (bagged) to dry off. The seed is small, of short life with no dormancy, and therefore is sown at once (nowadays it can be stored, very dry, in a deep freeze).

The Hawaiian solution method has been adopted, sometimes with minor variations, in many other countries. In Queensland, for example, a solution containing 50 p.p.m. of sulphurous acid (as sulphur dioxide) and 85 p.p.m. of phosphoric acid has proved satisfactory.

The solution method in 'lantern' batteries

A combination of the 'lantern' and preserving solution methods is now in general use in Barbados, Florida, Hawaii, India, Mauritius and many other countries. The male and female parents are placed in pails of preserving solution and are covered with lanterns made either of closely woven, pollen-proof cloth (Fig. 4.13) or polythene (Fig. 4.12), great care being taken not to damage the fragile panicles and arrow stalks. Thereafter crossing takes place, the female arrows ripen, the seed is dried and sown and selection is made in the usual manner.

The breeding and selection of new varieties

There are three stages in cane-breeding which involve choice: the selection of varieties to be introduced from other areas, of the parents to be used in crosses and of seedlings from the resultant plant population. All are fraught with difficulty and pose many hard statistical and practical problems. Acceptable decisions concerning the first and second choice can be made only by breeders or geneticists. In making the third choice, however, the specialists are helped by experienced field officers. It is this stage of the selection of new varieties which will be discussed.

Hundreds of seedlings are produced from a single cross, every one of which is a new variety of possible commercial worth, How are they to be assessed? There is neither sufficient land nor money for each to be raised and tested against standard varieties in controlled experiments. Therefore, individual seedlings of unproven merit are chosen from the trays or flats in which they have been sown, and the rest are discarded. The elect are then raised and planted as single stools in first-year trials. Some are rejected because of poor juice quality (Brix as determined by hand refractometer), while others

Fig. 4.12 Polythene lanterns used for crossing marcotted arrows in India (From Stevenson 1965)

proceed to second-year trials because their yield is high and they have other desired qualities. It is in making selection from a single stool that the art, experience or some innate instinct of the field officer causes him to suggest for further trial some varieties which fail to meet measured criteria such as yield and number of stems. In Barbados the skill of the late C. B. Foster in this respect (there

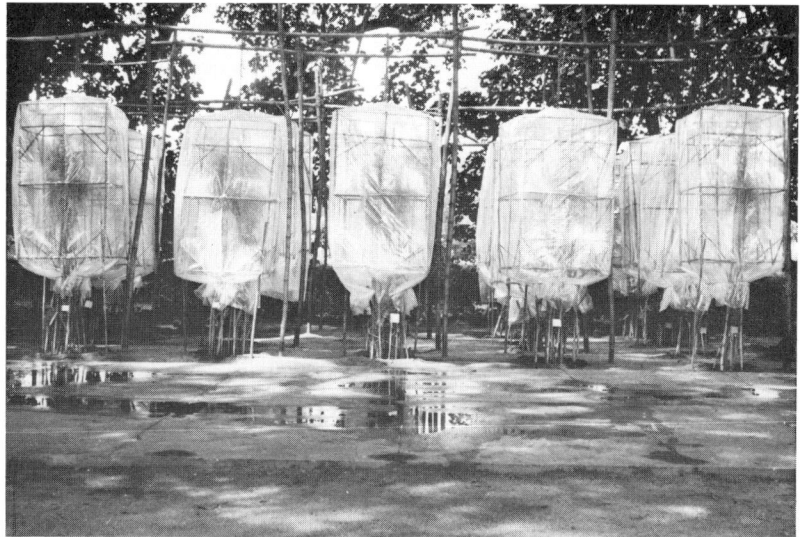

Fig. 4.13 Lantern batteries in Barbados (the skirt of one has been raised to show the arrangement of the arrows) (From Stevenson 1965)

must be others like him elsewhere) was legendary. He saved from the scrap-heap several varieties (including B 37161) which later were planted extensively on a commercial scale.

When uniformity trials are carried out to determine the appropriate size of plot for use in precise experiments, each stool is weighed. Although the stools are of the same variety, and grown under identical conditions, the differences in yield are far greater than might be expected. Therefore the successful choice between different varieties, based on the growth of a single stool, is a tribute to the flair, as well as to the knowledge, of all concerned.

Bunch planting

In Hawaii, Mangelsdorf (1946, 1953) devised a method of planting bunches of between five and fifteen seedlings, rather than individuals. Bunch planting has been adopted in several other countries. Critics of this method suggest that the conditions of selection favour canes which are vigorous, do not readily form tillers and have a low sugar content. The reply is that all the commercial varieties now being grown in Hawaii were chosen by bunch planting, have large stools and are rich in sucrose. Several experiments were carried out in Australia to discover whether bunch planting is superior to the standard practice of single planting. Skinner (1971) reported that, in general, the results were inconclusive and did not justify a change in method.

The melting-pot

After many years of controlled breeding in which both male and female parentage is known a new technique, the melting-pot, in which only the identity of the female parent is certain, was devised in Hawaii by Mangelsdorf and his co-workers. The melting-pot, and its underlying principles, were described by Warner (1954b). It is essentially a polycross involving carefully selected female (male sterile) and male varieties. Random pollination is ensured by scattering arrows of each variety throughout the breeding area, and by moving them frequently. Seed is collected for germination and evaluation from both male and female arrows. The advantage of the melting-pot is that many parental combinations can be explored at minimum expense; its disadvantage is that the male lineage is unknown.

Although biparental crosses continue to be made in Hawaii, the concept of melting-pot crosses aroused considerable interest and differences of opinion among cane breeders and geneticists. Stephenson (1965) gave a balanced account of these arguments in a general discussion on the philosophy of cane breeding.

At the end of a review of procedures in selection, Skinner (1971) succinctly described the position as follows: 'Considerable progress in both the theory and practice of sugarcane selection has been made during the last 20 years. However, practical cane breeding today is very much an art as well as a science. Some of the principles discussed are firmly based on experimental evidence. Others have no scientific basis, . . .' It is difficult to disagree with this conclusion.

Breeding, selection and trials
D. I. T. Walker and N. W. Simmonds.

Mr Walker is Director of the WICSCBS, and has over twenty years' experience of practical cane breeding. Dr Simmonds has been consultant to the station since 1961; he is in the Edinburgh School of Agriculture and is Honorary Professor in the university.

Introduction

The literature of sugar-cane breeding is scattered and no single authoritative account is available. Stevenson's (1965) book covers the history of breeding and the making of crosses but not selection procedures, which represent a very large part of the effort. The volumes of *Proceedings* of ISSCT congresses (triennially 1924–38 and 1950–present) contain numerous papers on breeding while the ISSCT *Sugar-cane Breeders' Newsletter* (1956–present) is devoted to informal reports of current work. Useful review references on cane breeding are: Alexander (1973); Anon (1971); Arceneaux (1965, 1968); Brandes & Sartoris (1936); van Breemen *et al.* (1965); Coleman (1968); Daniels (1965); Empig (1974) Hogarth (1978); Lalouette (1968); Loh & Wu (1965); Mangelsdorf (1946, 1953); Rao (1965); Roach (1971); Skinner (1971); Smith (1965); Symposium (1962); Urata & Warner (1959); Wagenaar (1959); Warner (1954a, b).

Sugar-canes are wind-pollinated outbreeders; they are highly polypoid, many are aneuploid and regular Mendelian segregations are neither expected nor observed. Being outbreeders and clonally propagated, they are highly heterozygous and intolerant of inbreeding. Crosses between clones therefore display great variability and it is among such F_1 progenies that cane breeders seek new varieties.

As explained under the evolution of the crop, parents with $2n$ ~ 100 chromosomes for a commercial crossing programme will generally be two or more generations away from the cross between the wild species. In constitution they have 85–95 per cent noble and 15 per cent or fewer *S. spontaneum* chromosomes. About 40 per cent are genetically male sterile and perhaps a further 10 per cent are poor pollen producers. The varieties maintained for crossing in any one station consist mainly of locally adapted varieties, supplemented by foreign varieties adapted to presumptively similar conditions elsewhere and by others which have special characters (erectness, disease resistance, etc.) which the breeder wishes to introduce into his programme. With the progress of time, all programmes accumulate a pool of parents, some of them the best local varieties, some near-varieties that were not quite good enough to be grown commercially, some foreign stocks and so forth. In all programmes the pool is constantly changing, new entries displacing discarded ones, though some breeders are less inclined than others to discard older but successful parents in favour of recently selected clones of unknown combining ability. This reflects in their strategy of breeding (Fig. 4.14).

Scale of programmes in various countries

Most major industries operate breeding programmes as parts of their research efforts. Table 4.4 shows the order of magnitude of some of these programmes.

In the subtropics the major constraints to breeding have been unsuitable natural daylengths (many varieties fail to flower) and low temperatures (causing failure to produce fertile pollen). These have been largely overcome by photoperiod and temperature control (Brett 1950; Chilton *et al.*, 1965; James 1969 and such subtropical programmes can now have a wider genetic base than formerly.

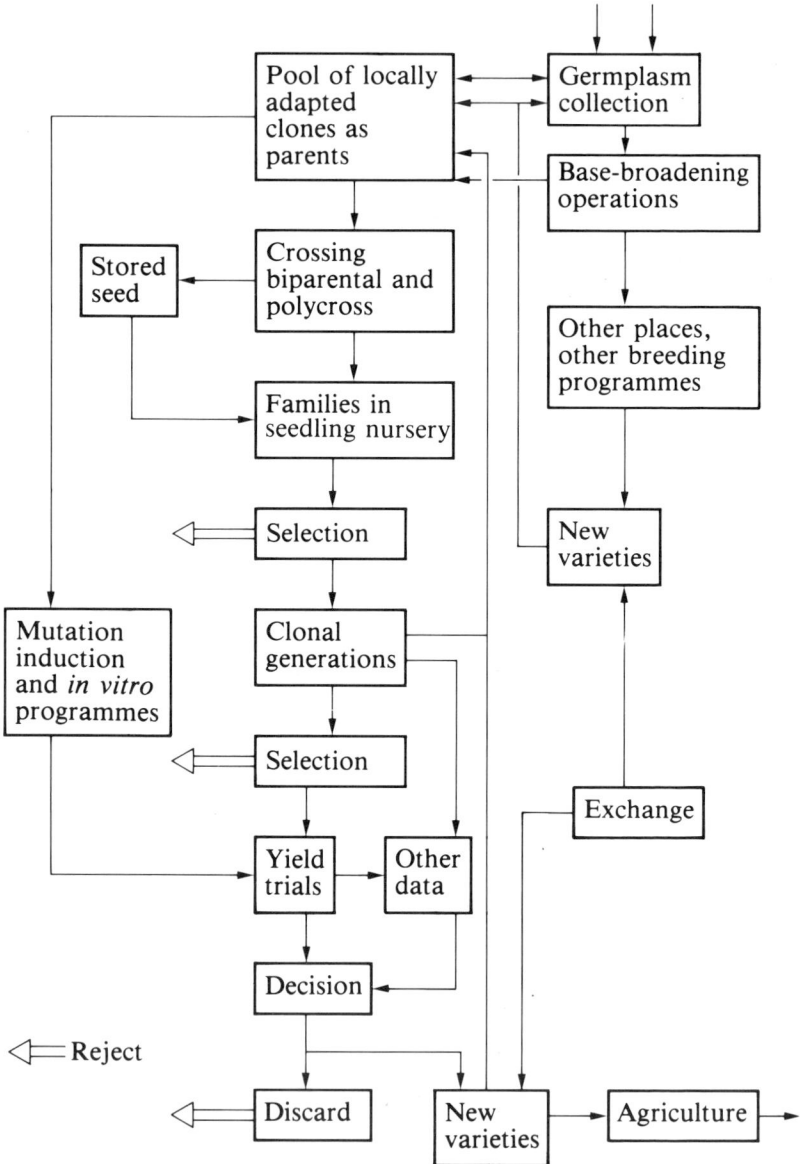

Fig. 4.14 The general pattern of cane breeding

Breeding, selection and trials 127

Table 4.4 Information on some sugar-cane breeding programmes (based on scattered literature relating to the 1970s)

Country	Call sign*	Annual breeding effort		Substations for stage 1	Sugar production served (m tonnes)
		Crosses incl. polycrosses	Seedlings (m.)		
Tropical					
Barbados and Caribbean	B, BJ, BR, BT, DB	1,700	0.65	10	8.0
Other Caribbean	C, CR, D, PR				
Fiji	LF	600	0.025	1	0.3
Hawaii	H	4,000	1.0	4	1.0
Mexico	Mex	700	0.2	8	3.0
Philippines	Phil	1,000	0.1	2	2.6
Mixed					
Australia	Q	700	0.075	4	3.0
Brazil	RB, SP, IAC	1,200	4.1	9	8.0
India	Co	110	0.25	8	6.0
Subtropical					
Argentina	NA, TUC	150	0.1	2	1.4
Mauritius	M	480	0.06	2	0.7
Réunion	R	150	0.08	1	0.25
South Africa	N	600	0.2	6	2.2
Taiwan	F	400	0.5	9	0.8
US mainland	CP	350	0.2	4	1.5

* Internationally agreed letter-code that precedes the number for variety identification (examples in Table 4.5).

Absolute size of programme plays a major part in the likelihood of success, and it can be seen in Table 4.5 that quite small industries raise large numbers of seedlings. Conversely, some larger industries have relatively small programmes. The principal motive for a local crossing programme is the belief (plausible though difficult to prove) that adaptation to local conditions is vital and that selection in a different environment may lose such adapted varieties at an early stage in the testing. A common compromise is for a breeding station to distribute seed to a number of substations (as is done in Hawaii and in the Caribbean). A less satisfactory compromise is to decentralize the testing of the largest possible number of clones at an early stage. A number of industries, some of them technically very successful (e.g. Iran), do not support breeding programmes but have depended on occasional importations of commercial varieties from

Table 4.5 *Widely and narrowly adapted sugar-canes (an asterisk· indicates where the variety is still important)*

Identity	Source	Places
		(a) *Some widely adapted varieties grown in countries distant from parent industries where they were originally selected*
BH 10(12)	Barbados	Caribbean, Mauritius, Mexico
B 37161	Barbados	Caribbean, Venezuela
B 37172	Barbados	Caribbean, Mauritius, Philippines
B 4362	Barbados	North Caribbean·, Central America·, Malagasy Rep., Mexico·
B 49119	Barbados	Caribbean, Central America, Venezuela·
B 51129	Barbados	North Caribbean·, Mauritius·, Réunion·
Co 281	Coimbatore	Brazil, Egypt, South Africa
Co 290	Coimbatore	Brazil, North India, Louisiana, Mexico
Co 421	Coimbatore	Central Africa, India
CP 48–103	Florida	Louisana, Iran·
H 32–8560	Hawaii	Hawaii, Peru, Philippines, Puerto Rico
H 37–1933	Hawaii	Hawaii, Peru
NCO 310	S. Africa	Australia·, Central Africa·, Egypt·, Iran·, Louisiana·, South Africa·, Sudan, Taiwan
PINDAR	Australia	Australia·, Fiji, Malagasy Rep.·
POJ 2878	Java	Colombia·, Cuba, Java, Mexico
RAGNAR	Australia	Ecuador·, Fiji·
		(b) *Narrowly adapted canes important in only one place*
B 62163	Barbados	Barbados·
CB 41–76	Brazil	Brazil·
Cl 41–223	Florida	Florida
CP 44–101	Florida	Louisiana
CP 65–357	Florida	Louisiana·
F 160	Taiwan	Taiwan·
H 50–7209	Hawaii	Hawaii·
M 134–32	Mauritius	Mauritius
PT 43–52	Taiwan	Taiwan

other parts of the world. Indeed, many varieties are widely adapted: Table 4.5 lists some of the world's major varieties that are or have been grown in more than one industry. Many of these are historic clones resulting from the first major thrust of interspecific cane breeding in the 1920s; very widely grown canes are unlikely to be common in the future owing to the much larger and more widely dispersed breeding effort.

Aims of breeding

All plant-breeding programmes have to do with numerous, even very numerous, characters and cane is no exception. The main ones are listed in Table 4.6.

The ultimate object of cane breeding is economical yield of sugar. Cane yield (Y) and sucrose content (Q), sustained over ratoons, are

Table 4.6 *Characters for selection in sugar-cane*

Character	Observations
Cane yield, Y	Cane number (related to tillering) × cane weight (related to length and thickness). Low repeatability. Estimated only in proper trials
Sugar content, Q	Of cane, but can be approximated by Brix measurements of juice. Low-middling repeatability and some early selection pressure can be exerted
Fibre content	Conventional factory requires fibre within 11–15% range. Extremes approximated by hardness testing in field; otherwise laboratory
Sugar yield, YQ	Economic sugar yield, better approximated by an economic index $Y(Q-m)$ (Simmonds 1979) where m is in range 4–6
Habit	Erect, non-lodging habit is favoured, especially for mechanized harvesting. Visual selection at high Y; lodging is generally very dependent on cane length. Probably unattainable in crops > 16 months
Ratooning	Sustained yield over crops, a very important economic factor for most industries; many environmental components
Disease resistance	Necessary for all. Major diseases affecting economic yield over a typical cycle are as follows: *Fungi:* Smut, red rot, rust, downy mildew, leaf scorch. *Bacteria:* Ratoon stunting, leaf scald, gummosis. *Viruses:* Fiji disease, grassy shoot, mosaic.
Insect resistance	Few known and have been little selected for but objective tests are under development (e.g. Coburn & Hensley 1971).

Table 4.6   *(continued)*

Character	Observations	
Miscellaneous	*1. Flowering:*	Depending largely on latitude (i.e. daylength pattern) and maximum temperatures, flowering may be frequent and damaging because it checks growth (Rao 1977). Highly repeatable at a given latitude and readily selected against.
	2. Spines, etc:	Sharp leaf edges and sheath hairs are very undesirable in canes that are to be cut by hand without burning.
	3. Brittleness:	Varieties that shatter into short pieces in mechanical harvesting add to costs or suffer lost yield. This character is of considerable importance to all industries using mechanized harvesting.
	4. Herbicide tolerance:	In industries such as Hawaii that depend on heavy use of chemicals for weed control every variety is evaluated for tolerance of normal agronomic practice.

therefore paramount. Habit of plant is important in most countries for economic harvest either by hand or by machine. Trashiness is important to industries that harvest without burning while, in a few industries, the trash and tops have real importance as animal feed, fuel and thatch. Disease resistance matters only in so far as it affects economic yield but, nevertheless, a satisfactory level of resistance is of paramount importance, since other means of control are usually not feasible or are too expensive.

Strategy of breeding

The most widely used crossing pattern is, as it virtually always is in plant breeding for quantitative characters, one of 'generation-wise assortative mating'; the better current parents are successively displaced by the best of their progeny. A few programmes, however, depend heavily on old 'proven' parents and even repeat specific crosses at the expense of the generation-wise approach (see Walker 1962, for discussion).

Paired crosses (in which both male and female parents are known) and polycrosses (female parents known, open-pollinated by diverse males) are both used. Programmes differ in emphasis from predominantly biparental to predominantly polycross. The former provides better information, but the latter is cheaper and allows quick exploitation of very diverse parents and also random crossing among a chosen group of parents, even if all shed pollen. Each approach has its uses and a judicious combination will no doubt remain the normal procedure. The polycross philosophy was largely developed by the Hawaiian workers (Warner 1953a) and has been adopted by other countries in recent years, notably Brazil, Barbados and Australia.

Whatever kind of cross is used, the breeder's objective is to make families having high mean performance and sufficient genetic variance to permit effective selection. Such families arise from combination of parents each having a high general combining ability (GCA) plus the unpredictable bonus of a favourable genetic interaction or specific combining ability (SCA) (Simmonds 1979, sects 4.6, 4.7, 5.6). The GCA can be estimated from a sufficient number of paired crosses and/or from polycrosses; SCA is unpredictable. The practical procedure is therefore to examine smallish samples of diverse families and then exploit more intensively those that give evidence of better-than-average performance (implying high SCA). Empirical exploration defines the best parents and combinations, not a priori judgement: the net must be widely cast.

Having said this, experience suggests that genetic variance for sugar content is essentially additive, so that somewhat greater confidence can be attached to the rejection of low-Q parents than of low-Y parents. This, in turn, often leads a programme towards some disassortative mating e.g. crossing high Y : low Q with low Y : high Q. An analysis of these trends is needed.

Selection schemes

The essential problem of selection is to start with very numerous seedlings (in the range 10^4–10^6) and reduce these efficiently to the one or few best clones contained therein. As numbers of genotypes decline, so clonal individuals of the survivors increase. The process of selection is based on trials in three to six distinct stages. Single unreplicated seedlings constitute stage 1; single or twin-row unreplicated plots are common for stage 2; and larger plots, replicated or not, for stage 3. Stage 4 generally consists of a series of replicated trials using larger plots with discard (guard) rows, approved experimental designs and statistical analysis for their interpretation. Stage

5 or 6 represents the multiplication of selected varieties for release to farmers. However, there are wide variations in detail and no one system could claim to be the best for all industries, due both to the varying economic weighting of different characters and to the varying environments within each industry. Two contrasting systems are shown in Table 4.7.

A further factor which must influence selection rate is the clonal repeatability (or broad-sense heritability) of the observed characters. Most characters of economic importance in sugar-cane have a rather low repeatability (George 1959, 1962; Ladd *et al.*, 1974; Mariotti 1972a, b, 1974; Rao *et al.* 1967; Skinner 1965) both between plants within a field (partly competition between plants – Skinner 1961) and across sites and years; but it is generally agreed that, when assayed at a suitable stage of development, sugar content has a higher repeatability and lower genotype–environment interaction (GE) than cane yield. Hence sugar content can be more confidently used as a criterion for selection in the earlier stages of trial. Commonly this is estimated using the hand refractometer to obtain the Brix of the juice in stages 1 and 2, while more detailed laboratory analysis of the cane is undertaken only on the clones reaching advanced stages. In turn, some of the components of cane yield have higher repeatability than others: most studies indicate a descending sequence of thickness, cane number per unit area and cane length. If the repeatabilites of components of yield are known, various selection schemes can be examined for their relative efficiencies. High efficiency would result from a combination of greater advance in mean from stage to stage with least effort/cost in terms of land and manpower for recording information. In these terms, an optimum intensity for each stage, considering a defined set of characters each time, can be determined. James and Miller (1975) report such a study for sugar-cane over six stages of selection. This indicated that selection intensity could justifiably be lower in Stage 1 (6 per cent) than in the stage 2 clonal plots (11 per cent).

A scheme which overcomes some of the problems of the early stages of selection is being used in Fiji. Called the 'mass stool population' (Daniels *et al.* 1971), this technique can handle 500–1000 clones at several sites, uses objective selection criteria (particularly Brix and cane number) that can be observed and applied by relatively unskilled staff and allows considerable flexibility. For example, the best clones can be advanced to stage 4 after the second harvest while borderline clones repeat the earlier stage.

The place of disease testing varies with the pathogen concerned. Resistance can sometimes be estimated simply by observing naturally infected plots in conditions which favour infection (e.g. rust disease in the presence of the very susceptible B 4362), but it often requires special laboratory or field tests under controlled inoculation

Table 4.7 *Two sugar-cane selection schemes compared*

	Hawaii	Year	Guyana	Year
Stage 1	1–1.5 m. total seedlings in bunches of 10 at 10 substations *Selected on visual characters at 12 months*	0	75,000 total seedlings as singles at one site *Brix at 12 months, selected in 7-month ratoons*	0
Stage 2	50,000–70,000 at same station; unnumbered stools 3.5 ft apart *Selected on Brix, etc. at 12 months*	1	7,500 rows of 18 ft at same site *Brix at 12 months, selected in 7-month ratoons*	$1\frac{1}{2}$
Stage 3	5,000–7,000 *Numbered* exchanged between stations, stools 3.5 ft apart; propagated to 14 ft rows on substations 3 *Selected on Brix, etc. at 12 months*	2	750 6 rows of 18 ft at same site *Brix, etc. at 12 months, selected in 7-month ratoons* *Numbered* propagation and reserve plots on station	3
Stage 4	Ca 800 30 × 24 ft plots, replicated on substations *Selected on weight and cane analysis at 12 months*	4	75 30 × 36 ft replicated one site on plantation *3 crops recorded for yield and cane analysis* + reserve propagation plots	$4\frac{1}{2} + 5\frac{1}{2}$
Stage 5	200–300 10 × 20 ft plots replicated on plantations in relevant zone, *full yield data at 12 months for 2 crops*	5+6	30 × 36 ft replicated one site, *3 crops recorded for yield and cane analysis*	$6\frac{1}{2} + 7\frac{1}{2}$
Stage 6	40 × 40 ft plots replicated on plantations in relevant zone *full data after normal 2-year crop* + substation increase plots	7+8		

Notes: The Hawaiian programme starts with a much larger population of seedlings, and emphasizes adaptation to the various environments, represented by 10 substations. Plantation trials running for the full 2-year crop are not planted before stage 6. The Guyana programme, serving a smaller and less varied industry, concentrates on ratooning ability and the replicated tests, though run on plantations, are few in number.

(e.g. cane smut and Fiji disease). Few programmes now emphasize inoculation of all seedlings with, for example, mosaic (virus), since most have reduced disease prevalence in their industries. But it is common practice to inoculate and record rather carefully those clones reaching stages 4 or 5, to guard against widespread planting of a susceptible clone which could in turn lead to resurgence of a disease that was otherwise under control. There is a large literature on sugar-cane diseases, much of it in the *Proceedings* of ISSCT congresses and in the ISSCT *Sugar-cane Pathologists' Newsletter* (1968–present). Martin *et al.* (1960) is the leading textbook.

Though at present outside the main thrust of cane breeding, mention should be made of mutation breeding and tissue culture techniques. In attempts to remove one undesirable characteristic from otherwise good commercial varieties, cuttings have been irradiated and mutations sought. Some favourable mutants have been reported, for example: reduced flowering (Rao 1972), resistance to red rot (Rao *et al.* 1966) and absence of sheath spines (Jagathesan & Sreenivasan 1970). However, associated changes affecting yield tend to be produced simultaneously and, in practice, mutation breeding leads to sets of clones that need complete re-evaluation in yield trials. Tissue culture and subsequent differentiation of subclones has been successfully achieved in Hawaii, Fiji, Taiwan and Argentina. Subclones differing in chromosome number and several other characteristics have been raised. The most promising line of work so far reported has been the isolation of subclones of Pindar resistant to Fiji disease (Krishnamurthi 1977). These techniques will doubtless continue to be developed but are unlikely to supersede conventional sexual recombination in sugar-cane breeding.

A basic necessity of all trials is that they should contain the current standard commercial varieties. In interpretation, emphasis must be given to the rank of the variety relative to the standards, environment by environment. Though each such comparison is made with a particular environment in mind, collective interpretation of all trial results by the use of regressions is often instructive and, in this way, trial results can be combined more effectively to help in the final decision as to whether to extend a new variety commercially (Ruschel 1977; Simmond 1979, sect. 6.2). Always, there is the assumption that trials truly predict agricultural performance. One test of this assumption is, unfortunately, not very encouraging (Walker & Simmonds 1980).

At this final stage, the breeder usually hands over to the farmer clones which he is confident are at least as good as existing standards as to sugar content Q and have superior yield in at least some environments. They must also have acceptable levels of resistance to disease. The farmer completes the process of testing by observing, on the field scale, such important aspects as labour acceptance, machine performance, ratoon survival, weed control requirements and herbicide tolerance, irrigation response and maturity patterns. These are characters which are but poorly tested in conventional variety trials but can make a great difference to economic value.

Chapter 5

Cultivation and planting

For many years land was cleared by axe and saw; it was a tedious business and highly labour intensive. There was a rapid transformation when tractor-mounted bulldozers and rippers were developed: they work quickly and require few (but highly skilled) operators and maintenance crews. Trees are uprooted, pushed into windrows with the rest of the vegetation and burnt when sufficiently dry. The cleared area can be land formed to provide adequate drainage, or given a precise grade to facilitate irrigation.

When preparatory cultivation was carried out by hand, using hoes or forks, or by animal-drawn implements, the heavier soils could be worked only when they were much too moist to allow a satisfactory tilth to be imparted to them. This changed *ca.* 1900 when shares carried on cables and drawn across the fields by steam-engines (Fowler steam ploughs) were introduced into several countries. In the 1930s steam-ploughs were replaced to some extent by gyrotillers. However both gave way, a few years later, to crawler and wheel tractors. The first tractors were driven by petrol or paraffin engines and later ones by diesel engines. They draw implements mounted on tool bars, controlled by hydraulically operated linkages, and thereby eliminate wasteful headlands.

Systems of cultivation

Although several different agricultural systems were developed when cane was first planted on a large scale generally under rain-fed conditions, they had two things in common: they were devised to meet local climatic conditions and were dependent upon an abundant supply of cheap labour. In dry areas, in the absence of irrigation, soil and soil moisture conservation were of supreme importance, while in wet low-lying areas adequate drainage was the main requirement.

Cane holes

In Barbados, Mauritius, Réunion, the drier parts of Jamaica and on sloping land in Antigua, cane used to be planted in holes, a practice which has now been abandoned but which was carried out in Barbados until recent years. (90 per cent of the Barbados crop was planted in cane holes in 1962, but only 10 per cent in 1972). The holes were 0.61 m (2 ft) square and 17.5 cm (7 in) deep, and most commonly spaced at intervals of 1.52 m (5 ft) by 1.52 m. Those shown in the well-known eighteenth-century print *Cane-Holeing in Antigua* (Plate 13 of Vol. 1 of Deerr's *History of Sugar* 1949), appear to be much larger. A similar illustration, together with a description of the operation as carried out in Jamaica, was given by Craton & Walvin (1970). Two two-eyed setts were planted in each hole. The cane grew well because the holes retained moisture and, on sloping land which had not been protected with a mulch, received soil particles washed in by heavy rain.

Cambered beds

Adequate drainage of low-lying flat land in areas of high rainfall was achieved in many countries by growing cane on cambered beds separated by deep drains. The beds were of varying width, perhaps most frequently 6.1 m (20 ft), and the drains 0.6 m (2 ft) wide and 0.45 m (1.5 ft) deep. For many years the camber was created manually by throwing soil with spade or fork from the drains to the centres of the beds. Later, when tractors replaced hand labour in preparatory tillage, the camber was maintained by mouldboard ploughs, or discs, travelling along the beds and turning their furrow-slices to the centres (gathering). If the camber became too pronounced, the mouldboards or discs were replaced by chisels in this first tillage operation. The full sequence of cultivation operations on soils which for the most part are heavy clays was:

1. Plough (or chisel) to uproot old cane stools and to break the soil into large clods.
2. Harrow, after an interval of ten days or so, to produce soil of good tilth.
3. Reharrow, if the tilth is still unsatisfactory.
4. Ditch, using a suitable tool, to reopen the drains.
5. Make planting furrows 1.52 m (5 ft) apart.

All cultivation took place in dry weather. Where irrigation was practised, cane was then laid in the furrows (which ran along and not across the beds), covered with soil by breaking the banks, and the cambered shape of the beds was restored. In unirrigated areas it was necessary to await the first showers of the wet season before

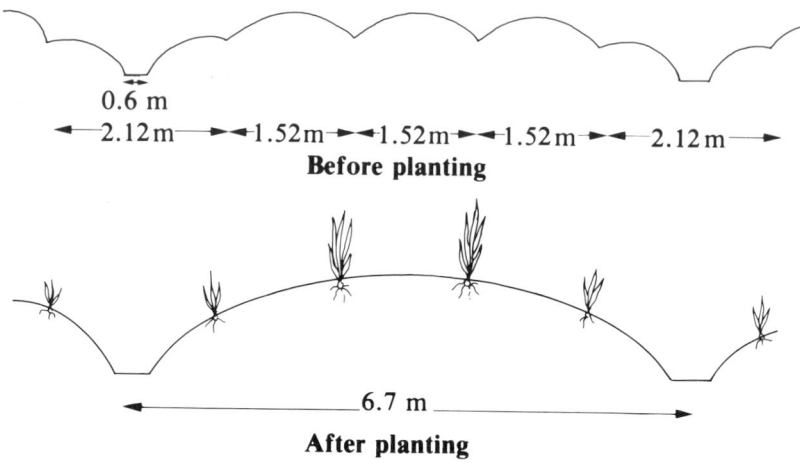

Before planting

0.6 m

←2.12m→ ←1.52m→ ←1.52m→ ←1.52m→ ←2.12m→

After planting

6.7 m

Fig. 5.1 A cross-section of a typical cambered bed

newly prepared fields could be planted. Cross-sections of a typical cambered bed, before and after planting, are shown in Fig. 5.1.

In theory the chisels (harrows) created bands of cultivated soil, 0.45 m (1.5 ft) deep, of the same curvature as the cambered surfaces. Excess rain (or overhead irrigation water) would percolate through this layer and, because of the camber, enter the drains through their sides. In practice there was considerable soil erosion, especially at the sides of the beds, and frequent clearing of the drains was necessary. Because of this, and also because the outside rows of cane were 0.6 m further apart than the centre rows (Fig. 5.1), their growth was restricted. The cambered bed system was never entirely satisfactory.

The system in Guyana

In Guyana sugar-cane is grown on narrow strips of land along the coast and bordering the mouths of the Demerara, Berbice and Courantyne rivers. The whole of the cultivated area lies within 13 km of the Atlantic Ocean, and much is below sea-level. The mean annual rainfall of the sugar belt is 2.34 m (92.31 in), and there are two distinct dry seasons. Near the coast the soils are heavy and saline (frontland clays) but their salinity decreases markedly further inland. Similar heavy soils, though not saline, occur close to the rivers (riverain clays). More distant from the coast and from the rivers, the frontland and riverain clays give way to highly acidic peat soil (*pegasse*). Beyond the cultivated area are vast tracts, known as swamp reservoirs, in which water from the interior is collected in conservancies for use in transport and irrigation on the estates during the dry seasons. Water from creeks is also used for the same purposes.

It is only by means of a complex system of drains, dykes and canals that arable farming is possible. Guyana was once a colony of Holland and the very existence of its sugar industry is a tribute to the Dutch genius for reclaiming and cultivating low-lying areas. A wide, well-built sea-wall protects the coastal land from inundation and substantial banks safeguard the riverside fields. However, drainage water must be discharged, and this is done either by pumps or, where possible, by sluice gates (*kokers*) which are opened at low tide. The drainage canals (sidelines) pass along the ends of the fields and receive water from the infield drains. At the opposite ends, and at a higher level, are the transport and irrigation canals (middle-walks) whose offshoots (cross-canals) establish field boundaries. Each field is surrounded by a dam bed, 11.28 m (37 ft) wide, which prevents water from the higher level canals from flowing into the field (except when required to do so) and irrigation water from the field into the drainage canal. Within the field are cambered beds, of varying width but usually 7.32 m (24 ft) from centre to centre, separated by drains. The beds may run either from middlewalk to sideline (English layout) or from cross-canal to cross-canal (Dutch layout). Cane is planted in rows, 1.83 cm (6 ft) apart, which run across the beds. There has been no claim that drainage takes place other than by water flowing over the cambered surfaces. Typical English and Dutch layouts are illustrated in Fig. 5.2. The sequence of cultivation operations carried out when a field is to be replanted is as follows:

1. Plough with discs to uproot the old cane stools and to restore the camber of the beds.
2. Harrow with discs to improve the tilth of the soil.
3. Reopen the drains with a mechanical digger.
4. Move the soil from the drains to the centres of the beds to complete their camber.
5. Harrow with tines or chisels to break up clods and to smooth the surface of the beds.

The field may then be submerged to a depth of 0.3–0.45 m for a period varying from three to six months, after which the water is drained off and cane is planted.

Flood fallowing

The benefits of flood fallowing, as this submergence is called, were recognized many years ago. For example John Justus Deeges, in a petition for a grant of land dated 7 March, 1827, stated that plantation Ruimzeght, having been in cultivation upwards of fifty years, had lost much of its original fertility; therefore it should be rested

Transport and irrigation canal (middlewalk) supplied from a
swamp conservancy or a creek

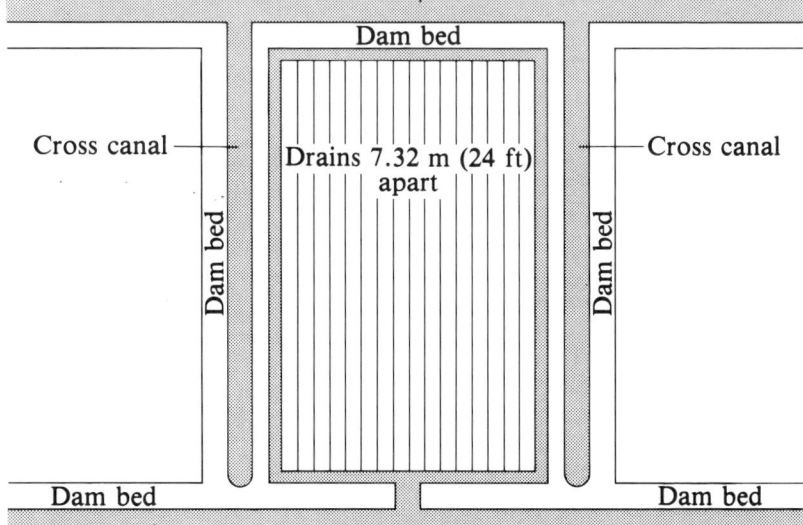

Drainage canal (sideline) to a pumping station or a sluice gate
English layout

Transport and irrigation canal (middlewalk) supplied from a
swamp conservancy or a creek

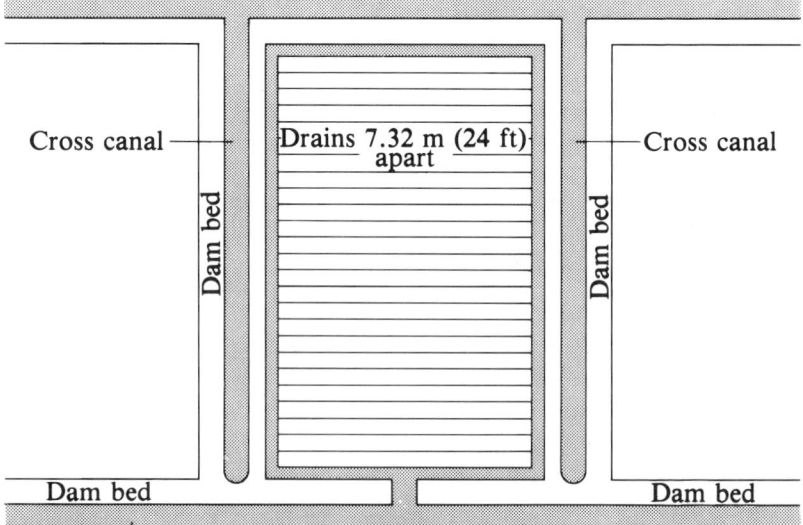

Drainage canal (sideline) to a pumping station or a sluice gate
Dutch layout

Fig. 5.2 The layout of cambered beds in Guyana

and flooded with fresh water in order that it might again become productive.

Flood fallowing improves the texture of the soil, which becomes more friable and eliminates dry land weeds; and also causes its nitrogen content to increase in three months by 13.5 kg/ha (the equivalent of 1 cwt sulphate of ammonia per acre). It has been suggested that the soil crumbs are protected by a layer of ferrous ions, developed by the reducing conditions induced by flood fallowing, which become oxidized to the ferric state when the water is removed; and that this is responsible for the improvement in tilth. Whatever the reason, flood fallowing causes an increase in yield of 40 per cent over a three- or four-year crop cycle. Flood fallowing is not practised on *pegasse* soils because of their porous nature.

Florida and Mozambique

Difficulties similar to those of Guyana occur elsewhere: for example in Florida and Mozambique.

In Florida cane is grown on land which has been reclaimed from the Everglade swamps by the installation of an extensive and well-managed drainage system. The 'muck' soil which has been recovered has a high organic matter content, in some places in excess of 60 per cent, and its fertilizer requirement for successful crop production is unusual if not unique.

The difficulties of combining flood protection with the provision of an adequate drainage system have also been faced in Mozambique, where cane is grown on vertisols in the delta of the Zambezi River protected by 200 km of embankments. The problems of in field drainage have not yet been satisfactorily resolved, though regional drainage is adequate.

Louisiana banks

In Louisiana the successful growth of cane on low-lying flat land, with a high water table and under heavy rainfall, was achieved by the development of a system of ridge and furrow cultivation in fields shaped in the form of 'turtle-backs'. The ridges on which the cane is grown, 0.45 m (1.5 ft) in height, are spaced at intervals of 1.8 m (6 ft). Each of the furrows acts as a drain. Water flows from the furrows into slightly deeper quarter drains, 20 m or more apart, which are at right angles to the ridges, and is then discharged (aided by the 'turtle-back' shape) at the sides of the fields into deeper ditches which run parallel with the ridges. The field drains lead into area drainage canals (Fig. 5.3).

The great merit of the Louisiana bank system (as the ridges and furrows are called) is that it allows all stages of cane production to

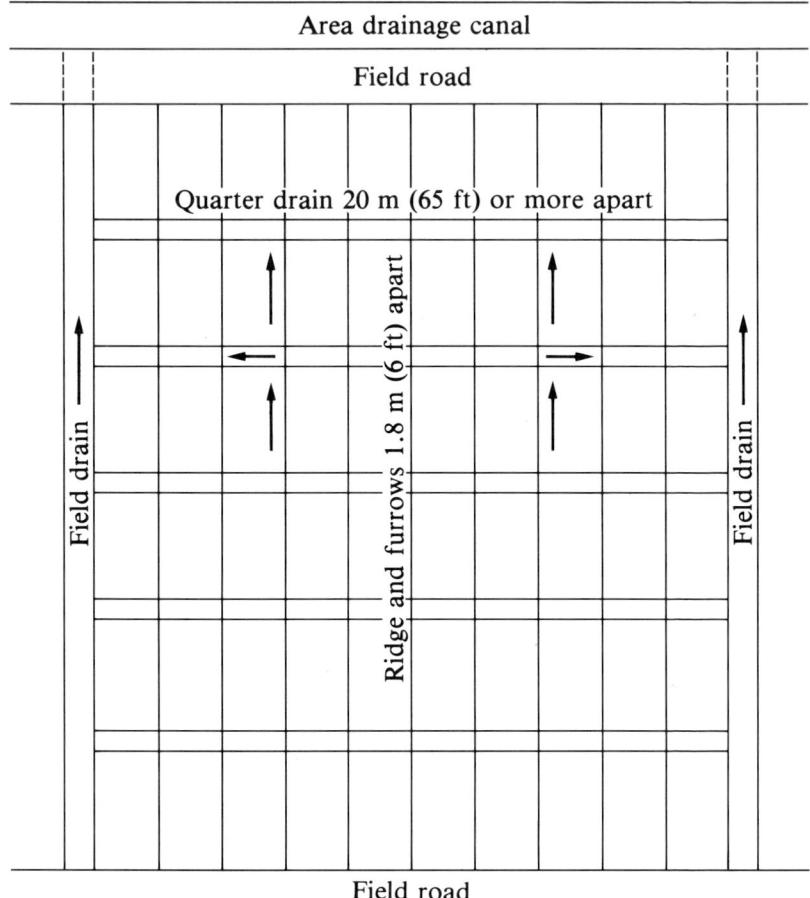

Fig. 5.3 A modified Louisiana bank layout

be mechanized. Land preparation, planting, the application of fertilizers and herbicides, the maintenance of the ridges, and reaping, can be carried out by wheel tractors which straddle the cane rows and draw suitable implements. A unique feature of Louisiana is the use of high-clearance tractors: they haul discs in fields of established cane to increase the height of eroded banks to the required level, and at the same time destroy weeds. In doing this, or any other cultivation operation, the quarter drains are filled and must quickly be reopened.

Cultivation on sloping land

On sloping land the emphasis is on soil and soil moisture conservation. There is at least one method of planting which is effective in this regard. After having imparted a good tilth to the soil, ridges and furrows are made along the contour (Fig. 5.4). Cane is placed in the furrows and covered with soil by breaking alternate ridges either mechanically or by hand. The unbroken ridges retain moisture and prevent soil erosion until well-formed lines of cane shoots serve the same purpose. They are then broken and a level surface is restored (Fig. 5.5).

This is of special value where planting is still a manual operation. Inevitably soil is pulled downhill to cover the cane by those using

Fig. 5.4 Sloping land contoured for planting (Photograph: M. S. Blackburn)

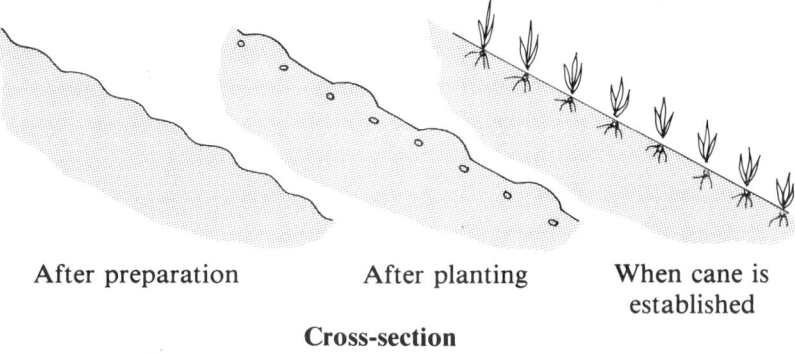

After preparation   After planting   When cane is
                                                        established

Cross-section

Fig. 5.5 Soil and moisture conservation on sloping land

hoes or similar implements, and thereby accelarates erosion. On the other hand, if a ridge is split to cover two adjacent rows, half of its soil must be moved upwards and half downwards. The same effect is achieved, of course, if covering is done by machine.

For similar reasons share and disc ploughs should never be used on steeply sloping land: their furrow-slices are always turned downhill. Repeated ploughing with discs has caused the removal of topsoil from large areas of shallow, undulating *rendzina* soils in the Naparima district of Trinidad. To rehabilitate the land the crowns of most of the hills are now planted with teak (*Tectona grandis* L.). All preparatory tillage operations on such land should be carried out by chisels or tines.

In the past the degree of slope has restricted the use of mechanical harvesters, and therefore has been a major factor in deciding whether new land should be brought into cultivation and whether old land should continue to be planted with cane. For example Lee & van Groeningen (1973) found that one machine, the Cameco Cost Cutter, could not operate on slopes greater than 8 per cent. More recently (1979), in Trinidad, the Class combine harvester, which can operate across the cane rows, on the smooth bank and furrow-free slopes present, has operated effectively, with the main limitation, in respect of the slopes which can be harvested, being the ability of the transport equipment to cope with the degree of slope present, rather than the stability of the harvester.

The measurement of sloping and undulating land

Surveyors' plans depict and measure land in a fixed horizontal plane; which is as it should be because considerable changes to the surface can be made by bulldozers, landslips and other means. Nevertheless, it is the unmeasured surface of the soil which must be cultivated and, if it is sloping or undulating, the surveyors' figures are little more than rough underestimates of its area. In these circumstances precise cost control of agricultural operations is not easy. There is another problem: the expression of yield. Should it be in terms of the plan area or of the surface area? The relationship between the two and the degree of the slope is not linear: if the slope is 25 per cent, the surface area is increased by roughly 3 per cent; but if it is 50 per cent, the increase is more than 12 per cent.

Where land has long been in cultivation and piece-work has been measured in terms of cane holes, or rods 3 m (10 ft) in length, the surface area of each field is well known. For example, at the most common spacing of 1.52 m (5 ft) ×,1.52 m, there are 4,304 cane holes to the hectare 1,742 holes/acre). Figures for measurement in rods are equally well known. On flat land there are 489 rods/ha (198 rods/acre) of the standard cambered bed illustrated in Fig. 7.1;

and on hilly land, where drains are not required, 539 rods (of four rows of cane)/ha (218 rods/acre).

In order to avoid confusion and to minimize the risk of financial wrongdoing, the surface areas, not the plan areas, of sloping and undulating fields should be used for all management-control purposes. This also is the practice of those who grow orchard and tree crops on slopes even steeper than those planted with cane; but for them the counting of trees or bushes, planted at known spacings, is not difficult. It is only when land is bought or sold that the transactions must be based on surveyors' plans.

The trend towards uniformity to meet the requirements of mechanization

In 1911 Noel Deerr wrote: 'The cane is grown under so many diverse conditions that no general sketch of its husbandry is possible.' While this is still true to some extent, in recent years there has been a trend towards uniformity. Spurred by the need to reduce production costs, especially those incurred in reaping by hand, modifications of the Louisiana bank system, so well suited to mechanical cultivation and reaping, has replaced (or is replacing) cambered beds and other forms of layout throughout the cane-growing world. For example in 1962 a directive was issued in Trinidad that the cambered beds of all fields to be replanted must be converted to 'Caroni ridges', a local adaptation of the Louisiana system. The method of doing this, and the results achieved, were described by Donawa (1973). Even the enormous difficulties of drainage and irrigation in Guyana have been overcome and 10 per cent of the area replanted each year is being converted to a type of ridge and furrow cultivation which, it is hoped, will allow the cane to be reaped (by mechanical harvesters (Young-Kong & Jackson 1976).

It has been suggested that cultivation systems have been altered to meet the needs of mechanization, rather than vice versa; and to a large extent this is true. Nevertheless, modern agricultural practices are better *per se* than those which they have replaced. In particular, land-forming eliminates irregularities within the field; and uniform layout leads to uniform drainage.

Exceptions to this general trend occur in China, India, Indonesia, Pakistan, Egypt, Taiwan and parts of Africa and South America where traditional methods are still used. In several of these areas landholdings are small and change to modern methods of husbandry, were this desirable, would not be easy. The position in Taiwan was described by Liu (1969) as follows:

The average size of farm in Taiwan is only about 1 hectare. Because of water control and labour supply, this small piece of land ordinarily is planted in many kinds of crops, each according to his own convenience. This random planting of many crops in a small area makes many improvements almost impossible; for instance, mechanization, irrigation or drainage.

Preparatory tillage operations

Turner (1945), working in the West Indies, attempted to classify soils according to their tillage and drainage requirements. He recognized four groups:

1. Soils which drain freely and have a permanent natural tilth. They require neither to be drained nor tilled.
2. Soils which gradually lose their tilth during the crop cycle. They require to be cultivated and to have some field drains to remove surplus water.
3. Soils which need to be drained but not cultivated.
4. Soils of poor natural tilth which require deep cultivation and an elaborate layout, such as cambered beds, to allow adequate drainage.

Examples of group 1 are rare as, indeed, are those of group 3; and it is difficult to determine whether some soils should be placed in group 2 or in group 4. Whatever the merit of Turner's classification, his work caused cane growers throughout the world to think deeply about the purpose of tillage and to reassess the value of their practices.

Tilth

The factors which determine whether a field shall be replanted or allowed to ratoon are discussed in Chapter 10. Briefly they are:

1. The need to create good tilth in the soil so that the plant might more easily form a large root system and thereby increase its yield.
2. The need to replace an established variety with one which has better growth characteristics: yield, including the ability to ratoon; and habit, with special reference to the use of mechanical harvesters.
3. The need to re-establish the plant population after severe damage has been caused by reaping in wet weather.
4. The need to replace susceptible with resistant or tolerant varieties when outbreaks of disease occur. One of the latest examples of this has occurred in the Caribbean countries of Guyana, Jamaica, St Kitts and Trinidad (and also Barbados) where the variety HJ 5741, which occupied 26,500 ha, has been attacked by smut disease. It is now being replaced, as quickly as possible, mainly

with the 'moderately resistant' B 41227 (Young-Kong 1976). Subsequently the variety B 4362 has been severely attacked by rust, in Cuba in particular, and has also had to be replaced.

However, it is the need to impart tilth, to put the land in 'good heart' which is usually the main consideration.

Most of the soils on which cane is grown are placed in Turner's groups 2 and 4: they should be ploughed and, at least to some extent, they should also be drained. Good husbandry requires that their structure should be such that they can easily be penetrated by roots to a depth of roughly 0.46 m (18 in) which, as shown in Fig. 2.7, is sufficient to accommodate all except Evans's rare rope systems.

Tilth is created mechanically by dry-season cultivation. There are many ways of doing this, each suited to local conditions of climate, field layout and method of reaping. Nevertheless they have several things in common:

1. In the first operation the old cane stools must be uprooted and destroyed by the passage of chisels, discs or shape ploughs. In doing this large clods of earth are formed. The few stools which survive should be dug up. M. Clarke (1979) stated that volunteer stools of B 49119 (susceptible to the disease) were the most significant source of smut in infection in Barbados.

2. The second operation is to harrow with chisels, discs or tines and thereby to form a seed bed. Any clods which remain, however small they may be, will disintegrate with the first rains, or when irrigation water is applied.

   Turner suggested that there is a close analogy between the effect of desiccation on soil in the tropics and subtropics, and of frost in colder climates. Both are essential, in their different circumstances, to the formation of good tilth by clod fracture. When particles of dry soil containing a proportion of swelling clay are moistened in hot climates, the colloids in the clay expand to such an extent that the aggregates into which they have been formed on drying are shattered; and in cold climates the expansion which takes place when water is transformed into ice acts in a similar manner.

   Nevertheless, sometimes it is necessary to reharrow, especially when heavy clays are being cultivated; but the first passage of the harrows must always be to a greater depth than the ploughed layer of soil (to prevent the formation of an impermeable 'pan'). Considerable experience and judgement are required in the timing of these operations if a satisfactory tilth is to be created.

3. Having cultivated the land, planting furrows are then made by the passage of V-shaped double mouldboards, variously called middle-busters, banking bodies and furrowing bodies; or by discs.

The spacing of the rows

Many experiments have been carried out in several countries to determine the optimum spacing of the rows, with special reference to yield and having regard to the cost of the seed-cane. With few exceptions they have shown that there is little to be gained by reducing the intervals between the rows to less than 1.5 m; moreover, spacings of 1.5–1.8 mm allow the easy passage of wheel tractors in cultivation operations, before and after planting, and also when the cane is being reaped. Hebert *et al.* (1967) reported that 1.8 m (6 ft) spacing was still the most economic in Louisiana, but qualified their conclusion by stating that one of the reasons for this was: 'The cost of the additional seed-cane that would be required and of adapting present cultivation implements to narrow row spacing would not be justified . . .' Thompson & du Toit (1967) found that, while spacings at intervals closer than 1.4 m (4.5 ft) might give higher yields of cane in South Africa, this was more than offset by reduction in quality. In the discussion which followed the reading of their paper, Kenning described experiments carried out in Argentina which showed that the optimum spacing was 1.6 m. Previous work by Cross had suggested that equal quantities of cane and sugar were obtained at any spacing between 0.9 m (3 ft) and 2.2 m (7 ft). The cultivation and harvesting equipment, developed for mechanized cane production, has been based on a 1.5 m spacing, which gives a satisfactory balance between facility for mechanical operations and yield losses which can occur at wider row spacings.

Considerable success is now being achieved in several countries by a new method (pineapple planting) in which two rows 0.9 m (3 ft) apart are made simultaneously and are spaced at intervals of 1.8 m (6 ft). This layout facilitates other operations, especially drip irrigation.

Some of the spacings which have been used at various times in different areas are given in Table 5.1.

Planting

Cane is then placed in the furrows, either by hand or machine, and is covered with soil.

Planting by hand

Before furrows were made, the usual method of planting by hand was to dibble two- or three-eyed setts (in the industry 'eye' is a synonym of 'bud'), using mattock-like tools, either into cane holes or into the land at set intervals along the lines of accepted patterns. Each seed-piece produced an identifiable stool. Nowadays this prac-

Table 5.1   *Some examples of the spacing of cane rows*

Distance		Country	Source
0.9–1.5 m	3–5 ft	India	Clarke (1959)
1.2–1.35 m	4–4 ft 6 in	Puerto Rico	Samuels & Capo (1954)
1.3–1.4 m	4–4 ft 6 in	Guadeloupe	Deerr (1911, quoting Bonâme)
1.2–1.5 m	4–5 ft	Java	Deerr (1911, quoting Prinsen Geerligs)
1.2–1.5 m	4–5 ft	Brazil	Barnes (1974)
1.4 m	4 ft 6 in	Swaziland	Humbert (1975)
1.4 m	4 ft 9 in	Australia	King *et al.* (1965)
1.5 m	5 ft	Mauritius	Clarke (1962)
1.5 m	5 ft	Trinidad	Donawa (1973)
1.5 m	5 ft	South Africa	Barnes (1974)
1.8 m	occasionally 6 ft		
1.2–1.8 m	4–6 ft	Jamaica	Innes (1954)
1.6 m	5 ft 3 in	Argentina	Kenning (1967)
1.65 m	5 ft 6 in	Belize	Seaton (1969)
1.65 m	5 ft 6 in	Cuba	Deerr (1911 ,quoting Reynoso)
1.8 m	6 ft	Louisiana	Hebert *et al.* (1967)

tice is still carried out only on small holdings to fill gaps (supply) in fields which are seldom replanted, or where there has been germination failure.

Where furrows are made, and planting is still by hand, whole canes, slightly overlapping, are laid in them. The stems are then cut *in situ* into pieces of varying length (appropriately three-eyed setts) by slashing them with a cutlass (machete). The reason for doing this is that sugar-cane exhibits top dominance. When the top bud of a stem grows, it is stimulated by hormone-like substances (auxins) which, at the same time, retard the development of the lower buds in increasing degree from top to bottom; so much that the lower ones remain dormant. Therefore, if whole stems are planted there are large gaps in the rows of the young shoots which arise from them. These must be filled, at considerable expense, with supplies. However, if the stems are chopped into three-eyed setts before being covered, all the buds can germinate to form continuous rows of uniform growth and supplies are often unnecessary (Fig. 5.6).

The setts are then covered with soil by breaking the banks by hand (using hoes) or by machine and, where furrow irrigation is used, the centre of the cane banks is left standing to assist with the control of irrigation water flows.

Planting by machine

Machines are replacing manual labour in most agricultural oper-

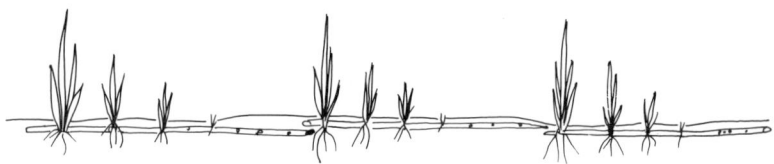

Where the stems have not been cut

Where the stems have been cut

Fig. 5.6 The effect of top dominance on germination

ations, and planting is no exception. The implements used, of varying degrees of complexity, are usually hauled by wheel tractors. In the simplest, setts carried in bins are fed by chute into the furrows and then covered with soil by discs. Others have attachments which open the furrows, chop whole stems into setts, place the setts in the furrows, apply bands of fertilizer parallel with and on each side of the row, and finally consolidate the soil with light rollers. In Australia, where it is necessary to take precautions against pineapple disease (Ch. 7) the machines may be fitted with a series of nozzles by which the setts are sprayed with organo-mercury compounds before being planted. Alternatively, they may have tanks containing the fungicide through which the setts must pass. A typical planting machine is illustrated in Fig. 5.7.

Planting material

If the seed-cane is of medium girth, with nodes roughly 15 cm apart, the quantity required to plant fields which are laid out in rows 1.5 m (5 ft) apart is 3.5–4.0 tonnes/ha. Higher rates (6–8 tonnes/ha) are essential when planting is carried out by machine because the setts are then placed in the furrows with less precision.

In Louisiana, where severe frosts can occur, precautions used to be taken to preserve seed-cane from the previous year's harvest. It was buried in the ground, a practice called 'windrowing', or protected from the cold weather in *matelas* until warm weather

Fig. 5.7 The Toft two-row planting machine. Bundaburg, Queensland, Australia (Photograph: Toft Bros)

returned. Now the setts are planted at a depth of 8 cm (3 in) on the tops of banks in the autumn (September–October) and the soil is consolidated by rolling. The planting density is 4.5 tonnes/ha. In spring (March–April), as the soil becomes warm, the tops of the banks are cut off to a width of 38 cm (15 in); thereafter the banks are raised by discs as the cane grows.

Frosts also occur in Argentina, where the setts are planted in shallow furrows, also in the autumn, and covered with soil to a depth of 15 cm (6 in). When warmer weather arrives most of the soil is removed. Thereafter the young shoots are earthed-up until the banks are roughly 1 m (3 ft) high. Because of the altitude of 300 m (1,000 ft), the frosts at Tucuman are more severe than at Jujuy and Salta, the other provinces of Argentina in which sugar cane is a major crop.

When planning reaping strategy, it is important to ensure that the material reserved as planting material should be young (older stems have hard basal buds which do not germinate quickly), of the required variety, and located close to the fields which are to be replanted. Generous allowance should be made for the accidental loss of nurseries by fire or by any other calamity.

Under irrigated conditions, where it is possible to control the age and nutritional status of the seed-cane, it is customary to arrange for seed-cane of six or seven months of age to be available at planting time. It has also been found that an extra application of nitrogen, of around 100–150 kg/ha, can improve seed-cane quality, giving better germination and leaf growth, and result in increased cane yields. Since the quality of seed-cane, and its freedom from diseases such as smut, is of such importance, frequent use is made of a series of nurseries, starting off with a small area of specially selected seed, which is carefully tended and rogued and, perhaps, heat treated to control ratoon stunting disease. Progressively larger nurseries are then established with the time of planting scheduled to provide seed-cane of the appropriate age for field-scale planting.

Where there is no danger of frost and irrigation water is available, or sufficient rain has fallen during the harvest period to allow planting to take place in unirrigated areas, the tops of young plant cane about to be reaped provided excellent seed-cane. Their eyes are soft and germinate readily. Quite apart from this, there is a financial advantage, if 'crop' rather than 'date' accounting is practised, in using tops as planting material. The tops carry only the cost of their cutting and transport as a charge to the newly planted field; whereas the full cost of growing the cane (in addition to that of cutting and transport) must be transferred, instead of being written off, when whole stems are used.

Stems used as planting material must be of the highest possible quality, and in particular their eyes must be undamaged. Injury to the eyes is most frequently caused by pests and by rough handling. Therefore, nurseries should be examined at regular intervals to make sure that the cane is not infested with *Diatraea* spp., *Chilo* spp. or other stem borers (Ch. 6). Planting material must also be free from disease and, where necessary, protected from infection. Of the diseases disseminated by seed-cane, ratoon stunting disease (RSD) and smut are by far the most serious. Where RSD has been detected, planting material is treated either with hot water or hot air, before being used, in order to kill the pathogen. The setts are treated with a fungicide in Australia to prevent them from being destroyed by the fungus *Ceratocystis paradoxa*, the causal agent of pineapple disease. Similar measures are taken in Brazil, Mauritius, South Africa and Taiwan (Ch. 7).

The consequence of using inferior planting material in Louisiana, twenty-five years ago, was described by Hebert (1956). Twenty per cent of the crop was used as seed-cane each year. Whole canes were chosen from old ratoon fields and less productive areas, and the amount of 'whacking or segmentation' of the stems varied from estate to estate. In spite of planting 'two running stalks and a lap' (6–12 tonnes/ha, according to the thickness of the cane used), there were large gaps in the rows (top dominance?) and 'it was not uncommon to see fields overrun with weeds'.

Supplying

In the rows the setts are placed end to end with a slight overlap. The buds give rise to primary shoots, the basal buds of which produce secondary shoots, the buds of the secondary shoots form tertiary shoots and so on (Ch. 2). More shoots are usually formed than can be sustained, and several wither and die, resulting in a stalk population characteristic of the variety, soil and climatic conditions. Significant increases in yield, well beyond the quantity of additional seed-cane required, can result from increasing the quantity of seed used and 'double planting' of poor tillering varieties can be well worth while, with the yield benefits extending with the ratoon crops.

Since cane is now grown in continuous rows, it is impossible to define a single stool in the field. Information which formerly was provided in terms of stools – for example the numbers of stems or the intensity of infestation by pests – should now refer instead to 1 m (or 1 yd) of a cane row.

With good seed material and under good soil moisture conditions, or where irrigation water is available, supplying should not be required and the level of supplying necessary is generally in the inverse to the effectiveness of the supervision given to the planting operation. Gaps of more than 0.6–0.9 m (2–3 ft) should be filled with setts at the earliest opportunity, to ensure that the shoots are not shaded out by the more advanced growth of the original setts.

Variations from the general pattern

Variations of this sequence of operations are practised in some important, mainly subtropical, cane-growing countries: Australia, Egypt, Fiji, Louisiana, South Africa, Taiwan and parts of Mauritius.

The problem of rocks in Mauritius

Mauritius is unique: of volcanic origin, the land is strewn with basalt rocks, the remains of lava flows of the late Tertiary and early Pleistocene periods. The clearing of these rocks to expose the underlying soil for cane cultivation was described by Evans (1962). When such land is replanted, the trash is burnt, the stools destroyed by rotary hoes or light harrows; and then sub-soilers, drawn by heavy tractors, are passed through to expose the stones. These are removed by bulldozers or by hand and are stacked in pyramid shapes or as walls 0.91 m (3 ft) wide by 1.22 m (4 ft) high between rows of cane. The spacing of the walls can be so close that they separate only two rows of cane, but ten to twenty row intervals are more usual. Further preparatory cultivation is unnecessary.

Cover crop

Forty or fifty years ago legumes were grown in many countries as cover crops between successive cane cycles, and then turned under to improve the quality of the soil with the special purpose of increasing its humus content. The cow-pea (*V. unguiculata* L. Wolf syn. *V. sinensis*) was grown in Cuba, the velvet bean (*Stizolobium aterrimum* L.) in Fiji, sunn hemp or sanai (*Crotalaria juncea* L.) in India and *Melilotus indica* L. in Louisiana. In Trinidad woolly pyrol or black gram (*Vigna phaseolus* syn. *V. mungo* L.), cow-pea, sunn hemp and jackbean (*Canavalia ensiformis* L.) were chosen from thirty-two species of legumes which had been introduced for this purpose by 1930.

The practice was discontinued in the 1930s and 1940s everywhere except in Fiji, Florida, Louisiana, South Africa and Taiwan in all of which, except Fiji, cane is grown under subtropical conditions.

In Fiji the crop distribution in 1956 was one-third of the area in plant cane, one-third in ratoon and one-third in short-term fallow sown to a leguminous green crop (South Pacific Enterprise 1956).

In Australia, it was recommended that old stools be destroyed by rotary hoes where the soil has a good natural tilth; and on other soils by disc ploughs. Legume seeds should then be broadcast or planted in rows, followed by spike-harrowing and rolling.

The velvet bean (var. Somerset) was widely planted until 1957, when it was heavily attacked by a wilt disease caused by the fungus *Phytophthora drechsleri*. It was replaced in affected areas by the cow-pea cultivars Poona and Giant which were susceptible to *Fusarium* wilt, nematode damage and attack by *P. vignae*. They were succeeded by Reeves's Q 1582. A breeding programme to produce disease- and insect-resistant cow-peas was started at the Queensland Bureau of Sugar Experiment Stations in 1955. Early commercial successes were Meringa, a cross between Malabar and Buff, high yielding and resistant to *Phytophthora* wilt, and Mulgrave, whose parents are Havana and Black Eye 5. On some of the northern soils, mung bean (woolly pyrol) is grown (Bieske 1965).

When the cover crop is ready to be turned in, the land should be ploughed more deeply by discs and the furrow slices not inverted but left in almost vertical bands to encourage aerobic decomposition of the legumes, which is achieved in roughly three weeks. In soils of good natural crumb structure no further tillage is necessary; else-where, in more intractable soils, one or more disc ploughings, followed by harrowing, are made according to the judgement of the planter. For this purpose spike harrows have been replaced every-where by disc harrows.

The use of cover crops in South Africa was described by Pearson (1954). Sunn hemp and the Mauritian bean, a cultivar of the velvet bean, were most widely grown. More recently kenaf (*Hibiscus sabdariffa* L.) has been used successfully for the same purpose.

Sesbania spectabilis L. is grown in Florida.

Despite all that has been written on this subject, King (1960) reported that in subtropical as well as in tropical countries, the organic matter content of virgin soils brought into cultivation rapidly falls to a constant level, and that the practice of conserving crop residues, credited in some countries with yield increases, had caused no measurable response in an experiment of twenty-five years' duration in Queensland. Moreover, in the discussion which followed the presentation of this paper, Summerville stated that in most cane fields in Queensland monoculture, without the planting of cover crops, had been practised for many, many years.

Suffice it to say that the practice of growing of cover crops between cane cycles has declined rapidly in recent years, and probably will continue to do so.

Choice of variety

The first requirement of a variety is that it should not be susceptible to the major diseases prevalent in the area in which it is planted. As described in Chapter 4, varieties released by cane-breeding stations for general cultivation are resistant or tolerant to most of the diseases known to occur in the countries which they serve. Suitable precautions are taken to prevent damage by those diseases against which the varieties do not have in-built resistance. Thereafter the main criteria for selection are the yield of sugar and the ability to ratoon. Mention has been made of growth habit in relation to the use of mechanical harvesters; but there are several other factors which must be considered.

In theory, early ripening varieties should be grown to ensure that, when they are reaped at the beginning of the crop season, the juice is of high quality and the *rendement* good; and, provided that the harvest can proceed in an orderly manner, this should be done. However, if the reaping programme cannot be controlled and is determined largely by such occurrences as fires of 'unknown' origin, it may be unwise in such circumstances to plant early ripening varieties. Many of them, for example B 4362, quickly dry out and their stems become pithy and cork-like if their reaping is delayed.

Another character which might deter the grower from planting a variety is the hardness of its rind. It was for this reason that B 49119, known as 'iron cane' in Jamaica, was not more widely grown in some Caribbean countries where cane is still being cut by hand. In other countries, where reaping is increasingly being mechanized, this is relatively unimportant: the cutting devices of harvesters can easily be changed or sharpened.

The sudden and calamitous failure of the Bourbon cane in Brazil, Mauritius and several other countries, mentioned in Chapter 1, took place many years ago. Nevertheless it is still remembered, and many growers feel that it is unwise to plant one variety, however good, to the exclusion of all others. This opinion was reinforced when the area under B 34104 was rapidly expanded in Guyana, Jamaica and Trinidad in the 1940s; and equally rapidly reduced when this new variety succumbed to leaf scald disease in Guyana, to mosaic disease in Jamaica and showed little ability to recover from froghopper damage in Trinidad. Mention has already been made of the recent failure of HJ 5741 in these same countries (and also in Martinique and St Kitts), caused by an outbreak of smut disease. For all that, the sugar industries of several countries relied successfully for many years on a single variety: South Africa on NCo 310, Mauritius on M 134/32 (Fig. 5.8) and Colombia amongst others on POJ 2878. Moreover the replacement of NCo 310 and M 134/32 was caused not by poor performance but by the availability of even better varieties; while POJ 2878, after fifty years, still is widely grown in Belize and Colombia.

The lesson to be learned is that a new variety should be introduced gradually; and only after successful performance over many years should it be widely planted.

Time of planting

In order to achieve the maximum yield of sugar per arable hectare per annum in areas unaffected by frost, fields should be planted as

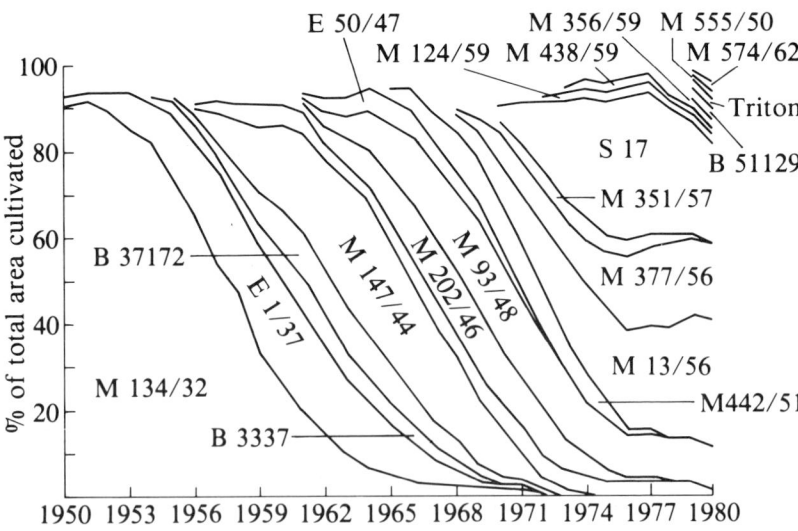

Fig. 5.8 Varietal trends 1950–80, Mauritius (Sugar Industry Research Institute, Mauritius)

quickly as possible after preparatory cultivation operations have been completed. To achieve this in the irrigated areas of Hawaii, reaping and planting often take place simultaneously in adjacent blocks of land. Where irrigation is not practised, planting must be delayed until the onset of the rains; but then should be carried out with dispatch. In many tropical countries the timing of the onset of the wet season can be predicted within narrow limits, and its arrival is often preceded by heavy showers. Provided that the showers cause water to penetrate the soil to a depth of, say, 5 cm (2 in), cane tops should be laid in the furrows and the banks broken. The newly exposed surface soil may be dry but that which surrounds the planting material will be wet. Invariably germination is excellent.

The principle that as little time as possible should be lost between successive cane cycles is not universally accepted. As has been discussed earlier, in some countries, especially in the subtropics, it is felt that monoculture adversely affects fertility. Therefore it has been advocated that a leguminous crop (green manure) should be grown and incorporated with the soil to improve its structure before the next cane cycle is started. In Jamaica, Chinloy & Hogg (1969) found that rotation with Pangola grass (*Digitaria decumbens* Stent) arrested the decline in yield of plant cane on one estate, probably because the population of harmful root organisms (nematodes) was reduced during the period under lea; but this is not a common occurrence. The relation between sugar-cane and soil fertility is discussed in Chapter 3.

The growing of food for local consumption is combined with cane cultivation in some countries (Barbados, Mauritius, India, Indonesia, Pakistan and Taiwan, among others). For example in Barbados in 1979 it is stipulated by law that not less than 12 per cent of the arable area of every landholder shall be planted with vegetables. The area under cane and the time of its planting must be arranged to meet this requirement, with consequent loss of sugar production, although where cane is interplanted with vegetables, particularly beans, which do not form a large canopy and which only occupy the ground for a short time, there may be little or no loss of cane yield.

Yield

One of the most significant indications of the success of a grower is the productivity of his land expressed as the yield of cane (or sugar) per hectare per annum. The yield per hectare reaped, though often quoted, is of little importance unless related to the age of the cane at harvest. For example a yield of 170 tonnes cane/ha reaped, if the average age at harvest is twenty-four months, indicates lower productivity than 90 tonnes cane/ha if annual cropping is practised.

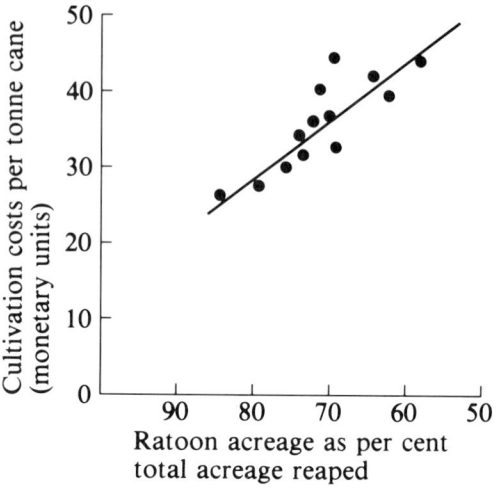

Fig. 5.9 The relationship between production costs and ability to ratoon

That is not all: the level of productivity itself must be judged by the cost of its achievement. The many considerations which determine the size of the replanting programme, year by year, and therefore the proportion of younger, higher-yielding fields, are discussed in Chapter 10. In general, provided that the required level of production is maintained, the higher the ratio of ratoon to plant cane, the lower will be the average cost of production. This is illustrated in Fig. 5.9. The data refer to one year's operation of thirteen estates which were under the same management.

Ratoons and ratooning

After a newly planted cane field has been harvested, the bud and root primordia of the rootstocks develop when ecological conditions are favourable and give rise to a 'stubble' or 'ratoon' crop. In exceptional cases as many as twenty ratoons may be reaped before the field is replanted. As the new shoots grow and develop roots, the old roots die and decompose. Thus each successive crop is sustained by water and nutrients absorbed by its own new root system. However, as time goes by the soil surface can lose its structure by the beating action of rain and become compacted by the passage of wheeled equipment: it loses the tilth so carefully and expensively imparted to it by preparatory cultivation. Consequently, other factors being constant, the formation of an adequate root system becomes increasingly difficult for each successive ratoon, and yields decline (Table 5.2).

Table 5.2   *The yield of sugar-cane in Barbados in 1976*

Category	Area (ha)	Yield (tonnes cane per ha)
Plant cane	2,221	70.07
First ratoons	2,217	62.98
Second ratoons	2,070	60.84
Third ratoons	2,390	58.91
Fourth ratoons	1,096	56.09
Fifth and older ratoons	377	56.29
Total	10,371	61.75

Source: Adapted from Clarke *et al.* (1976).

By and large the establishment cost of plant cane is very much higher than the cultivation cost of ratoon cane, a difference which has been increased by the widespread cultivation of modern vigorous hybrid varieties, and the increasing realization that, in many instances, the usual cultivation practices such as subsoiling can actually cause a loss in yield, particularly when irrigation is not applied promptly after the cultivation is completed. Indeed the high cost of establishing plant cane may be likened to and sometimes is accounted a capital investment, with the yield of cheap ratoon cane as dividends.

In some countries the maximum number of ratoons permitted to be grown, or the proportion of the assigned area planted with cane allowed to be reaped each year, is determined by statute or by an inflexible system of crop rotation. Reference has been made to legislation to this end passed in Barbados and Queensland, and to the patterns of agriculture in Java and Taiwan. Elsewhere, in deciding the size of the planting and replanting programme, and therefore the area to be left to ratoon again, the grower gives serious thought to many factors of which the most important are:

1. The balance between the lower yields of cheaper old ratoon cane and the higher yields of more expensive younger cane, taking special note of the effect of output on the incidence of fixed charges.
2. The condition of the sugar market, bearing in mind the need to fulfil quotas fixed by international agreement; and to take up the shortfalls of other producers thereby staking a claim, based on performance, for preferential treatment in future negotiations.
3. The social consequences, in some countries, of having little or no work for field employees in the period immediately preceding the start of grinding operations if the area in young plant cane is unusually small. At that time of year agricultural activity is

confined to newly planted fields: the others are well grown, have covered in and await harvest.

4. Technical considerations, especially the wish to increase the area planted with a superior new variety as quickly as possible, or to combat an outbreak of disease by planting resistant varieties. An example of the former was the spectacular expansion of B 37161 in Barbados from 8 per cent by area in 1942 to 96 per cent in 1949; and of the latter, the rapid replacement of B 34104, stricken with leaf scald disease in Guyana in 1950. In countries where the froghopper is a serious pest the population of eggs in diapause, which enables the insect to survive from one year to the next, increases with the age of the ratoon; but the eggs are killed by desiccation during cultivation preparatory to replanting. Therefore, if adequate control of the pest cannot be assured, it is unwise to have a large proportion of the acreage in old ratoon cane.

Comparable data for 1969 concerning three groups of estates in two different countries are given in Tables 5.3 and 5.4. They illustrate the principles set out above. They also show that, where it is very dry, the cost of irrigation can be of paramount importance. For all these reasons there is wide variation in the number of ratoon crops taken in different areas. Where advanced technology is used the number is usually within the range of three to nine.

Table 5.3 *A comparison of the cultivation costs of plant and ratoon cane, 1969*

	Cultivation costs (£)		
	Frome Jamaica	*Monymusk Jamaica*	*Caroni Trinidad*
Plant cane per ha	187.55	134.67	143.81
Ratoon cane per ha	68.47	91.25	40.20
Plant cane per tonne	2.16	1.30	2.43
Ratoon cane per tonne	0.81	1.22	0.56
Average per tonne cane, including overheads, maintenance, research and depreciation	2.02	2.48	1.66
Average per tonne 96° sugar	21.68	29.62	16.68

Notes
1. At Monymusk irrigation was responsible for one-half of the cultivation cost of plant cane and two-thirds of that of ratoon cane.
2. At Caroni froghopper control accounted for 30 per cent of the cultivation cost of ratoon cane but for only 1 per cent of that of plant cane.
3. There were considerable differences between soil conditions and the scales of wages paid at each group of estates.
Source: Adapted from Tate & Lyle (1969).

Table 5.4 *Area of plant and ratoon cane harvested in 1969 (ha)*

Category	Frome Jamaica	Monymusk Jamaica	Caroni Trinidad
Plant cane	493	689	2,182
Ratoon cane	4,332	5,291	18,134
Percentage plant cane	10.2	11.5	10.3

Source: Adapted from Tate and Lyle (1969).

In certain circumstances it is difficult to define the crop cycle. For example in Cuba, many years ago, the land was planted after having been imperfectly cleared: tree roots and stones were left *in situ*. If a cane stool died, it was replaced; but the field as a whole was not replanted until twenty years or more had elapsed. Other problems of definition occur where reaping frequently takes place in wet weather. Stools are uprooted and the damage can be made good either by replanting the whole field or by renovating its most badly affected parts. When push-rake harvesting was first used on estates on the windward coast of Hawaii Island an arbitrary figure of 30 per cent was chosen to decide which course should be taken. Immediately below this limit slightly less than 30 per cent of some fields would be plant cane at the next harvest, and roughly 70 per cent ratoon cane of various ages. In which categories should they be placed? Similar decisions must be made frequently in the Cauca Valley of Colombia where, at one estate in 1972, replanting accounted for 80 per cent of the total cost of ratoon cultivation: 2,500 of 3,100 pesos/ha (Holguin 1973).

Weed control in sugar-cane

Weeds have been described as plants which are out of place, and there is certainly no room for them in efficient cane growing. When they are allowed to grow without restraint, as in abandoned fields, they quickly smother and destroy sugar-cane.

For many years the control of weeds was achieved by labour-intensive means; gangs of people, armed with hoes or forks, moved across fields and uprooted the weeds. This method is still practised by small growers whose holdings are worked by family labour.

Mechanical weed control

Cultivation in dry weather, prior to replanting, should cause all weeds which are propagated vegetatively to be destroyed by desiccation. Deep cultivation of the cane fields of Trinidad during certain parts of the dry season reduced the invasion of perennial grass weeds (Blackburn *et al*, 1952), but due to the differences in seasons, this method was not so effective in Jamaica.

In some countries, for example Australia, mechanical weed control is still preferred to the use of herbicides. Discs, drawn by wheeled tractors in earth-forming (moulding) operations, also destroy weeds, and, with good timing, the two functions may be carried out simultaneously. However, it is essential that the weeds be destroyed when they are young and not allowed to establish deep roots. Therefore, occasionally discs may have to be used to control weeds when there is no need to mould the cane rows. However, mechanical cultivation may set back the cane crop by root pruning.

Chemical weed control

Chemical weed control was first attempted in the 1920s and 1930s but the substances used, sodium chlorate and sodium arsenite, are so dangerous – sodium chlorate because of its inflammability, and sodium arsenite because of its mammalian toxicity – that they were little used. Then came the development of the hormone-type herbicides 2,4-D and MCPA, and with them the beginning of a new era in weed control. These herbicides were the first used to destroy weeds in established cane fields. As their use was limited to non-woody broad-leaved weeds, and not all of these, oil and pentachlorophenol (PCP) were frequently added as a contact herbicide. This technique, however, only killed established weeds and it is the newly germinating plant cane that suffers most from weed competition. In Hawaii they developed a method of spraying 2,4-D immediately after planting. A single treatment controlled the weeds until the cane closed in when the leaf canopy restricted further weed growth.

With the development of the triazine herbicides (atrazine, ametryne, etc.) and the ureas (e.g. diuron) which were relatively insoluble, these could be sprayed after planting the setts but before the weeds germinated and a much wider range of weeds including many annual grasses could be controlled. If broad-leaved weeds were also a problem 2,4-D was sometimes added.

It is also possible to apply these herbicides after the cane has germinated and on ratoon crops. Where the cane trash was left as a mulch this restricted weed growth in the mulched areas, though creeping weeds and some strong-growing grasses could often grow through the mulch. Dalapon was the first effective grass-killer to be developed, but this damaged many varieties of cane and could not be sprayed over the crop. However, it gives good control of the annual grasses *Digitaria* and *Setaria* in Mauritius if the spray is directed away from the cane. Paraquat will knock back the grasses, but will also scorch the cane severely if sprayed on the green leaves. Paraquat, and more recently glyphosate, have been widely used to clean up a grass infestation before either the plant cane germinates or the ratoon cane develops after cutting. Glyphosate will also kill nutgrass (*Cyperus* spp.). Asulam has been used selectively to kill grass weeds, particularly wild sorghum (*Sorghum halepense*), but also *Digitaria* spp., *Rottboellia* and many annual grasses. Frequently it is used in a mixture with Actril to control a range of grasses and broad-leaved weeds in established cane.

The current recommendations for cane growers in South Africa are summarized in Table 5.5. This illustrates the flexibility available to cane planters, and the wide range of mixtures that can be used to grow sugar-cane without weeds. Similar local recommendations are available in cane-growing regions.

Many of these herbicides, particularly the insoluble ones such as atrazine and diuron, when applied to existing weed growth act more

Table 5.5   *Herbicide recommendations in South Africa, 1982*

Herbicide	Rate per ha	Weeds controlled	Comments
1. *True pre-emergence: short term* (4–5 weeks)			
2, 4-D	4 litres	B/L some G	Beware of drift on to other plants
MCPA	7 litres	B/L some G	
2. *True pre-emergence: long term* (6–12 weeks)			
Dual + atrazine*	2.75 litres + 2 litres FW	B/L, G	Under favourable conditions Ce well controlled
Lasso + atrazine* (EC 384)	5 litres + 2 litres FW	B/L, G	Under favourable conditions Ce well controlled
Destun + atrazine	5 kg + 2 litres FW	B/L, G, Ce	Corrosive to metal. Do not leave mixture in tank or pipelines. Some activity on Cr. Use on soils of less than 35% clay only
Velpar + diuron	0.5 kg + 2.5 kg	B/L, G	*Only* on ratoon cane, *not* on plant. Under favourable conditions Ce well controlled. Add directly to tank. Do not make a slurry
Sencor + diuron	2.0 kg + 2.0 kg	B/L, G	Under favourable conditions Ce well controlled

Table 5.5 (continued)

Herbicide	Rate per ha	Weeds controlled	Comments
3. *Late pre-emergence: long term* (6–12 weeks)			
Dual + atrazine + Gramoxone	2.75 litres + 2 litres FW + 1.5 litres	B/L, G, Ce, Gr	Spray before cane leaves unfurl OR as a directed spray
Lasso (EC 384) + atrazine + Gramoxone	5 litres + 2 litres + 2 litres	B/L, G, Ce, Cr	Spray before cane leaves unfurl OR as a directed spray
Velpar + diuron	0.5 kg + 2.5 kg	B/L, G, Ce	Controls Ce better after emergence; good on *Panicum maximum*
Sencor + diuron	2 kg + 2 kg	B/L, G, Ce	Controls Ce better after emergence. One of the least phytotoxic of all recommendations
4. *Post-emergence: short term* (5–6 weeks)			
Diuron + 2, 4-D + S	2.5 kg + 2 litres	B/L, G, Ce, Cr	Good under moist soil conditions
Diuron + MCPA + S	2.5 kg + 4 litres	B/L, G, Ce, Cr	Spray before tillering of grasses
Diuron + Actril DS	2.5 kg + 1.25 litres	B/L, G, Ce, Cr	Direct spray away from cane foliage
Diuron + Certrol DS	2.5 kg + 1.25 litres	B/L, G, Ce, Cr	
Ametryne + 2, 4-D + S	3 litres + 2 litres	B/L, G, Ce, Cr	Good under moist or dry conditions
Ametryne + MCPA S[†]	3 litres + 4 litres	B/L, G, Ce, Cr	Spray before tillering of grasses
Ametryne + Actril DS[‡]	3 litres + 1.25 litres	B/L, G, Ce, Cr	Direct spray away from cane foliage
5. *Post-emergence: long term* (8–12 weeks)			
Sencor + diuron	2 kg + 2 kg	B/L, G, Ce	
Sencor + ametryne	2 kg + 3 litres	B/L, G, Ce	Use on soils of greater than 25% clay only
Velpar + diuron	0.5 kg + 2.5 kg	B/L, G, Ce	Good on *Panicum maximum*. Only for ratoon, not for plant cane

[*] Atrazine: where clay content of soil is greater than 35 per cent 3 litres should be used. Atrazine is also available as a WP which should be applied at 1.25 kg/ha, or 2.0 kg/ha if the clay content of the soil is greater than 35 per cent.
[†] S = surfactant.
[‡] Use on soils of greater than 25 per cent clay only.

Key: B/L = broadleaf weeds; Ce = *Cyperus esculentus*; G = grasses; Cr = *Cyperus rotundus*.

Source: *A Herbicide Guide for Cane Growers*, South Africa.

effectively if 1 or 2 per cent of a self-emulsifying refined mineral oil such as Actipron® is added.

Damage to cane: phytotoxicity

- Herbicides sprayed before the cane emerges are far less damaging than those sprayed post-emergence.
- Older cane is more susceptible to damage from chemicals than is younger cane.
- Post-emergence applications should always be directed so as to spray as little cane leaf area as possible, e.g. use drop arms and flood jets in the inter-row.
- Poorly grown cane suffering from 'wet feet', nematode damage or nutrient deficiency, is more susceptible to damage than is well-grown cane. Some cane varieties appear to be more susceptible to damage than others.
- When these varieties are grown, particular care should be taken to direct post-emergence sprays away from the crop foliage, or preferably good pre-emergence sprays should be used.
- Hot, humid conditions increase the likelihood of damage to cane.

Ratoon cane

Ratoon cane develops ground cover more quickly than plant cane and suffers less from competition, particularly from *Cyperus* spp. which do not grow very tall. Cheaper, short-term treatments are often satisfactory, but care must be taken to prevent serious competitive weeds, which are difficult to control, becoming established and competing with successive ratoon crops.

Field borders and railways

Chemical weed control is also used to protect field borders, main drains and railway tracks, not only to keep them clean but also to prevent them becoming a source of weed invasion into the cane fields. Dalapon and 2,4-D are the traditional herbicides used to clean up such areas. Paraquat may also be added, used alone or with diuron. More recently glyphosate has been used where the weed growth is varied and deep rooted, but this is currently a more expensive treatment.

While field drains and borders are adequately treated using knap-sack sprayers, railways are sprayed with a boom sprayer attached to a tank drawn by a locomotive.

Application

At first herbicides were applied by knapsack sprayers whose lances had been fitted with double T-jet nozzles. The asymmetric orifices of the nozzles allowed even distribution of the liquid over the soil surface. Knapsack sprayers have been replaced in large-scale operations by tractor-mounted boom sprayers or aerial application, but are still widely used by small cane growers and for special problems.

Damage to crops in adjoining fields and gardens

The herbicides used in sugar-cane, particularly 2,4-D, can be phytotoxic to several other plantation and garden crops. The small droplets in the spray can drift with the wind, and where 2,4-D or mecoprop ester is used, these may volatilize and move downwind for quite a distance. Even though the ester is described as 'nonvolatile' like the iso-octyl ester frequently used, there are often more volatile esters present as impurities which can cause damage. Consequently, all herbicides should be used with great caution in boundary fields and on roads and railways which pass, by right of way or private arrangement, through land not under the complete control of the cane grower. If expensive claims for damage are to be avoided, sensitive areas such as these should be sprayed only on calm days, or when a favourable wind is blowing, using non-volatile products.

Changes in flora

There is considerable variation between weeds in their reaction to pre-emergence herbicides. Although few seeds germinate, some adapt more readily than others to their new environment, by selection or mutation. For example corn grass (*Rottboellia exaltata* L.) for many years the most noxious weed in Trinidad was eradicated but soon replaced by fowlfoot grass (*Eleusine indica* Gaerth.). A survey of weed flora in fifty fields, representative of the several soil types on which sugar-cane is grown, was made before herbicides were used on a large scale in that country. Twenty years later, Goberdahn (1973) examined the same fields, using the author's notebooks as a guide, and found that whereas nineteen species (all dicotyledons) were no longer present in the fields, two grasses, *Andropogon annualatus* Forsch. and *Panicum maximum* Jacq., with *E. indica*, had increased in status and were abundant and often dominant, and two others (*Brachiaria platyphyllia* Nash and *Ischaemum rugosum* Salisb.) occurred frequently and often were abundant. No broad-leaved weed had become important and the cause of their decline must be attributed to the long use of substituted phenoxyacetic acids.

Destruction of cane prior to replanting

Glyphosate and more recently Fusilade have been used to kill established cane stools and grass weeds prior to replanting.

Maintenance of soil fertility

Three aspects of sugar-cane agronomy have provoked considerable discussion and controversy. They are the abandonment of the use of organic manures, especially in tropical countries: the long-term effect on soil fertility of monoculture; and the similar consequences of pre-harvest burning.

Artificial fertilizers and organic manures

Perhaps the most striking development in sugar-cane between the two world wars was the replacement of farmyard (pen) manure by artificial fertilizers. Although it was shown in many countries that this gave rise to equally good yields on all soils, and better on nutrient-deficient ones, and although it was manifestly uneconomic to keep animals, replaced by tractors in all other operations, for the sole purpose of making pen manure, nevertheless in some quarters the change was regarded with suspicion and disapproval. Sir Albert Howard went further: he ascribed all the medical, economic and agricultural ills of the West Indies to it. Follett-Smith (1943) showed that this extreme view could not be supported by data concerning the sugar industries of Guyana and Barbados. Statistics given later by Saint (1953) and Hudson (1973) confirm this conclusion. Yet all misgivings concerning possible soil deterioration cannot be dismissed out of hand. The effect would not be immediate and would be reflected, other things being equal, by a gradual diminution in yield which could well be masked by increases caused by other factors, such as the planting of new and better varieties. By and large the soils of Barbados are extremely shallow *rendzinas*, often less than 1 m in total depth, and overlying coral rock. Nevertheless there is no evidence, as yet, that they have been impoverished by lack of pen manure.

In some subtropical countries agricultural practices are different from those in the tropics. Attempts have been and still are made to maintain or to increase the humus content of their sugar-cane soils while, at the same time, full use is made of artificial fertilizers. Cover crops, usually legumes, are grown between cane cycles and then ploughed in; and trash, bagasse and other by-products are returned to the soil. Whatever the organic material, nitrogen is required to facilitate its decomposition by bacteria, a process so timed that cane nutrition does not suffer.

Reference has been made also to the enigma of Queensland, where in 1959 the recommendation of the Bureau of Experiment Stations to grow cover crops was carried out by very few cane growers. Indeed in a discussion on yield decline, with special reference to ratoon stunting disease, King (1960) stated that, year in year out, the organic matter content in the surface horizons of soil under cane remained constant. He reported that in an experiment of twenty-five years' duration: 'Neither the protective trash cover on the soil nor the incorporation of the residue in the ploughed zone has resulted in a larger crop due to moisture conservation in a low-rainfall area or increased yield due to soil improvement.'

Similarly in Hawaii the advice of Eckhart, given in 1909, that the cane cycles be interspersed with fallow periods during which leguminous crops should be grown, was not accepted. Instead a programme of intensive mineral fertilization was followed, and the yields increased and continue to do so. The achievements of the Hawaiian sugar industry attracted considerable attention. Agee (1931) gave a thought-provoking analysis in detail of these practices to his Hawaiian colleagues, and Moir (1932) to an international gathering. Agee's conclusion was that, crop after crop, a mass of roots formed in part by the fertilizers each year decayed and thereby increased the reserves of soil organic matter. This probably was the safeguard which '. . . holds our agriculture on a sound footing . . .'. Doubtless the excellent physical properties and great depth of the red volcanic soils on which much of Hawaii's cane is grown also helped to achieve these satisfactory results.

Agee's supposition was supported by crop-rotation studies carried out in Uganda where Martin (1944) found that fertility, measured by the proportion of water-stable crumbs in the soil, could be maintained if a three-year period under elephant grass (*Pennisetum purpureum* Schum and Thonn.) followed three years under exhaustive cash and subsistence crops. Organic manures failed to improve soil structure, being oxidized before they could be decomposed to colloidal dimensions. Legumes were equally unsuccessful. Grasses, on the other hand, were particularly effective. Martin attributed this to their widely dispersed, stringy root systems, as opposed to the taproots of legumes. Sugar-cane has a root system similar to that of elephant grass.

Effect of monoculture of cane

Many sugar-cane-growing areas have used the same land for sugar-cane for many years without any obvious deleterious effect: as sugar-cane is a perennial grass which has additional characteristics such as some value in controlling erosion, this is to be expected. Maclean (1975) compared land which had been under sugar-cane for ninety years with adjacent uncultivated land; chemical and physical characteristics were very similar, except for some compaction and lower porosity in the inter-rows of the cane, which is to be expected in this industry which relies heavily on mechanized inter-row cultivation.

Pre-harvest burning

In none of those countries in which pre-harvest burning has been practised for many years has it caused sugar production demonstrably to be reduced. On the contrary, the application of modern technology has caused spectacular and sustained increases in yield in all of them: Australia, Guyana, Hawaii, Louisiana and Trinidad to 1969, after which other factors caused production to decline.

By contrast, recent economic pressure caused pre-harvest burning to be brought into general use in St Kitts and Barbados, and the effect of so doing is causing much concern (Table 5.6). After three years of controlled burning in St Kitts, it was reported in 1974 that yields and the ability to ratoon had been seriously reduced (WISA 1974). Therefore the practice was discontinued. Likewise in Barbados, where a closely reasoned examination of the causes of declining yields in that country led to the conclusion that pre-harvest burning was an important factor and that, unless drastic and concerted action was taken, deterioration would continue (Hudson 1973). In 1975 the President of the Barbados Sugar Producers' Association attributed the fall in production to poor rainfall (30,600 tonnes), reduced acreage (31,600 tonnes) and burning (24,400 tonnes). Therefore in that year permission to carry out pre-harvest burning was withheld. Nevertheless, as described in Chapter 8, fires of 'unknown origin' occur frequently in both countries.

The conservation of moisture is crucial to successful cane production in Barbados and St Kitts, and it might well be that the

Table 5.6 *Annual production of sugar in Barbados and St Kitts 1965–80 (tonnes tel quel)*

Year	Barbados	St Kitts
1965	206,963	39,155
1966	181,551	38,445
1967	211,862	39,232
1968	167,999	35,470
1969	144,952	36,000
1970	160,420	27,265
1971	140,451	25,449
1972	116,500	26,406
1973	120,839	24,463
1974	112,680	26,732
1975	101,967	25,855
1976	106,486	36,460
1977	119,836	42,794
1978	103,785	40,899
1979	117,110	40,745
1980	135,493	36,609

Source: Adapted from WISA (inc.) Annual Reports and ISO Yearbooks.

loss by fire of the mulching effect of trash has indeed contributed largely to the decline of the industry in those islands. For all that, sugar production in Barbados was roughly 200,000 (1965) and m.p.t. 100,000 (1975) tonnes when fires of 'unknown origin' already were a serious menace; similarly in St Kitts, where the corresponding figures are 40,000 and 25,000 tonnes (Table 5.6). Perhaps lower standards of husbandry and changes in cultivation practices made during a period of financial and political difficulty, their effect so difficult to quantify, have played a greater part than is generally recognized in causing the reduction in yield. Be that as it may, sugar-cane has been grown on the same land in these countries for roughly 300 years and future production trends, with special reference to the incidence of fires, will be watched with great interest.

Chapter 6

Pests

In most countries where sugar-cane has been widely grown as a plantation crop the ravages of pests have at one time or another threatened the very existence of the industry. Fortunately, means have been found either to contain or to overcome these menaces. Spectacular successes have been achieved by biological control; and in more recent years, after the discovery of the remarkable insecticidal properties of HCH and DDT, by the use of synthetic organic chemicals. In this description reference is made not only to pests which still demand attention but also to some which have ceased to be of economic importance.

Each geographical region has its own distinctive cane insect fauna; and those insects which damage the crop are either generalized plant feeders, or may be restricted to grasses, or to *Saccharum* species, their hybrids and near relatives. None appears to be confined to sugar-cane. Their natural range has in many cases been extended by unintentional passage from one country to another.

Most regions also have indigenous rodent pests which damage cane in a wide range of ecological conditions. More important than these in some countries, however, are the black and brown rats, both thought to have originated in Asia Minor and the Orient.

Nematodes can cause serious losses, especially where cane is grown on light sandy soils.

Distribution

Many regional lists of the pests of sugar-cane or accounts of its economic entomology have been published. Some are little more than comprehensive catalogues; in others casual and unimportant species are excluded, assessments are made of each insect's effect on the plant's well-being, and control measures are discussed. A list of the most important accounts is given in Table 6.1.

Table 6.1 *Lists of sugar-cane pests or accounts of sugar-cane entomoloy*

Region	Reference	Title	Content
Mexico	Van Zwaluwenburg (1951)	The insects affecting sugar-cane in Mexico	Full economic description
	Flores & Ruano (1961)	Principles lagas la cana de azucar en Mexico	Full economic description
	Flores (1972)	Plagas de la cana de azucar en Mexico	Historical review with list of species
Continental USA	Ingram *et al.* (1951a)	Insect pests of sugar-cane in continental United States	Full economic description
Florida	Ingram *et al.* (1939)	Sugar-cane pests in Florida	Full economic description
Abaco Island, The Bahamas	Beg & Bennett (1973)	Insects associated with sugar-cane on Abaco Island, The Bahamas	Full economic description
Cuba	Scaramuzza & Barry (1960)	A list of insects and animals affecting sugar-cane in Cuba	List of species
The Dominican Republic	Martorell *et al.* (1973)	Preliminary investigations on the sugar-cane insects of the Dominican Republic	Full economic description
Puerto Rico	Martorell & Medina-Gaud (1967)	Status of important pests in Puerto Rico and their control	Full economic description of major pests only
Lesser Antilles and Trinidad	Box (1954)	A preliminary list of the insects affecting sugar-cane in the Lesser Antilles and Trinidad	List of species
Venezuela	Box & Guagliumi (1954)	The insects affecting sugar-cane in Venezuela	List of species
	Guagliumi (1960)	Actual situation of entomology of sugar-cane in Venezuela	List of species with notes on economic importance
	Guagliumi (1962)	Las Plagas de la cana de azucar en Venezuela	Full economic description
Colombia	Potes (1954)	The sugar-cane entomology of the Cauca Valley of Colombia	List of species with notes on economic

The Americas	Bates (1967a)	Pest control in sugar-cane in the Americas	General review
Hawaii	Pemberton (1951)	The present status of insect pests of sugar-cane in Hawaii	General review
	Bianchi (1960a)	Entomological changes in the sugar cane field of Hawaii: 1930 to 1959	General review
Ryukyu Islands	Takara & Azuma (1969)	Important insect pests affecting sugar-cane and the problems on their control in the Ryukyu Islands	List of species with notes on economic importance
The Pacific (including Australia, Hawaii, Fiji, Taiwan, Java and the Philippines) Papua New Guinea	Pemberton (1963)	Insect pests affecting sugar-cane plantations within the Pacific	List of species with pest status
	Szent-Ivany & Ardley (1963)	Insects of *Saccharum* spp. in the territory of Papua and New Guinea Further records of insects collected from *Saccharum officinarum* in the territory of Papua and New Guinea, with notes on their potential as pest species	List of species with pest status
	Bourke (1969)		List of species with pest status
India	Butani (1961)	Annotated list of insects on sugar-cane in India	List of species
Mauritius	Jepson & Moutia (1939)	The progress of applied entomology in Mauritius during the years 1933 to 1938, with reference to insects of sugar-cane	Full economic description
	Williams & Moutia (1954)	Some aspects of sugar-cane entomology in Mauritius	Full economic description
South Africa	Carnegie (1971)	Our more important cane insects in the South African sugar industry	Economic description
	Dick (1951)	Sugar-cane entomology in Natal South Africa	Full economic description
General	Box (1953)	List of sugar-cane insects	List of species with their distributions
General	Long & Hensley (1972)	Insect pests of sugar-cane	General account with economic description

Major pests

The major pests of sugar-cane are easily identified. They are moth borers, froghoppers, white grubs and rodents. Other pests cause limited damage or appear infrequently; and some insects, notably leafhoppers and aphids, are vectors of virus diseases.

Moth borers

Moth borers are acknowledged to be the most destructive insect pests of cane. For convenience they are grouped as stem borers, and top and/or shoot borers. The distinction is somewhat arbitrary: most stem borers destroy shoots, although shoot borers do not attack stems. Typical damage caused by stem and top borers is illustrated in Fig. 6.1.

Stem borers

Stem borer eggs are deposited on the cane leaves; and shortly after hatching the larvae feed on the epidermal tissue of young leaves before burrowing into the stems, sometimes gaining entrance through the soft tissue of the bud primordia. They emerge subsequently as adults. All stages of the lift cycle are present throughout the year where climatic conditions do not preclude continuous breeding. Elsewhere, especially in subtropical countries, larvae become quiescent during the winter and carry the pest from one year to the next.

In young cane the larvae might act as shoot borers by destroying the growing-points, thereby causing the leaf spindles to turn brown, to die and become 'dead hearts'. In past years, when planting densities were low, the gaps so formed were supplied. The modern method of planting seed-cane in a continuous line in the row is such that many more tillers emerge than can be sustained; consequently some die for physiological reasons. Therefore damage caused by stem borers in this manner is relatively unimportant.

Significant losses are caused by larvae as they eat their way along older stems. Tissue is destroyed, the sucrose content of that which remains is reduced and access is given to pathogens, especially to the fungus *Glomerella tucumanensis* (Speg.) Arx and Muller, the causal organism of red rot disease. The extensive literature on the assessment of these losses has been reviewed by Metcalfe (1969). It has been estimated that the value of the sugar lost by moth-borer damage each year in the Americas in the mid-1960s was US$55 m. (Bates 1967). Exact quantification is not possible: but it is essential that reports should reflect accurately the effect of changes in agri-

Cane stem   Cane top

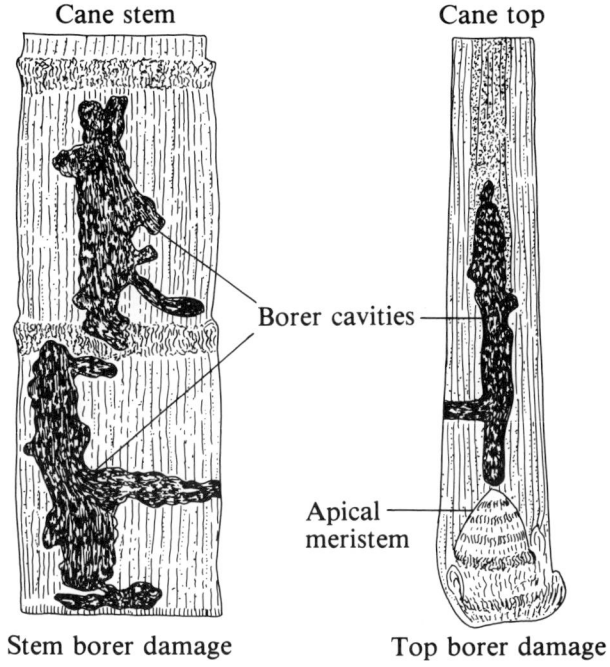

Borer cavities

Apical
meristem

Stem borer damage   Top borer damage

Fig. 6.1 The damage caused by stem and top borers (After Deerr 1911)

cultural technique, choice of variety and attempted control measures on the relative incidence of these pests.

The method of estimation most widely used is the proportion of joints bored expressed as a percentage of all millable joints (PJI). Cane may be examined in the field or at the factory; and the infestation measured by external damage or, after having been split longtitudinally, by internal damage.

As with rice, most of the important borers of the New World are species of *Diatraea* and those of the Old World species of *Chilo*. Indeed the comment that: 'the genus *Diatraea* is confined to the American continent; all records claiming it as a pest in Asia really refer to *Chilo*', made by Grist (1975) concerning rice is also true with respect to cane. For example, in grouping the Crambine moth-borers, Bleszynski (1969) classified both *Diatraea striatalis* Snellen, the striped stem borer of Java, and *Proceras sacchariphagus* (Bojer), a pest of sugar in Mauritius, as *Chilo sacchariphagus* (Bojer). Nevertheless the genera *Chilo* Zincken and *Diatraea* Guilding are so close that they form a compact monophyletic group and are kept distinct mainly for practical purposes.

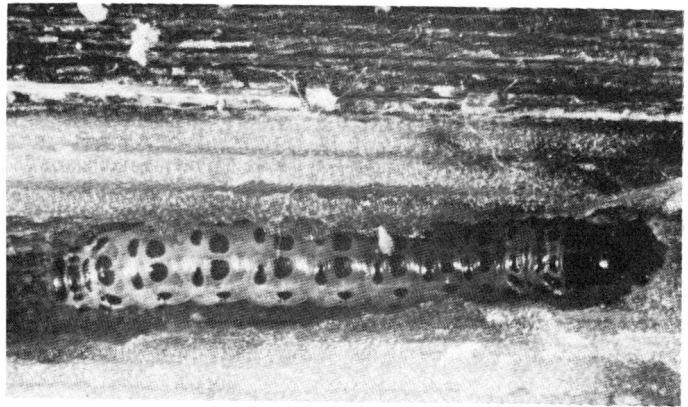

Fig. 6.2 Adult and larva of *Diatraea saccharalis* (Photo: J. R. Metcalfe)

The small moth borers, *Diatraea* spp. (Lepidoptera, Crambinae)

Of the twenty species of *Diatraea* which attack cane, *D. saccharalis* (Fabricius) is by far the most important. It causes damage wherever the crop is grown in the Americas: from Louisiana through Mexico, Colombia, Venezuela, Peru and Brazil to Argentina; and in most of the Caribbean islands. *Diatraea saccharalis* is also the most widespread and destructive insect in the developing sugar industry of the lower Rio Grande Valley of Texas (Fuchs *et al*. 1973).

Life history

Eggs of *D. saccharalis* are laid on the leaves in clusters of ten to thirty; they are 1 mm in length usually in three or four rows, partially overlapping one another in a uniform manner. The larvae, 2 mm in length on hatching, live on the leaves for ten days and are specially vulnerable to parasites, predators, insecticides and adverse weather conditions during this period. Thereafter they feed on the epidermal tissue of young leaves for about ten days, then enter the stems or shoots where they feed for a further four to five weeks and, after having prepared exit holes, pupate. The larvae are easily distinguished: their heads are black, whereas those of other species, except *D. rufescens* Box, which occurs in Bolivia, are yellow or brown. The average life of a larvae, having survived the first dangerous period, is thirty-three to thirty-five days; and that of a pupa, six days. The moths which emerge from the exit holes are inactive during the daytime, but can often be seen in large numbers in cane fields shortly after sunset (Fig. 6.2). The other species of *Diatraea* and those of *Chilo* which attack cane have similar life histories.

Diatraea bucksella Dyar and Heinrich is a pest in Colombia and Venezuela; *D. centrella* (Höschler) in Guyana and Trinidad; *D. considerata* Heinrich in Mexico and Venezuela; *D. dyari* Box in Argentina; *D. guatemalella* Sch. in Costa Rica; *D. impersonatella* (Walker) in Brazil, Guyana, Peru, Surinam, Trinidad and Venezuela; *D. magnificatella* Dyar in Mexico; *D. rosa* Heinrich in Venezuela; and *D. tabernella* Dyar in Costa Rica and Panama. In 1968, when 8,000 ha were planted with cane in an abortive attempt to establish a sugar industry on Abaco Island in the Bahamas, the most serious pests were *D. lineolata* (Walker) and *D. centrella*, the latter accidentally introduced in planting material from Guyana (Beg & Bennett 1973).

The level of infestation varies from year to year and from country to country. Control measures have been regarded as unnecessary where the PJI was of the order of 5. Such places are Jamaica (Chinloy 1955), Puerto Rico (Martorell & Medina-Gaud 1967). In attempting to elucidate the reasons for the low incidence of damage in Trinidad, Pickles (1947) stated that while both egg and larval parasites were of some importance, non-parasitic larval mortality of unknown origin accounted for 58.3 per cent of the population. He suggested that a proper understanding of this might provide a solution to the *Diatraea* problem throughout the Caribbean.

Biological control
Larval parasites

Where damage was greater, often 20 PJI, counter-measures were taken: biological control by the establishment of exotic larval parasites was attempted, in some areas with great success. Three Tachinids, all from the New World, were the most effective agents: *Lixophaga diatraeae* Townsend (the Cuban fly) in relatively dry areas, *Metagonistylum minense* Townsend (the Amazon fly) in wet areas and *Paratheresia claripalpis* Van der Wulp. (the Peruvian fly). A comprehensive review of work in this field was given by Bennett (1969).

Lixophaga was released and controlled *D. saccharalis* in Antigua and St Kitts; and has helped to do so in Guadeloupe. In Antigua in the mid-1930s damage was reduced from 14 to 6 PJI (Warren 1945). This level was maintained, with safeguard of annual liberations of *Lixophaga*, until the demise of the sugar industry in that island thirty years later. The results of a similar campaign in St Kitts were even more satisfactory. The PJI fell rapidly from 20 to 5; and with the exception of two years, has remained at this low level without the help of further releases. In those exceptional years, 1945 and 1946, the PJI rose to 11. Pickles (1946a) felt that unusual climatic conditions had caused the resurgence and confidently – and correctly – predicted less damage in future years. There was no reason to consider the introduction of new parasites and no need to boost the *Lixophaga* population. Bennett & Pschorn-Walcher (1969) confirm Pickles's assessment. Guadeloupe is unique in that all these three well-known larval parasites coexist there. Despite the presence of *Metagonistylum*, in 1947–48 the PJI was roughly 20. *Lixophaga* was then released and damage reduced to 10–15 PJI. Later, following the introduction of *Paratheresia* in 1954, the PJI fell to 3 (Simmonds 1960). Lemaire (1966) confirmed that the PJI in Guadeloupe was much the same several years later. Although *Lixophaga* is indigenous to Cuba, and is the most important parasite of *Diatraea* in that country, damage in the late 1950s was 12–15 PJI. (Scaramuzza 1960a). A campaign of annual releases to boost its population was started in 1961, and by 1965 the PJI had been reduced to 5 (Long 1969, quoting Scaramuzza).

In Guyana, St Lucia and Venezuela introductions of *Metagonistylum* were more effective than in Guadeloupe. Their complete success against *D. saccharalis* in Guyana has been described by Myers (1934) and Cleare (1939). By 1963 the only borers of consequence were *D. centrella* and *Castnia licoides* Boisduval; and the annual PJI during the period from 1954 to 1963 ranged from 6 to 8 (Bates 1967b). Similar results were achieved in St Lucia where liberations of *Metagonistylum* at Roseau and Cul-de-sac in 1934 caused a reduction from 20 to roughly 10 PJI. *Diatraea centrella*, unaffected by the parasite, was responsible for most of the residual damage (Box 1939). This degree of control was maintained until the sugar estates were converted to banana plantations, some twenty years later. In Venezuela *Metagonistylum* was effective against *D. bucksella* and *D. rosa* as well as *D. saccharalis*. Consequently, liberations made in the Aragua Valley in 1950 caused the PJI to fall from 16 (1947–52) to 6 (1956) and eventually to 2 (1965) (Bennett 1969, quoting Box).

Reference has been made to the partial success of *Paratheresia* when introduced into Guadeloupe. In Peru, where it is indigenous, annual releases made since 1953 in the Chicama, Santa Catalina, Lambayeque, Zana and Pativicla valleys caused a spectacular reduction in damage. For example the PJI at Paramonga Estate fell from 18 in 1953 to 4 in 1958 (Risco 1960). Similar releases were made in Bolivia. However, Bennett felt that these claims, and those made for *Lixophaga* in Cuba, would be more widely accepted if supported by additional data, especially on the ecology of natural populations.

Bennett also described and discussed the many unsuccessful attempts which have been made to establish Tachinid larval parasites elsewhere. In particular, he drew attention to the striking contrast between the control achieved by *Lixophaga* in St Kitts and its failure, under similar climatic conditions, in Barbados.

Egg parasites

There is one more important group of parasites to be considered: the wasps, *Trichogramma* spp. (Hymenoptera, Chalcidoidea) which attack moth-borer eggs in both the Old and New Worlds. Their influence in exerting a measure of natural control has long been recognized; and for many years it was thought that even greater benefit would accrue if the *Trichogramma* population were increased by mass liberations at critical times of the year. This method, first used in Guyana in 1921 (Cleare 1928), was adopted in several other cane-growing areas. Tucker (1951) reported that, as a consequence, the PJI in Barbados was reduced from 22 in the years 1930–33 to 12 in the years 1946–49; and equal success was said to have been achieved in Guyana, Louisiana, Mexico and Peru. Experimental release in Colombia, Florida and Puerto Rico, however, gave inconclusive results. The claims of success were not universally accepted and became the subject of considerable controversy. The procedure was re-examined, fell into disrepute and, except in Peru, has been abandoned throughout the Americas. Only the indigenous species, *T. fasciatum* Perkins, formerly known as *T. minutum* Riley, was involved in the foregoing. A full description of the *Trichogramma* : *Diatraea* relationship, and of these events, has been given by Metcalfe & Brenière (1969).

Cultural control

Certain cultural practices might help to reduce damage in those places where biological control has not yet been achieved. For example in higher latitudes, if crop residues are thoroughly burnt after harvest, protection from the cold for overwintering borers is removed and infestation in the following year thereby reduced (Ingram *et al.* 1951). This approach has been examined in detail by Charpentier & Mathes (1969).

Chemical control

Alleviation by cultural practices is at best a palliative and, in many places where effective biological control of heavy infestations has not yet been established, the use of insecticides has been recommended as a last resort. In the lower Rio Grande Valley in Texas, monocrotophos was the most effective of nine insecticides tested against *D. saccharalis* (Fuchs *et al.* 1973).

Chemical control has been and is practised intensively only in Louisiana where *D. saccharalis* survives the winter in hibernation. When warm weather returns, the first generation is usually small and erratic. Three of four broods, often overlapping, then follow until winter sets in. Sodium fluosilicate and cryolite were first used in the 1920s. They were succeeded by ryania which, in turn, gave way to endrin in 1958. In 1963 resistance to endrin was first noticed at Port Allen in Louisiana (Yadav *et al.* 1965). High levels of resistance were widespread by 1969 and endrin was replaced by Guthion, an organophosphorus insecticide, or, if the fields were to supply livestock feed, by carbaryl a substituted carbamate which does not accumulate in animal fat. There has also been a change in the methods used: granules proved superior to dust, then sprays replaced granules; and as techniques develop, lower volumes are being used (Long 1969). Long asserted that more insecticide was applied to sugar-cane in Louisiana than in any other part of the world. This may well have been true, though it is possible that the dubious honour might now be held by Australia, where insecticides are used to control white grubs; or to Trinidad, where the annual cost of froghopper control is roughly £400,000. However, there can be no doubt at all that the chemical control of *D. saccharalis* in Louisiana is an expensive operation.

Resistant varieties

In his discussion on the philosophy of sugar-cane breeding Stevenson (1965) stressed the need to enquire whether varieties do in fact differ in their intrinsic resistance to insect pests; and, if they do, the characters upon which such resistance depends. He then quoted the opinion of Mathes *et al.* (1954) that varietal susceptibility to damage by *Diatraea* does exist, and is due not to one but to many characters. In support of this Stevenson gave Tucker's reasons for regarding B 37161 as susceptible to borer attack (Tucker 1951). On the other hand he also referred to the findings of Blackburn (1950) in Trinidad, Chinloy (1955) in Jamaica and his own in St Kitts: that in small plots palatability rather than true susceptibility was being measured; that while B 37161 might be highly infested, in experiments where the insect had a choice of varieties on which to feed, when planted on a large scale its PJI reverted to the norm for the area. He concluded that: 'It is unlikely that agreement on these contentious questions will be reached between entomologists, agronomists and geneticists until a great deal more fundamental information is available.'

Mathes & Charpentier (1969) believed that control can be achieved by planting borer-resistant varieties, but the supporting examples do not bear close examination. In making the allegation that borer damage in Guadeloupe increased because of the replacement of the noble cane BH 10(12) with recently developed hybrids, Simmonds (1960) quoted opinions of the level of infestation for the former and PJI for the latter, not comparable data. The high level of borer damage at Woodford Lodge Estates Ltd in Trinidad was caused not by the extensive planting of B 37161, as claimed by Simmonds (1951), but by the repeated – almost obsessive – application of DDT in drift dusting against the froghopper, where fields were dusted as many as twenty-seven times in one season. The insecticide drifted before the wind resulted in heavy borer damage, which diminished progressively further to leeward. When in 1961 this estate was taken over, drift dusting with DDT was abandoned and the borer problem declined to negligible importance; as it had been throughout this period at Trinidad Sugar Estates Ltd nearby where B 37161 was the major variety grown, but drift dusting with DDT had not been practised. In Louisiana Long *et al.* (1960) found that treatment with DDT not only failed to control *D. saccharalis* but caused a significant increase in infestation. Therefore, these citations do not illustrate a general acceptance that borer populations, on a large scale, have been affected by the choice of cane variety. Nor does the experience of Parthasarathy (1950), who in Madras, found considerable variations in the incidence of *Chilo indicus* (Kapur) on different varieties; but that in large-scale cultivation this factor was unimportant.

Nevertheless, attempts to identify those characters which might cause resistance should be encouraged; for the planting of resistant varieties, so successful against diseases, would be the most satisfactory method of preventing damage by pests. Differences have been shown between varieties in their ability to recover from insect attack, for example the planting of M 134 32 in Mauritius was the most important factor in reducing the status of the white grub, *Clemora smithi* (Arrow) (Williams & Moutia, 1954).

Chilo spp. (Lepidoptera, Crambinae)

Borers are unimportant in Australia (Mungomery 1965) and, for many years, insect pests as a whole in South Africa (Dick 1951). The stem borers of the Old World are pests of sugar-cane only in Burma, China, India, Malaysia, Thailand and the islands of the Indian and Pacific oceans. With the exception of the top borer, *Scirpophaga nivella*, which is described later, they are all species of *Chilo*.

Chilo and *Diatraea* have similar life cycles and cause the same sort of damage, though that inflicted by *Chilo* spp. appears, on the whole, to be less severe. The most important species, *C. infuscatellus* Snellen, is a serious pest in India, Java, Pakistan, the Philippines and Taiwan. Other species, generally of less significance, are *C. auricilius* Dudgeon in India, *C. sacchariphagus* (Bojer) in Java, Madagascar, Mauritius and Réunion and *C. venosatus* Walker in Java, the Philippines and Taiwan. The nomenclature of Bleszynski (1969) has been used to described this confused group.

Few quantitative records of damage have been published, but apparently little damage is caused in Taiwan, Mauritius and Réunion. Chen & Hung (1969) reported that liberations of the native egg parasite, *Trichogramma australicum* Girling, caused the PJI at Chisan in Taiwan to fall from an average of 3.15 to 1.24 during the years 1957–60; and Liang (1970) found that the PJI caused by all borers on one variety between 1957 and 1966 was only 4.29 per cent. Likewise in Mauritius, where *C. sacchariphagus* does not seem to be considered a serious menace; and in Réunion, where it was estimated that the loss caused by this pest was 5 per cent Caresche 1962). However, in Madagascar in 1953, *C. sacchariphagus* was particularly injurious in the north-west of the island: almost all the stems were infested and the PJI approached 40 per cent. Less damage was caused in later years, probably because of the increasing effect of *T. australicum* (Caresche & Brenière, 1962). Nevertheless serious damage undoubtedly is caused by moth borers in Bangladesh, India and Pakistan; but in such a vast area the importance and incidence of individual species varies widely. Many methods of determining loss have been used: for example Khan & Krishnamurati Rao (1956) who separated the formation of 'dead hearts' from internodal damage, regarded the former as the more important.

Pradhan & Bhatia (1956) asserted that *C. infuscatellus* was to India what *D. saccharalis* was to the Americas; and was known for its depredations throughout the subcontinent. In 1960 it was regarded as one of only two pests of economic importance in Pakistan (Carl 1962).

Other species are not so widespread. *Chilo indicus* is a major pest in peninsular India, especially Madras; but occurs elsewhere only in small pockets in northern India and Bangladesh. Where its incidence is highest, the Tiruchirapalli, Tanjore and South Arcot districts of Madras, investigation showed that more than half the millable canes were infested. David & Kalra (1967) gave a comprehensive description of this species and the measures taken for its control.

Chilo tumidicostalis also is a regional rather than a subcontinental pest and its presence has been recorded in West Bengal, Bihar,

Bangladesh and Assam. It was of little more than academic interest until an outbreak occurred in 1956 in the Nadia district of West Bengal, causing 95 per cent of the stalks to be damaged, with a PJI of 25. Gupta & Avasthy (1960) described this attack, gave an illustrated life history of the species and reported on the success of mechanical means of control after attempts with insecticides had failed. The mechanical measures taken were:
1. The killing of moths in light traps.
2. The removal and destruction by burning of tops showing primary infection.
3. The collection of egg masses.
All are highly labour intensive and suitable only for small plantings tended by family units.

A different system of agriculture is being adopted by the sugar industry of Java: ratoons are being grown where before only plant cane had been taken. Whatever effect this might have on the incidence of pests, in 1961 it was found that in East Java, the main producing area, there were only two stem borers of importance: *C. sacchariphagus* and *C. auricilius* (Ghani & Williams 1963).

Problems with stem borers have not yet arisen in the developing plantations of Sumatra.

Control measures
Biological control
The population of *Chilo* spp. is controlled in varying degree by local egg and larval parasites. Several attempts have been made to introduce exotic species, especially *Lixophaga* and, as in the Americas, to boost the population of egg parasites, *Trichogramma* spp., with mass liberations. Although some of the New World Tachinids will parasitize *Chilo* larvae, their permanent establishment in the Old World has not yet been achieved. Releases of *Lixophaga* were made, with little success, in India, Mauritius, the Philippines and Taiwan. Following the failures of 1947 and 1948 in Mauritius, Williams & Moutia (1954) decided that an agent for the control of *C. sacchariphagus* should be sought in the island of its origin, Java. The Tachinid *Diatraeophaga striatalis* Townsend was collected and multiplied; and then released in India, Madagascar, Mauritius and Réunion, so far without success (Ghani & Williams, 1963; Brenière *et al.* 1966; Etienne, 1969).

Releases of indigenous *Trichogramma* spp. were made as follows:
T. nanum Zehntner in India.
T. australicum Girling in Madagascar, Mauritius and Taiwan.
T. evanescens Westwood in China.
In addition *T. fasciatum* was introduced into Madagascar and the Philippines and *T. evanescens* into Mauritius.
Successful control has been claimed only in Taiwan and in two states

of India: Bihar and Madras. Elsewhere, either the practice has been abandoned or data supporting claims of success are not available. Metcalfe & Brenière (1969), who have given an excellent description of the *Chilo : Trichogramma* relationship, urged that critical studies should be carried out in those parts of India and the Far East which, with Peru, are the only remaining areas where this method of control is still in use. They also drew attention to the pressing need to revise the nomenclature of *Trichogramma* species.

Chemical control

In India effective control of *C. infuscatellus* and other borers has been achieved in experiments by pouring an emulsion of 0.75 per cent gamma-HCH on to setts at the time of planting (Siddiqi *et al.* 1959). Newer insecticides offer better control where economic.

Resistant varieties

The considerations which affect the breeding and selection of varieties resistant to *Diatraea* apply equally to *Chilo*.

The large moth borer, *Castnia licoides* Boisduval (Lepidoptera, Castniidae)

Life history

Much has been written about the large or giant borer, *Castnia licoides*, which occurs in Guyana, the Lesser Antilles, Surinam, Trinidad and Venezuela. Its life history has been described by Skinner (1929). Eggs are laid singly at the bases of the canes between the trash and the stems. They are 1.5 mm in length, spindle shaped, with five longitudinal keels, and hatch in seven to fourteen days. The emerging larvae, 5 mm in length, eat downwards and then tunnel into the lower parts of the canes; orange and pink at first, their colour changes to pale cream after the first moult. They prepare exit holes 75–100 mm above ground and then retire to the rootstock to pupate. The moths are large and fly by day. Damage is caused by the larvae and, although their depredations are similar to those of *Diatraea* and *Chilo*, they take the form of much wider tunnels confined to the lower internodes and rootstocks. When the crop is reaped, surviving larvae hide in the rootstocks whence they, and their progeny, infest the next crop. Consequently the older the ratoon the more likely it is to be attacked by *Castnia*.

Skinner referred to outbreaks in South Trinidad in 1927 and 1928; and in nine fields examined in detail found that 18.3 per cent of the stems had been bored. Whatever the importance of *Castnia* in Trinidad in former years, since 1950 – and possibly for much longer – it has been negligible. Surveys during crops 1950–52 showed the damage caused from this pest was so slight that its incidence ceased to be recorded (Blackburn 1951). In 1967, of the stems selected from 128 ratoon fields for inspection for borer damage in general, only 2.0 per cent had been affected by *Castnia* (Fewkes & Buxo 1969), and in 1968 2.1 per cent (Buxo 1968).

In Guyana, however, serious sporadic outbreaks have occurred. Bates (1967b) referred to those which took place in 1911 and 1921; and more recently during the five-year period 1961–65. In 1963 the incidence varied widely: although the average stem infestation in 145 fields was 4.6 per cent, for individual fields it was as high as 37.0 per cent. Considerable secondary damage was caused by other insects and the stem rot fungus, *Giberella fugikori* (Sawada) Ito. Bates considered that the factors which had contributed to this upsurge were:

1. Longer ratooning and therefore fewer fields ploughed out and flood-fallowed.
2. Prolonged dry weather.
3. The increased planting of B 47258, a variety particularly attractive to *Castnia*.

He found that *Castnia* larvae were killed by flooding affected fields for forty-eight hours. Adding various insecticides to the floodwater did not improve the treatment.

Shoot borers

Of the moth borers which damage cane only by killing young shoots, the most important are *Elasmopalpus lignosellus* in the New World and *Sesamia* spp. in the Old World.

The jumping borer or lesser corn stalk borer *Elasmopalpus lignosellus* (Zeller) (Lepidoptera, Phycitidae).

Elasmopalpus is a common pest of grasses and legumes in the tropics and subtropics of the Americas; and according to Box (1953) has been found in cane fields in Argentina, Cuba, Guyana, Louisiana, Mexico, Nicaragua, Peru and Puerto Rico. More recently damage caused by this pest has been reported in Venezuela (Kern 1956), Barbados, Jamaica and St Kitts (Bennett 1962) and Trinidad (Fewkes 1966). Fewkes gave a detailed description of the insect's life history. The larvae feed on young ratoon shoots, boring into them at or just below ground level. Having eaten the growing-point and surrounding soft tissue, each larva then attacks a fresh shoot and therefore destroys several during its development. Unlike other borers it then pupates in a silken tube outside the stem at soil level. Infestation is confined to young shoots, causing 'dead hearts'; older stems are not affected.

Under normal circumstances *Elasmopalpus* is not found in cane fields. Migratory moths and larvae, reared on surrounding vegetation, caused the outbreaks in Barbados, Jamaica, St Kitts and Trinidad. In Venezuela the affected fields previously had been sown to maize and insects lived on volunteer plants before attacking the cane.

Both Bennett and Fewkes noted the correlation between borer infestation and the burning of trash before and after reaping; and Bennett suggested that this might provide an olfactory attraction to gravid females. Fewkes emphasized that significant damage occurred in Trinidad only where cane was under stress from prolonged drought; in conditions of adequate moisture the loss of even half the shoots might be of little importance. This type of stress is prevalent especially in two-year plants and standover cane, but absent in ratoons and one-year plants; and becomes widespread and more severe when reaping programmes are disorganized by fires of unknown origin.

The pink borer and related species – *Sesamia inferens* (Walker) and other *Sesamia* spp. (Lepidoptera, Agrotidae)

Although many species and subspecies of *Sesamia* cause serious damage to cereals in the Old World, only four are pests of sugarcane. They are: *S. calamistis* Hampson, in Africa and the Mascarenes; *S. inferens* in the Orient; *S. uniformis* Dudgeon in northern India, Pakistan and the Philippines and *S. grisescens* Warr. in New Guinea. A comprehensive review of the literature has been made by Rao & Nagaraja (1969).

There is a remarkable similarity between *Sesamia* in the Old World and *Elasmopalpus* in the New World. *Sesamia* eggs have not been found in cane fields and infestation is by migrant larvae from nearby grasses. They enter shoots at ground level, eat young tissue and destroy the growing-points, thereby causing the formation of characteristic 'dead hearts'. They infrequently attack older cane (Venkatraman & Vasudeva Menon 1964). Having fed on several shoots, the mature larva forms a cocoon, either inside the shoot or in a shaded place adjacent to it, from which the moth emerges in due course. Life cycles of 66 and 102 days have been reported by Moutia (1954) and, with uninterrupted development, four or five generations occur each year.

The damage caused by *Sesamia*, like *Elasmopalpus*, depends on the ability of affected plants to compensate for 'dead hearts' by producing new shoots, which in turn depends upon the density of planting in newly established fields and the moisture and nutritional condition of the soil in fields of ratoon cane. Therefore it is difficult to assess with accuracy the status of *Sesamia* in those countries in which it is found. *Sesamia calamistis*, though widely distributed, appears to be only a minor pest of cane in Madagascar (Caresche & Brenière, 1962), Mauritius and Réunion (Williams & Mamet 1962), although slightly more serious on the windward side of Réunion (Caresche 1962). *Sesamia inferens*, also widely distributed, is a more important pest in India, where it is known as the pink borer (Gupta & Gupta 1959; Kumar & Kalra, 1965). *Sesamia uniformis*, of restricted occurrence, is a minor pest in northern India; and *S. grisescens*, confined to Papua and New Guinea, has been described as the most important pest of sugar-cane in that territory (Szent-Ivany & Ardley 1963).

Clean cultivation is the best means of preventing damage by *Sesamia*; in particular, grass weeds should be eliminated. Where labour is plentiful and inexpensive, and on small family holdings, it is possible to control incipient attacks by cutting out and destroying shoots which show early signs of infestation (Gupta & Gupta 1959). Nowhere has chemical control of these pests been carried out regularly in cane fields. Attempts to achieve biological control in Mauritius by the introduction of exotic larval parasites have had little success. Similar attempts are being made with egg parasites in the Philippines (Nickel 1977) and larval parasites in Taiwan (Chen & Hung, 1962).

The top borer – *Scirpophaga nivella* (Fabricus) (Lepidoptera, Pyralidae)

The top borer *Scirpophaga nivella* and its subspecies *S. nivella varia intacta* Snellen are second in importance only to *Chilo* spp. as pests of sugar-cane in the Old World. *Scirpophaga nivella* causes serious damage in Burma, China, India, Japan, Malaysia, Pakistan, Sri Lanka, the Philippines and Taiwan; and *S. nivella intacta* in Indonesia. A full description of the biology of *Scirpophaga*, the damage it causes and the control measures taken has been given by Avasthy (1969). Avasthy noted that, in due course, *Scripophaga* would probably be assigned to the newly erected genus, *Tryporyza*.

The moths are nocturnal and attracted to light. They lay masses of eggs, covered in scales, in overlapping rows on the undersides of leaves; but, on hatching, only one larva lives in each cane top. After being exposed for only three to four hours it gains protection by boring into the underside of the midrib of the youngest open leaf and then travels downwards through the leaf spindle and the joints in process of being formed. It then prepares an exit hole and pupates. The duration of its life cycle varies from twenty-five to eighty-three days. The number of generations in each year is difficult to determine: they overlap in all regions; and in subtropical India the larvae hibernate for three or four months.

The symptoms of attack by the top borer are parallel rows of shot holes in the leaves, caused earlier by larvae entering the leaf spindles; and 'dead hearts', surrounded by bunchy tops comprising the sprouting upper buds of infested stems. Canes attacked early are thin and stunted. In spite of winter hibernation, damage in the northern states of India and Pakistan is greater than in peninsular India, where development throughout the year is uninterrupted. Surveys showed that while damage at Coimbatore, Cuddalore, Peshawar and Rudrur was negligible, more than 20 per cent of the shoots were bored in Jullundur, Pusa, Rawalpindi and Shahjahanpur.

Exact quantification of the losses sustained is not easily made. Early maturing varieties suffer most, and damage increases progressively with each brood. Losses in India have been estimated to vary from 20 to 85 per cent by weight of millable cane, with a considerable reduction in the juice quality of that which remained (Avasthy 1969); 20–30 per cent in Taiwan (Lee & Pao 1962); and only 5 per cent in Java (Jepson 1954, citing Hazelhoff), though greater damage has been reported in recent years.

Mechanical control by the collection of egg masses and the removal of infested stems has been practised successfully for many years in India (Basheer 1959) and Taiwan (Lee & Pao 1962); and Hazelhoff (1932) reported that the cutting out of bored shoots in Java, called 'rogesan', reduced borer incidence by roughly 50 per cent. In the past insecticides have failed mainly because of the concealed habit of the borer larvae and the absence of distinct broods; but some success has been obtained in Java by injecting infected tops with the systemic insecticide, Furdan. Biological methods show promise: the introduction of *Isotima javensis* Rohwer from subtropical Uttar Pradesh has effectively controlled *Scirpophaga* in tropical Madras (Raja Rao 1964); and reports concerning its establishment in Taiwan (Chen & Hung, 1963) are encouraging.

Eldana saccharina (Walker)

Eldana borer is indigenous to Africa but does not occur on any other continent. It has long been known in West Africa where it lives on wild water grasses and sedges – tall *Cyperus* spp. In some parts of Africa it may cause serious damage to maize and sugar-cane.

The caterpillar stage of this moth is a very active, tough, brown leathery borer which wriggles actively when disturbed and can descend rapidly on a silken thread which it spins. It moves both backwards and forwards with equal ease. It is distinguishable from *Sesamia* by its brown to black colour; *Sesamia* is pinkish and less active.

The presence of *Eldana* is marked by frass on the outside of the cane stalks.

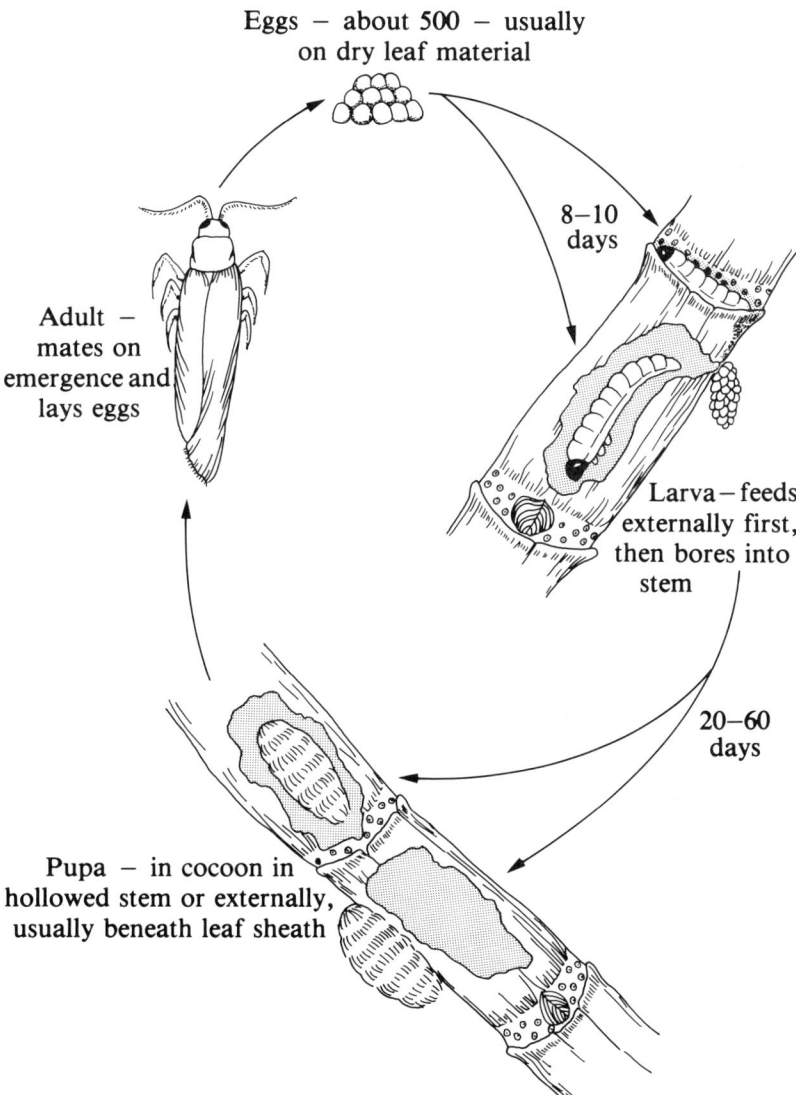

Fig. 6.3 Life cycle of *Eldana* (*Eldana. Know Your Enemy*. South African Sugar Association Experiment Station, Mount Edgecombe)

Life cycle

The life cycle is shown in Figure 6.3. The moth lives a few days. It does not feed, but mates, flies and lays its eggs preferably on dry leaf. The eggs hatch in about a week and the young larvae feed on the outside of the stalk around the node, after which it enters the stalk, feeding for the rest of its life inside the stalk. It pupates either in the hollow stalk or behind a leaf sheath. Breeding is continuous with a number of generations a year, but the time taken for each stage of the life cycle varies considerably under the influence of the climate and nutritional factors.

Control

The *Eldana* population increases with the age of the cane (see Fig 6.4). The following control measures are currently recommended in South Africa.

Recommended control measures are based mainly on management practices. Some natural control effected by such predators as ants is of great importance, but it is insufficient to prevent outbreaks, and there are as yet no useful insect parasites. The insect is well protected within the stalk and no effective insecticide has been identified. Indeed, the adverse effects of insecticides on ant populations must discourage their use.

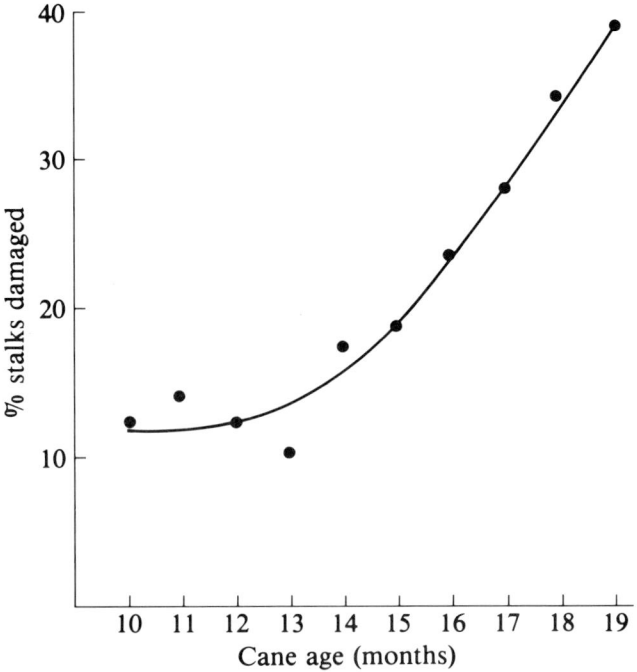

Fig. 6.4 The effect of age of cane on the incidence of *Eldana* (*Eldana. Know your enemy*. South African Sugar Association, Experimental Station, Mount Edgecombe)

A most important factor is cane age, because borer populations increase as the crop gets older.

Field hygiene is also important, because residues remaining in a field after harvest can serve as a source of further infestation. In areas where the pest is not yet well established, burning before harvesting will kill moths, eggs and young larvae and may be helpful in restricting populations.

The pest can be spread in seed-cane and it is important therefore that planting material should be free of the borer. Do not cut cane for seed unless it has been inspected for the presence of borer. If the use of infested seed is unavoidable, then follow one of these procedures:

1. Immerse the seed-cane in a suspension of insecticide for ten to fifteen minutes, in trials the following have proved to be effective:
   (i) Dieldrin or aldrin, prepared by mixing 2 g of the wettable powder with 1 litre water; or 6 ml of the liquid concentrate with 1 litre water. (These chemicals may soon cease to be marketed.)
   (ii) Fenitrothion at 2 ml of the liquid concentrate in 1 litre water.
   (iii) Phoxim at 2 ml of the liquid concentrate in 1 litre water.

2. Treat the seed-cane in a hot-water tank at 50 °C for thirty minutes.

3. Select stalks without signs of borer attack for planting, and either send bored stalks to the mill or burn them.

Any pieces of sugar-cane stalk remaining on the site after planting has been completed should be burnt.

To summarize, present control recommendations are:

(*a*) Cut cane as young as possible.

(*b*) Cut cane as low as possible.

(*c*) Leave a minimum of crop residues in the field. (This applies particularly to stalk material because it is the mature part of the stalks which contain most borers.)

(*d*) Ensure that planting material is borer-free.

(*Source: Eldana. Know Your Enemy*, South African Sugar Association, Experiment Station, Mount Edgecombe.)

It now seems unlikely that burning of an infested crop before harvesting is warranted unless the infestation is particularly severe. No control has been achieved in the field by spraying insecticides. A number of wasps have been found to parasitize *Eldana* eggs in the laboratory (Thompson 1981).

Froghoppers (Homoptera, Cercopoidae, Cercopidae and Aphrophoridae)

Froghoppers or spittle bugs occur throughout the tropics and subtropics. They seriously affect sugar production on the mainland of the New World in Mexico, Belize, Venezuela, and Brazil; but in the islands of the Caribbean only in Trinidad. One species is of minor importance in Guyana. In the Old World, with a single exception, they are of little consequence. The exception is *Clovia sarawakana* Lallmand, which occurs in north-western Malaysia (Tan & Johnson 1974). A classic report on these pests was made by Williams (1921) and up-to-date descriptions given by Fewkes (1969a, b).

The most destructive froghopper in Brazil is *Mahanarva indicata* Distant; but elsewhere the greatest damage is caused by species and subspecies of *Aneneolamia* (formerly *Tomaspis*): *A. postica jugata* (Fowler) in Belize, *A. postica postica* (Walker) in Mexico, *A. varia sontica* Fennah in Venezuela, *A. varia saccharina* (Distant) in Trinidad and *A. flavilatera flavilatera* (Urich) in Guyana. They are indigenous to the countries in which they occur and feed on a wide variety of grasses. Possibly when sugar-cane was introduced, and certainly when its cultivation was expanded, they adopted it as their main host plant. For example although Iturbe & Ruano (1963) noted that froghopper damage to cane in Mexico was first recorded in 1943, they felt that attacks could have taken place in previous years. Williams (1921) referred to such an attack in Vera Cruz in 1903, reported by Urich (1913), and of the high infestation found on a visit to that country in 1911 (Urich 1912). The production of sugar in Mexico increased from 100,000 tonnes in 1903 to 400,000 in 1943 and 1,750,000 in 1963. The awareness of froghopper blight in Trinidad arose in similar circumstances: it was in 1863, during the rapid expansion of the industry in the mid-nineteenth century, that atten-

Fig. 6.5 Some adult froghoppers (From Williams 1921)

a, b, c, d	*Aeneolamia varia saccharina* (Distant)	Trinidad
e	*A. flavilatera flavilatera* (Urich)	Guyana
f	*A. varia bodkini* (Williams)	Guyana, Venezuela
g, h	*A. varia carmodyi* (Kershaw)	Tobago
i	*A. varia propinqua* (Walker)	Venezuela
j	*A. varia bogotensis* (Distant)	Colombia
k	*A. lepidor* (Fowler)	Panama, Venezuela
l	No reference found	
m	*A. postica postica* (Walker)	Mexico
n	*Panabrus dominicanus* (Distant)	Dominica
o	*Zulia pubescens* (F)	Guyana
r, q	*Mahanarva tristis tristis* (Fowler)	Surinam
	Prosapia bicincta fraterna (Uhler)	Cuba
s	*Sphenorhina rubra sororia* (Germ.)	Brazil

N.B. m. and o. are no longer regarded as sugar-cane froghoppers

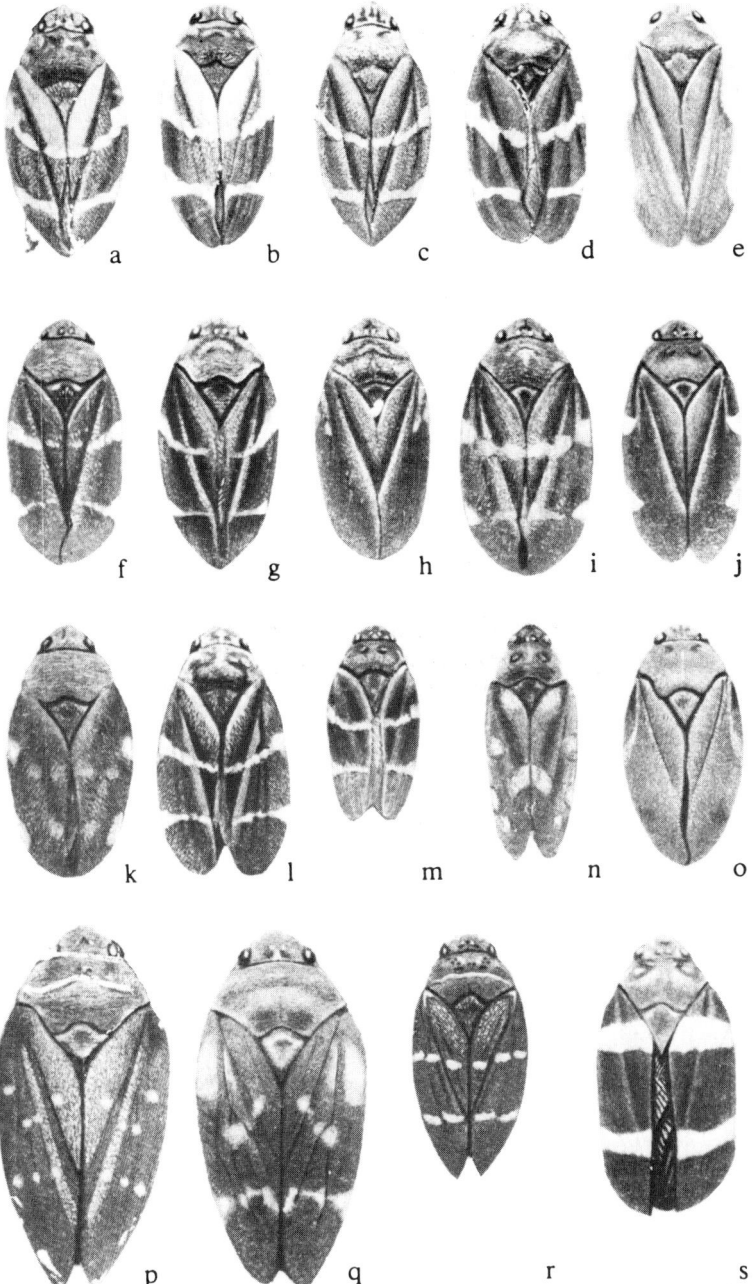

tion was first drawn to the connection between the damage and the insect (Cruger 1892). In the recent large development in Belize, newly planted cane was soon affected by invasions of adult frog-hoppers from the surrounding uncultivated bush (Fewkes & Buxo 1965); but damage from this pest had been reported as long ago as 1883 (Sir Daniel Morris, cited by Williams 1921). These observations are in accordance with the findings, made during investigations on varietal resistance, that froghoppers are attracted to sugar-cane; and that the attractant might be its leaf odour (Fennah 1939).

The life cycle

The eggs of *Aeneolamia* spp. are straw coloured and spindle shaped, 0.75 mm in length by 0.25 mm in diameter at their widest part, and are laid singly in the soil around the bases of the cane stems. A highly fecund insect might lay between 200 and 300 eggs, usually in three batches. Most of the eggs develop typical black hatching caps and nymphs emerge after an incubation period of fourteen days or more. The nymphs live on and under the soil surface in and around the stools, in masses of froth or spittle of their own manufacture and pass through five instars before the perfect state is reached. They feed on the roots of the canes but cause little damage. The adults are 7–9 mm in length and 3.5–4.5 mm in width across folded wings. In most species the forewings are dark brown, with two narrow transverse bands of lighter colour, but there are many excep-tions to this general pattern. The length of a life cycle is slightly less than two months and, according to the duration of the dry season, from two to four distinct generations occur each year. Other eggs go into diapause, survive the dry season and hatch shortly after the onset of the rains to give rise to the first brood of the next year. Such eggs are laid in increasing proportion by successive generations. The reasons for this must be deeply seated and reflect an inbred rhythm. Thus, although fourth brood adults of *A. varia saccharina* emerge in Trinidad in November in wet conditions, for some unknown reason most of the eggs which they lay go into diapause and there is no fifth brood. The entomogenous green muscardine fungus, *Metarrhizium anisopliae* Metchnikoff (Sorokin) is seen more frequently then than at other times of the year, but is not the chief cause of this behaviour. It is more likely that length of daylight, the composition of the cane leaves or excessive heat might influence the growth hormones and thereby induce diapause (Kirkpatrick 1957). The adults hide in the uppermost leaf axils during the day and emerge in late afternoon. They jump from leaf to leaf and plant to plant, hence the name froghopper, but migrate only for short distances.

There is a slight but nevertheless important difference in the modes of life of the Brazilian froghopper, *Mahanarva indicata* and

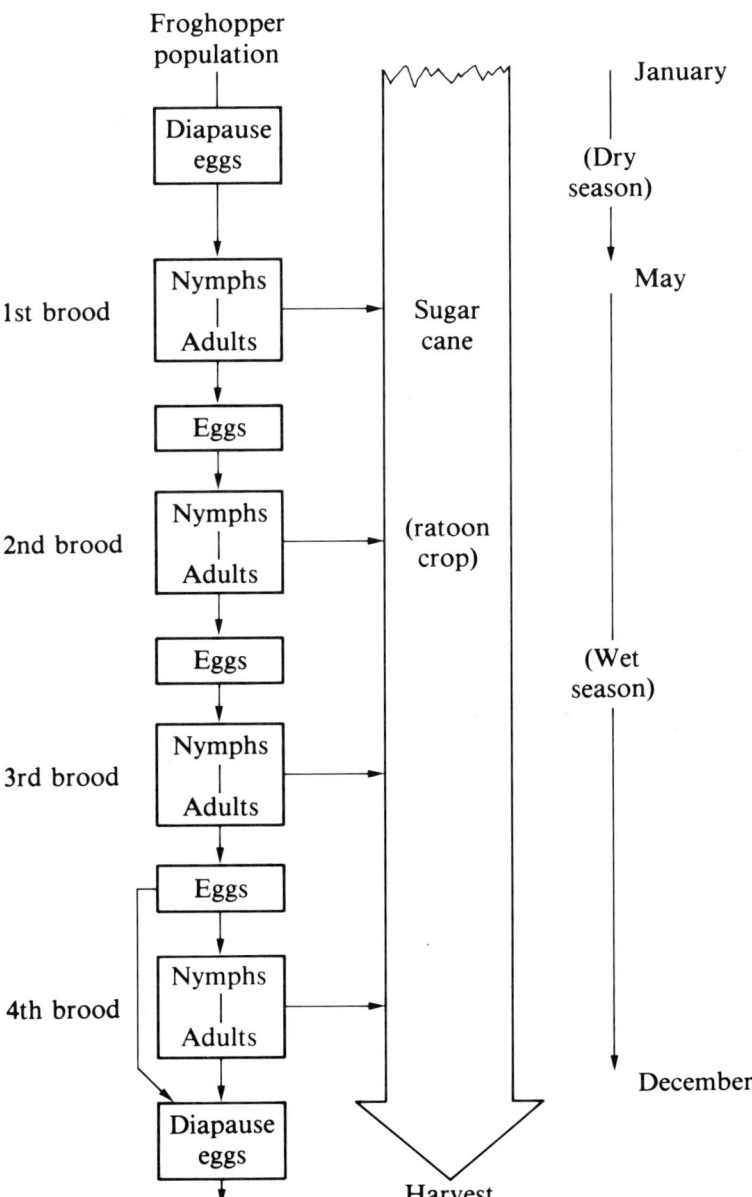

Fig. 6.6 The temporal relationship between a froghopper population and sugar-cane. (From Norton & Evans 1974)

the large froghopper, *Mahanarva tristis guppyi* (Urich), a sporadic pest in Trinidad. The eggs of these species are laid not in the ground but in the cane leaves, and the nymphs feed and live at the junctions of the leaf sheaths and laminae.

Damage

The adults of all species feed on the border parenchyma of the leaves and thereby cause minor physical damage. This takes the form of narrow strips, paler green than the adjoining tissue, which extend along the laminae from the points of insertion of the insects' stylets. Such marks are not easily noticed by the unpractised eye. Where the froghopper is a pest, necrosis develops around each feeding puncture and spreads longitudinally to form brown streaks which, if plentiful, coalesce and destroy the leaves (Fig. 6.7(a)). Growth is retarded; thin, stunted internodes are formed, and after a severe attack the smaller stems might be killed (Fig. 6.7(b)). The millable cane from affected fields, even after recovery, invariably has juice of low quality.

It has long been accepted that the agent or agents causing the destruction of leaf tissue are transmitted in the froghoppers' saliva

(a)

Fig. 6.7 (a) Streaks of froghopper blight caused by the feeding of *Aeneolamia varia saccharina* (Dist) (Photograph: B. H. Most) (b) Cane stool badly damaged by *A. varia saccharina* (dead leaves removed) (From Williams 1921)

when feeding takes place. Withycombe (1926) could find no disease organism in the saliva of *A. varia saccharina* and concluded that enzymes were responsible for the blight. Pickles (1937) agreed and suggested that the spread of necrosis after feeding was caused by

(b)

their continuing activity. Hagley (1967) identified the amino acids and enzyme groups present in the salivary glands and, after inoculating cane leaves with these substances, found that lipase produced lesions similar to froghopper damage.

The losses incurred vary from country to country. Iturbe & Ruano (1963) described *A. postica* as the second pest of sugar-cane in Mexico, but there is little doubt that the most severe damage has been caused by *A. varia saccharina* in Trinidad. The history of the Trinidad industry in the early twentieth century was one of foreclosures, bankruptcies and companies going into liquidation; and the most important factor in these failures was the froghopper. The yield of cane from blighted ratoon fields could be reduced by 50 per cent or more, the quality of the juice depressed by roughly 30 per cent and the stools so weakened that replanting was essential. Plant cane, by and large, was not affected. Eggs in diapause were killed by mechanical cultivation in dry weather, planting was delayed until the end of the froghopper season and the new crop reaped sixteen to eighteen months later. Reinfestation took place so quickly that only one and a half ratoons were taken: half of the area in first ratoons, and all of that in second ratoons, was ploughed out and replanted each year. Moreover, the small ratoon acreage retained in cultivation was neglected in varying degree from fear that money spent on it might not be recovered at harvest. In 1950, however, successful large-scale froghopper control was first achieved; during the next decade productivity in Trinidad increased from roughly 40 to 67 tonnes cane/arable ha per annum and a rapidly increasing number of ratoons were grown, with consequent economic benefit (Hanschell 1967). Progress was maintained until 1969, when heavy froghopper damage was in part responsible for a disastrous crop in 1970, and thereby gave a sharp reminder of the need for continuing vigilance.

Control measures
Biological control

Most of the early work on froghopper control was done on *A. varia saccharina* in Trinidad. Indigenous predators and parasites were multiplied and released, and natural enemies introduced. The most promising local agents were the egg parasite, *Anagrus urichi* Pickles (Pickles 1933); the Syrphid fly, *Salpingogaster nigra* Schiner (Guppy 1914), a predator of the nymphs; and the green muscardine fungus, *Metarrhizuim anisopliae*, a parasite of adults (Rorer 1910, 1911, 1913; Briton-Jones 1927). The Mexican bug, *Castolus plagiaticollis* Stal., a predator of the adults, was introduced from Vera Cruz (Urich 1914). None gave adequate control. The intensive but unsuccessful search for predators and parasites in tropical America during the period 1911–35 has been described by Williams (1921) and

Pickles (1942). It was felt later that other areas should be examined, and preliminary studies in East Africa have been made (Fewkes & Buxo 1965). The concentration of effort to find a solution to this problem made during the period 1925–34 is reflected in the carefully documented *Minutes and Proceedings of the Froghopper Investigation Committee* [later renamed the Sugar Cane Investigation Committee]. However, the only means of non-chemical control, the prevention of damage by frequent replanting, was uneconomic; and although varieties differed in their ability to recover from blight, none was resistant.

Chemical control

In theory the eradication of all first-brood nymphs should prevent any further attack, the potential progenitors of succeeding broods having been destroyed, but in practice this was and is not so. Although most of the eggs in diapause hatch roughly fourteen days after the onset of the rains, others remain dormant for longer periods. Therefore, to achieve nymph control, a chemical must have a long residual effect. The first insecticide recommended for this purpose was Cyanogas, a powder containing 50 per cent calcium cyanide (Hardy 1926); but its short period of four to six hours' lethal activity, and its high human toxicity, prevented it from being widely used.

Adult control was attempted by drift dusting. Ground limestone, impregnated with various insecticides, was released by tractor-drawn dusting machines travelling along the windward traces of infested fields in the evening, when a gentle breeze was blowing. The purpose was to cover the leaves with dust at a time when the adults were active. Pyrethrum and sabadilla killed quickly but had little residual value; therefore Pickles (1946a) recommended the use of a mixture of sabadilla with the slowly acting but persistent DDT. In large-scale experiments with DDT and HCH, Blackburn (1949) noted two defects: control was effective for only 90 m to leeward; and since the operation was taking place at the height of the wet season, the dust was quickly washed off the leaves and repeated applications were necessary.

Meanwhile Potter & Carrington (1947) found that gamma-HCH, diluted with ground limestone and applied as a root dust (Fig. 6.8) gave excellent and lasting nymph control. This was confirmed by Blackburn (1949), the most effective treatment being 125 kg of 0.65 per cent gamma-HCH/ha, but could not become estate practice immediately: the stools had to be cleaned of trash before the insecticide was applied, yet all the available labour was fully occupied in planting, and later in preventing the young plants from being smothered by weeds. The problem was solved by the widespread use of pre-emergence weed control with 2,4-D in newly planted fields; and

Fig. 6.8 Root dusting for froghopper nymph control in Trinidad sugar-cane (Photograph: K. Manchouk)

at last successful froghopper control was achieved (Blackburn *et al.* 1952). The methods developed in Trinidad were adopted elsewhere, with slight modification: in Mexico in 1952 (Hernandez & Flores 1956; Iturbe & Ruano 1963); in Venezuela in 1952 (Guagliumi 1954) in Guyana since 1950 (Bates 1957) and in Belize in the early 1960s (Fewkes & Buxo 1965).

Because the nymphs of *Mahanarva indicata* and *Mahanarva tristis guppyi* live not at ground level but in the axils of the upper leaves, control of the former in Brazil was by the aerial application of either 3 per cent gamma-HCH or 5 per cent carbaryl dust; and of the latter in Trinidad by the application of 0.65 per cent gamma-HCH dust, using knapsack blowers. Aerial applications were also made in some countries when adults appeared, either because nymph control had not been carried out, or because of invasion from surrounding areas. For this purpose a 3 per cent gamma-HCH dust was applied at 30 kg/ha in Mexico, and one containing either 25 per cent malathion or 5 per cent gamma-HCH in Belize. In Guyana concentrated formulations of gamma-HCH or dieldrin were applied in liquid form at the rate of 8.5 litres/ha.

There is a continuing search for new and more effective insecticides, but by and large the methods described are still used everywhere except in Trinidad. In 1954, nymphs of *Aeneolamia varia saccharina* in some areas of Trinidad were found to be resistant to gamma-HCH and other organo-chlorine insecticides. So attention was turned to organo-phosphorus insecticides. A 3 per cent malathion dust applied at 315 kg/ha proved effective against gamma-HCH-resistant nymphs, but had little residual value and repeated treatment of a rapidly increasing area was necessary to achieve control (Blackburn 1954). A 2 per cent carbophenothion (Trithion) dust used in the same way as malathion acted more slowly but was effective for the whole of the froghopper season (Neate 1958). Consecutively nymph control using either gamma-HCH or Trithion became standard estate practice for the next decade. In 1960 it was discovered that, where only organo-phosphorus insecticides had been used for four years, the froghopper nymphs had reverted to being susceptible to gamma-HCH (I. S. Clarke 1960).

Nymph control was costly: not only had insecticides to be bought, but the cleaning of stools by hand prior to their application became increasingly expensive as wages rose. Fewkes & Buxo (1969) developed the control of the adult stage by aerial application. Carbaryl (85 per cent sprayable) with sticker was chosen for first-brood control (at 2.8 kg in 28 litres/ha). For later broods the cheaper malathion (95 per cent) was applied at 1.4 litres/ha. On average, from three to four sprayings were required each year. This method was used on a large scale with complete success in 1967 and 1968, and replaced nymph control in northern Trinidad.

Thus, during a period of inflation, the cost of froghopper control at Trinidad's major estates fell from £13.8/ha in 1963 to £11.8 in 1968.

In 1969 and 1970, however, despite making nearly twice as many applications of insecticide as in 1967 and 1968, froghopper damage was severe. Evans and Buxo (1972) attributed this partly to a decline in the standard of management and partly to the apparent development of resistance by the insect. They pointed out that the size of sample from which population estimates were made had been reduced by 50 per cent and was inadequate; that there was an uncontrolled tendency of observers to report fictitious counts; that agriculturists failed to recognize the need to kill all first-brood adults, especially when their emergence was long drawn out; and that some adults had developed resistance to carbaryl.

So far there has been no report of resistance in froghoppers to gamma-HCH or carbaryl other than from Trinidad. More recently the synthetic pyrethroids have been used very successfully.

The mammalian toxicity of insecticides

Finally, reference must be made to the tragic lesson learnt in Trinidad in 1962 that insecticides of high mammalian toxicity should be used only when there are no alternatives; and supervision must then be as strict as possible. For general guidance, great care should be taken in the handling of a chemical which has an LD_{50} of less than 10; and when used its antidote, in many cases specific and difficult to obtain, should be readily available.

White grubs

The white grubs which are pests of sugar-cane are mostly larvae of the subfamily Melolonthidae but include one species of the Rutelidae, *Anomala orientalis* Waterhouse, and one Dynastid, *Alissonotum impressicolle* Arrow.

In Puerto Rico *Cnemarachis portoricensis* (Smyth) and *C. vandinei* (Smyth), in association with *Diaprepes abbreviatus* Linnaeus, comprise a major pest second in importance only to the moth borer, *Diatraea saccharalis* (Martorell & Medina-Gaud 1967). Elsewhere in the New World white grubs are of little consequence, except for sporadic outbreaks of *Clemora smithi* (Arrow) in the low-rainfall areas of Barbados: in 1913–25, 1930, 1931 and 1976 and in the Dominican Republic (Alam & Hudson 1976). It is in the rest of the world, especially in Australia and India, that they are a continuing menace to the industry. Their history and the control methods recommended for use in Australia have been described by Mungomery (1965). Avasthy (1967) has given a full report of the white grubs of India; and a comprehensive worldwide review has been made by Wilson (1969a, b).

Anomala orientalis in Hawaii and *C. smithi* in Mauritius, both introduced from other lands, have been controlled by natural enemies. Where indigenous species have long posed a serious problem, however, soil insecticides are providing a satisfactory solution. This applies to *Dermolepida albohirtum* (Waterhouse), *Pseudholophylla furfuracea* (Burmeister) and *Lepidiota frenchi* Blackburn in Queensland; *Leucopholis irrorata* Chevrolat in the Philippines; *C. portoricensis* and *C. vandinei* in Puerto Rico; *Holotrichia consanguinea* (Blanchard) and *H. serrata* Fabricius in India; *Cochliotis melolonthoides* (Gerstaecker) in Tanzania; *Hoplochelus rhizotrogoides* Blanchard in Madagascar; *Phyllophaga crinalis* Bates in Mexico; and *A. impressicolle* in India, Burma and Taiwan.

The life cycle

Eggs, varying in colour from pale white to cream, are laid in the soil, usually singly, and deeply enough to avoid death by desiccation: those of *A. orientalis* were found by Bianchi (1935) immediately beneath blankets of trash, but as deep as 45 cm in dry soil; and Avasthy (1967) noted eggs of *Holotrichia consanguinea* at vertical intervals in columns to a depth of 15 cm. The females of most species lay approximately thirty eggs each, often in a single batch, though those of *D. albohirtum* sometimes lay a second batch of the same number. *Holotrichia consanguinea* and *Clemora smithi* are more prolific: Bourne (1921) reported an average of 106 eggs per female for the latter. Obtuse-ovate when newly laid, the eggs approach spherical shape in roughly seven days as they mature. They range in size from 2.0 mm × 1.2 mm to 4.25 mm × 2.95 mm; and those of *C. portoricensis*, *Pseudholophylla furfuracea* and *Lepidiota frenchi* are encased each in its own pellet of earth. The period of incubation is influenced by temperature and varies from seven to twenty-five days (Table 6.2).

Table 6.2   *The incubation periods of the eggs of some white grubs*

Species	Incubation period (days)	Source
C. smithi	7–25	De Charmoy (1912)
H. serrata	8	Avasthy (1967)
H. consanguinea	8–18	Kalra & Kulrestha (1961)
A. orientalis	11–26	Van Zwaluwenburg (1937)
D. albohirtum	14	Mungomery (1965)
A. impressicolle	14–17	Avasthy (1967)
L. frenchi	16	Illingworth & Dodd (1921)

The grubs which emerge pass through three larval instars, meanwhile feeding on soil organic matter and cane roots. They are white in colour, broad and fleshy, and have a characteristic posture in the form of the letter C. The length of body of fully grown third-stage grubs varies from 30 mm (*A. orientalis* and *C. smithi*) to 50 mm (*D. albohirtum*).

White grubs have one or two-year life cycles according to the length of time spent in the larval stages. In the one-year species the first stage occupies roughly one month, the second a little longer and the third six months or more (Table 6.3). Pupation takes place in the soil at depths usually between 30 cm and 60 cm according to soil conditions; and the pupal period ranges from fourteen to thirty days. *Holotrichia consanguinea* is an exception: although its grub stages are much shorter, the adults compensate for this by remaining in the·

pupal cells for four months, making a one-year cycle.

The two-year species are pests only in Queensland: French's beetle (*L. frenchi*), the Childers beetle (*P. furfuracea*) and the Bundaberg beetle (*L. trichosterna. Lea*). The first grub stage of *L. frenchi* occupies two months, the second eight to nine months and the third six months in the upper soil, followed by five months at a depth of 35 cm in a pre-pupal condition. Its pupal period is one month. Although the life cycle of *P. furfuracea* occupies two years and is much the same as that of *L. frenchi*, it does not have such a distinct period of hibernation and the various grub stages overlap (Mungomery 1965).

Table 6.3   *The length of larval instars of some white grubs (days)*

Species	First stage	Second stage	Third stage	Source
One-year species				
D. albohirtum	30	35	182	Mungomery (1965)
C. portoricensis	32	50	169	Smyth (1917)
L. irrorata	49	67	224	Goseco, in Lopez (1931)
C. smithi	210–240 (all stages)			De Charmoy (1912)
	227 (all stages)			Bourne (1921)
H. consanguinea*	56 to 70 (all stages)			Avasthy (1967)
Two-year species				
L. frenchi[†]	61	240–270	180	Jarvis (1917)
				Illingworth, in Wilson (1969)

* Adults remain in the pupal cells for 166 days.
[†] Third-stage grubs of *L. frenchi* are active for 180 days and then spend 150 days in a pre-pupal condition.

When soil moisture and temperature are favourable the adult insects, cockchafer beetles, emerge. These beetles vary in colour from grey (*D. albohirtum*) to many shades of brown (*C. smithi* and *L. frenchi*). They also vary considerably in size, the beetles of *D. albohirtum* being roughly 33 mm long and 12 mm wide; while those of *P. furfuraceae* are 19 mm × 8 mm. After mating, the females return to lay eggs in the soil close to the cane roots. The beetles of all species except *A. impressicolle* cause virtually no damage to cane but feed on the leaves of nearby trees and shrubs.

Damage

Although first-and second-stage grubs of some species feed on cane roots, in general those of the third stage live longer, are extremely voracious and cause the greatest damage. They destroy the roots; and as they approach maturity eat the lower joints of the stems, and might burrow into them.

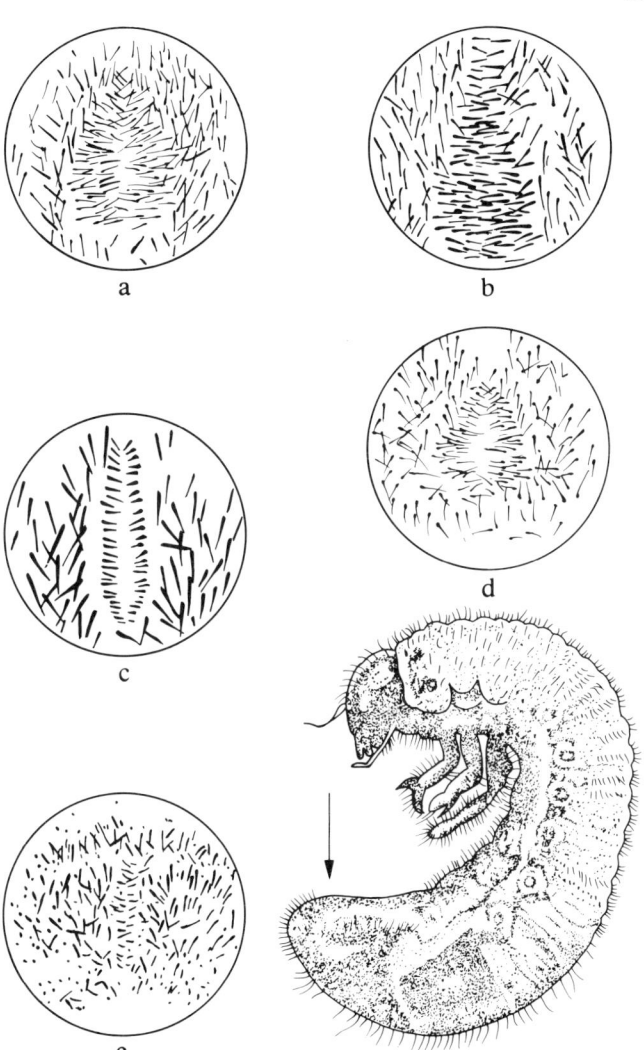

Fig. 6.9 (a) Means of identifying cane grub pests. Arrangements of hair groups on last body segment of grub at position shown by arrow: a. Frenchi grub; b. Consobrina grub; c. Grey-back grub; d. Childers grub; e. Bundaberg grub. (b) Stages in the life-cycle of the Childers cane beetle: a. Egg × 4. b. first-stage grub × 2; c. second-stage grub × 2; d. third-stage grub × 2; e. pupa × 2; f. beetle × 2; g. antennal club of female × 6; h. antennal club of male × 6 (From King *et al.* 1965)

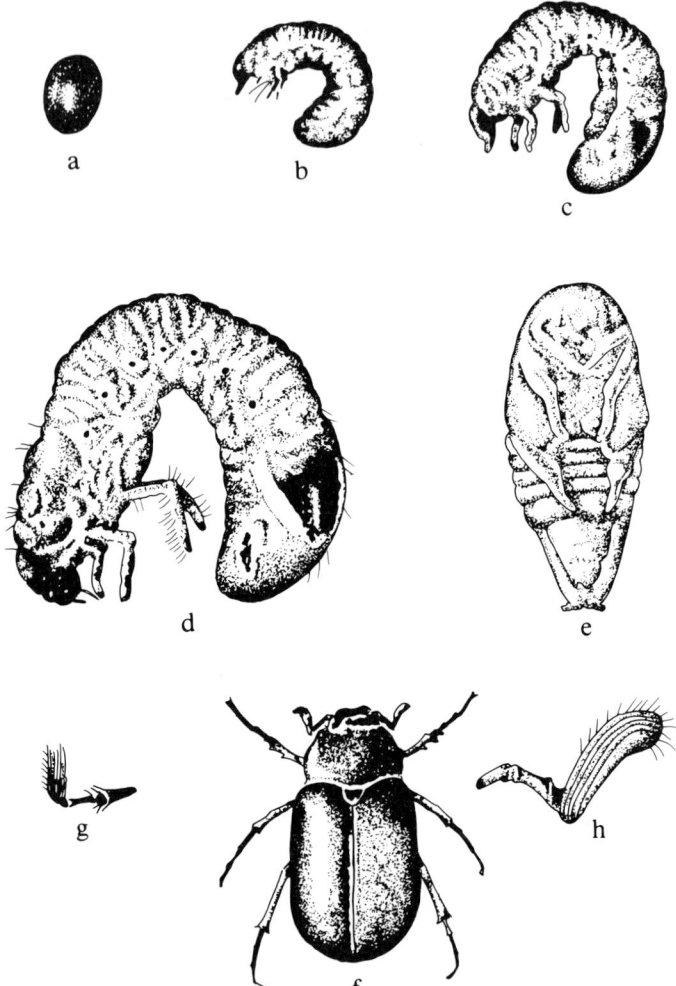

Fig. 6.9(b) (See also p. 205)

The first signs of damage to the crop are yellowing of the leaves and drooping of the inner leaf spindles. Later the stems deteriorate, the stools become loose and some overturn. The subsequent ratoon growth of damaged stools, even if uninfested, is usually weak.

The losses sustained vary according to the level of infestation from roughly 12 tonnes cane/ha (Wilson 1956) to the complete destruction of the crop (Avasthy 1967). Mungomery (1949) drew attention to the influence of environment on the severity of damage. One grub of *D. albohirtum* per stool was sufficient to cause serious damage to cane growing on relatively poor soil in a dry area of Queensland, three or four might have little effect in more fertile places.

Control

Many means of control have been advocated: the collection by hand of grubs and beetles, the destruction of the shrubs and trees on which the beetles feed, liming, changes in cultivation methods and in the time of planting. None was satisfactory. Adequate control has been achieved only by biological and chemical methods.

Biological control

In the New World lizards, frogs and toads have been effective in holding these pests in check, although sporadic outbreaks have been recorded. One such occurrence took place in Puerto Rico in the late 1870s after insectivorous lizards had been destroyed by the recently imported mongoose. It was halted by the introduction of the giant toad, *Bufo marinus* (Linnaeus). There was a recrudescence in the 1940s when the population of toads had been depleted by severe drought; thereafter supplementary chemical control was introduced (Wolcott 1936; Martorell & Medina-Gaud 1967). A similar relationship between white grubs and their predators exists in Barbados, and the severe attack in 1976 also is thought to have been caused by the effect of a succession of dry years on the numbers of toads. The use of insecticides had been advised (Alam & Hudson 1976).

Elsewhere successful biological control has been confined to Hawaii and Mauritius.

In the early years of this century, *ca.* 1908, *A. orientalis* was accidentally brought to Hawaii from Japan and *C. smithi* to Mauritius from Barbados. They caused considerable damage, but eventually were controlled by the introduction of parasites. *Campsomeris marginella* (Klug) *modesta* (Smith), a Scoliid wasp, was imported into Hawaii from the Philippines in 1916, quickly became established and successfully parasitized the grubs of *A. orientalis* (Muir 1917, 1919). Thereafter this pest was rarely seen until a severe though temporary outbreak took place in 1928. By 1934, however, *C. marginella modesta* had re-established effective control.

The first parasites established in Mauritius for the control of *Clemora smithi* were the wasps *Tiphia parallela* Smith, introduced from Barbados in 1913 and 1914, and *Campsomeris coelebs* Sic. from Madagascar in 1917 (de Charmoy 1917, 1923). Despite their activity it was estimated that *Clemora smithi* caused 20 per cent loss in yield in 1934; therefore further action was taken. Of the many parasites and predators released during the next few years, five Scoliid wasps were also established successfully and helped to give satisfactory control thereafter. Four were from Madagascar: *Campsomeris pilosella* Sauss., *C. erythrogaster* Dal., *C. lachesis* Sauss (and *Scolia carni-*

fax Sauss; and one from Java (*C. phalerata* Sauss). A full description of this work has been given by Jepson & Moutia (1939); and an account of its success by Williams & Moutia (1954) and Williams & Mamet (1962).

Resistant varieties

Although no variety is resistant to white grubs there are considerable and important differences between varieties in their ability, after an attack, to renew the root system by the development of dormant primordia. For example it is believed that the widespread planting of M 134/32 contributed even more than the establishment of parasitic wasps to the control of *Clemora smithi* in Mauritius (Williams & Moutia 1954); whereas the relatively weak root system of B 62163 caused its dramatic collapse when infested by the same pest in Barbados (Alam & Hudson 1976).

Chemical control

The current (1981) recommendations for grub control in Queensland issued by the Bureau of Sugar Experiment Stations are shown in Fig. 6.10.

If the land preparation has taken place in cold weather, insects in a pre-pupal condition, lying dormant below the depth of ploughing, will have escaped destruction and the HCH dressing is incorporated more deeply into the soil. Toxic effects on germinating setts are avoided by using lindane instead of the crude material, but there is some loss of residual protection (Mungomery 1965).

Rodents

Although many rodents attack sugar-cane, in general only rats are regarded as serious pests. Most of them are members of the family Cricetidae, the New World rats; or of the Muridae, the Old World rats. The ubiquitous Norway or brown rat, *Rattus rattus rattus* (Linnaeus) and the Alexandrine rat, *Rattus rattus alexandrinus* (Geoffroy), whose origins have been the subject of speculation for many years, have displaced indigenous species in many countries to which they have migrated or accidentally have been introduced. For example in Jamaica the aggressiveness of these rats and the predatory mongoose, *Herpestes javanicus auropunctatus*, specially imported from India, have caused the disappearance of the cane piece rat, *Oryzomis antillarum* (Metcalfe & Thomas 1966).

In Florida, Guyana, Hawaii and Mexico, where the rodent menace is most serious, control has been achieved by the use of poisoned bait in extensive campaigns carried out each year. An historical description of the black and brown rats, with special reference to Puerto Rico, has been given by Martorell *et al.* (1967); and an exhaustive survey of rodents in sugar-cane by Bates (1969).

Plant cane application
(for control in plant cane and three ratoons)

- Apply 130 kg/ha HCH 20 to sides of half-open drill.
- Place leading tynes to spread HCH across to centre.
- Place outer tynes to cover HCH with at least 5 cm of soil.

Plant cane

Ratoon cane application
(for control in 4th and subsequent ratoons)

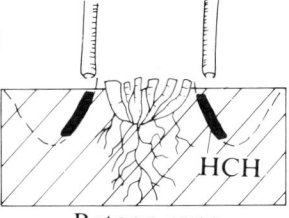

- Apply just after harvest of 3rd ratoons.
- Cut away from stool with discs or boards and apply HCH 20 to the cut face.

Ratoon cane

To protect 4th ratoon only − Apply 50 kg/ha HCH 20.
To protect 4th and 5th ratoons − Apply 85 kg/ha HCH 20.
To protect 4th, 5th and 6th ratoons − Apply 100 kg/ha HCH 20.

Warning:
HCH is a toxic chemical. Responsibility in relation to personal safety and the environment is essential. Follow manufacturers' recommendations and use protective clothing and respirators. Dispose of containers correctly.

Fig. 6.10 Grub control (Central district, Queensland, Australia)

Rodents were described by Collado & Ruano (1963) as the main pests of sugar-cane in Mexico, the most important of seven species being the sugar-cane rat, *Sigmodon hispidus* Say, of which there were two subspecies: *S. hispidus major* and *S. hispidus toltecus*; the Texan mouse, *Peromyscus leucopus texanus* Woodhouse; and the grey mouse, *Peromyscus boylii levipes* Merriam. Several species, including *R. norvegicus*, were also found in Guyana where the cane field rat, *Holochilus brasiliensis*, was so voracious that it was regarded as a major pest (Bates 1963). Bianchi (1960) reported that four species or subspecies caused great but unspecified losses in Hawaii. They were *R. rattus rattus*, *R. rattus alexandrinus*, *R. norvegicus* and *Rattus hawaiiensis* Stone. According to Doty (1960) the cotton rat, *Sigmoden hispidus littorales* was the most destructive pest in Florida, where the rice rat, *Oryzomis palustris coloratus*, and the Everglades water rat, *Meofiber alleni nigrescens*, also damaged cane. Doty's assessment was confirmed by Samol (1972), who stated that the roof rat, *R. rattus*, had increased in importance since 1963. Takano & Kondo (1939) regarded *Bandicota indica* as the most serious pest in southern Taiwan. Other rats and mice which attacked cane in Taiwan were described as *R. norvegicus*, *Rattus losea* Swinhoe, *Apodemus agrarius ningpoensis* Swinhoe and *Mus formosanus* Kuroda. Although rodents are relatively unimportant pests in Queensland, Redhead (1972) stated that 90 per cent of the damage took place in the area north of Townsville and was caused almost entirely by two indigenous Murid species: *Rattus contatus* Thomas and *Melomys littoralis* Lönberg. Similarly, *R. norvegicus* and *R. rattus* are minor pests in Mauritius, but Rajabalee (1969, 1970) found that control measures against them were taken at most of the estates in the humid zone of that country. Recently it was reported that the rat *Nesokia indica* and the mouse *Mus musculus* Linnaeus caused serious damage at Haft Tappeh in Iran (Azizi & Sund 1972).

Life cycle and biology

The life cycle, morphology and general biology of the species named above have been described by the authors cited. Although there is considerable variation between them, they have characteristics in common which affect their status as pests of sugar-cane:
1. They are highly fecund, mature quickly and have short gestation periods.
2. They cannot live on sugar-cane alone: their diet must contain protein.
3. The presence of potential nesting sites inside or close to fields predisposes cane to attack.

Favoured nesting places are rank vegetation on the banks of watercourses, uncultivated land, villages and lodged cane. Consequently, most of the injury in Hawaii, for example, occurs in high-rainfall areas, in fields near heavily vegetated gullies and during the second year of two-year crops; whereas younger, upright cane is unaffected (Hilton *et al.* 1972). Metcalfe & Thomas (1966) found a similar distribution of damage in Jamaica; and Takano & Kondo (1939) described cane fields close to waste areas, cemeteries, banks of rivers and aqueducts as most at risk in Taiwan. In Barbados rodents are not regarded as serious pests, though lodged cane may be eaten by rats; but in Trinidad, where annual cropping is practised and most of the stems are upright at harvest, such damage is unknown.

Damage

Rodents feed mostly by night. They eat into the basal internodes of stems and cause physical damage by the formation of characteristic scallop- or canoe-shaped cavities which, like stem-borer tunnels, allow the entry of pathogens. Affected canes often break and become a complete loss. An entirely different type of damage, the destruction of apical growing-points, is caused by two climbing species: the rat *H. brasiliensis* in Guyana (Bates 1969) and the mouse *P. leucopus texanus* in north-western Mexico (Collado & Ruano 1963). It has been suggested that this activity is associated with nutritive requirements during periods of reproduction.

The incidence of outbreaks is irregular, as is their severity. Major epidemics in Guyana, such as those of 1935, 1951, 1958 and 1959, were caused by the migration of rats from the savannahs to the cane fields and were associated with climatic extremes. If the savannahs are either flooded or very dry, with resultant bush fires, competition for food and nesting places forces many rats to move to the adjacent sugar estates (Bates 1963). Rats with their fur burning may also spread fire from burning fields to adjacent cane fields. Mungomery (1965) illustrated the uneven distribution of damage in Queensland where total losses of cane on the small farms in rat-infested areas varied from less than 1 tonne to 50 tonnes, which is typical of other areas, and makes it impossible to estimate the loss. Collado & Ruano (1963) estimated that, in Mexico, the most heavily infested plantations suffered between 8 and 11 per cent loss of cane in the early 1950s and control measures were required in 21 per cent of the area planted, since when there had been some reduction in damage. On this basis the overall loss, when the report was made, was roughly 1 per cent. Bates (1963) estimated that the loss of sugar caused by the epidemic of 1958 and 1959 in Guyana was at least 12,750 tonnes, 1.8 per cent of the total of 703,443 tonnes produced in those years. A similar level of damage, 1.7 per cent, was sustained in Puerto Rico in 1961: Martorell *et al.* (1967) calculated that 17,527 tonnes of sugar were lost but 1,000,559 tonnes, raw value, were produced. Mungomery (1965) estimated that in recent years annual losses in Queensland varied between 20,000 and 30,000 tonnes of cane, or 0.2–0.3 per cent of the total crop. Metcalfe & Thomas (1966) reported that damage by rats in 1965 amounted to 5.4 per cent of the potential yield of ten fields at one estate in Jamaica. It was estimated by Takano & Kondo (1939) that, in spite of killing 2,083,700 rats and mice during the years 1933 and 1934, the annual loss suffered from the depredations of these pests in Taiwan was roughly 7,000 tonnes of cane. Finally, Bianchi (1960) felt that although losses in Hawaii could not be estimated accurately from the meagre data accumulated over the years, the money spent on rat control, US $300,000 per annum, would give greater returns if more wisely spent. Twenty years earlier, Pemberton (1939) stated that rat control was the largest biological problem facing sugar-cane growers in Hawaii and that the damage by rats on one plantation 'amounted to about US $300,000'.

Whatever the damage they cause to cane, rodents are a menace to man and animals. Doty (1960) described them as a veritable reservoir of disease, referred to their connection with bubonic plague and gave the following examples of rodent-borne diseases endemic in the Florida Everglades: tularemia, trichinosis, anthrax and leptospirosis (Weil's disease). The need to control cane-field rats in order to prevent the transmission of Weil's disease was also stressed by Mungomery (1965).

Many methods of control have been attempted: changing an environment which provides ideal nesting sites, the rejection of palatable varieties, the introduction of predators, hunting, trapping and the use of fire. All have contributed to some extent but, as stated earlier, the most effective means is by the use of poisoned bait.

Resistant varieties

Although under some conditions in Queensland all varieties may be damaged by rats, certain characteristics confer a degree of protection. They are a thick barrel, hard rind and erect habit (Mungomery 1965). Similar observations have been made in other countries. In Guyana *H. brasiliensis* showed a liking for soft, low-fibre varieties (Bates 1963); and in Jamaica *R. norvegicus* and *R. rattus rattus* preferred to eat thin rather than thick canes (Metcalfe & Thomas 1966). Much earlier, Takano & Kondo (1939) noted that in Taiwan POJ 2725, which had soft, inclined stems, suffered heavier damage than the harder, upright POJ 2878 and F 108.

Control measures
Biological control

Rodents are attacked by various mammals, birds and reptiles although only the mongoose, *Herpestes* spp., has been introduced into cane-growing areas in large numbers for their control. Successful at first in several West Indian islands and Hawaii, later it turned to easier prey, both wild and domestic, and is no longer effective.

Rodents are also susceptible to disease, and the possibility of achieving control by the release of bacteria such as *Bacillus typhi murium* Loeffler has been examined, especially in Java and Taiwan. Before such methods could be adopted the risk to public health would require careful consideration; however, all the early experimental attempts in Taiwan failed (Takano & Kondo 1939). Strains of *Salmonella enteriditis*, with greater specificity to rodents, have been reported; and their use in cane fields appears to be worthy of consideration (Bates 1969).

Fire

The only pest of cane in Barbados mentioned by Ligon (1657) were rats. To kill them fire was set along the periphery of each infested field and, burning inwards, prevented their escape. His description was as follows:

. . . (the overseers) have only this recompence, which is by burning an army of the main enemies to their profit, Rats, which do infinite harm in the Island, by gnawing the Canes, which presently after will rot, and become unserviceable in the work of Sugar. And that they may do this justice the more severely, they begin to make their fire at the out-sides of that land of Canes they mean to burn, and so drive them to the middle, where at last the fire comes, and burns them all; and this great execution they put often in practice, without Assizes or Sessions . . .

The burning of cane fields prior to harvest continues to be an important control measure in many countries, but is not carried out primarily for that purpose. It was considered by Doty (1960) to have contributed more than any other single factor to the control of rodents in Florida. Nevertheless, additional measures are taken there and in other places where pre-harvest, and in some places also post-harvest burning are standard practices. Nowhere in modern times has burning been adopted solely for rodent control.

Chemical control

The chemical control of rodents is unlike that of other pests of sugarcane: most techniques entail the use of bait; and in recent years coumarin derivatives, acting as anticoagulants, have been the most successful agents. Formerly the fumigant Cyanogas (calcium cyanide) was used, especially in Taiwan, against species which lived and nested in tunnels; but for general purposes poisoned bait was either broadcast throughout the fields, restricted to the perimeters to prevent migration from nearby waste land and villages, or placed in strategically located containers.

Control by the distribution of bait containing crude pastes of phosphorus or arsenic was first attempted in the late nineteenth century. These were replaced by sodium fluoride, strychnine and red squill, a preparation made from the lily, *Urginea maritima*; but by the late 1920s they, in turn, had been succeeded by barium carbonate in Australia and thallium sulphate in Hawaii and Java. Zinc phosphide, antimony thirourea and sodium fluoroacetate were added to the list in the 1930s and 1940s; and endrin in the 1950s. Zinc phosphide is still used in Hawaii and in 1970 became the first rodenticide to be granted federal approval for broadcast application over food crops in the USA (Hilton *et al*, 1972). In contrast, sodium fluoroacetate at a strength of 1 : 1,000 is effective against rats but highly toxic to humans. Moreover, no antidote to it is known. Consequently the Queensland Department of Health has been reluctant to sanction its use so long as less dangerous and equally efficient chemicals are available (Mungomery 1965). In the late 1940s recognition of the anticoagulant properties of a substituted coumarin, renamed warfarin, led to its use, and also to that of related compounds, in rodent control. There is considerable variation in the relative susceptibility of *Rattus* spp. to warfarin: a concentration in the bait of 5 : 1,000 is considered adequate for *R. norvegicus*, whereas 25 : 1,000 is recommended for the control of *R. rattus*. Bates (1963) noticed in 1961 that on several plantations in Guyana *H. brasiliensis* had developed resistance to warfarin, and concluded that endrin pellets had played a major role in the successful control campaigns of the two previous years.

Rajabalee (1970) reported on the acceptance by rats in Mauritius of waxed blocks containing anticoagulants. In Queensland, Mungomery (1965) recommended the distribution in dry weather – to prevent decay and the formation of mould – of small cellophane sachets smeared with fresh linseed oil, containing 3.5 g of wheat, treated with thallium sulphate, placed at 9 m intervals along each seventh row of damaged cane. The more expensive warfarin was reserved to control stubborn foci of infestation where rats were resistant to other chemicals.

Aerial spreading of bait is now standard practice in many countries. Bates (1969) has illustrated some of many forms of preparation: devices to make the bait waterproof, 'dog biscuits' used in Hawaii, waxed blocks in Mauritius and pellets in Guyana. He warned that it would be unwise to adopt without question baits and techniques which were successful elsewhere. Local species and strains of rodents, and different ecological conditions, might require them to be modified.

Miscellaneous pests

There are many more pests of minor or local importance. The elephant, hippopotamus and wild pig occasionally damage cane in Africa; the wallaby and white cockatoo in Australia; and the monkey in Barbados. Termites might eat setts at the time of planting (Harris 1969); mealybugs can affect the manufacture of syrup (Ingram *et al.* 1951b) and the growth of cane (Dick 1969b); and scale insects occasionally cause appreciable losses (Rao & Sankaran 1969; Fewkes 1972).

Others of greater interest are nematodes, the West Indian cane fly, army worms, the leafhopper and, where the industry is being expanded, new pests of increasing importance.

Nematodes

Many species of nematode have been recorded from the roots and the rhizosphere of sugar-cane. Most are either root-knot nematodes, *Meloidogyne* spp., or lesion nematodes, *Pratylenchus* spp. The damage which they cause is difficult to quantify and their economic importance is therefore not known. Interest in them has been spasmodic.

By and large attempts to kill nematodes have been confined to the testing, as soil fumigants, of halogenated hydrocarbons such as D-D, EDB and DBCP. Holtzmann & Wismer (1967) described trials carried out with these chemicals in Hawaii, but found that the increase in yield of sugar from treated plots was insufficient to justify their use. Dick (1969a), in reviewing the results of experiments carried out over the years, reported that in economic terms fumigants were rarely successful; and that, in general, large-scale trials gave inconsistent or disappointing results. Consequently the commercial use of nematocides in cane fields has been negligible. Nevertheless, worthwhile responses to their use has been recorded at Chirundu in Zambia (Thompson 1977); and experiments with them were continued in South Africa, where nematodes were known to depress the growth of cane on sandy soils. It is now recommended that in such areas the organo-phosphorus granular nematocide, aldicarb (Temik), should be applied to plant cane in the furrow at 30 kg (15 per cent a.i.)/ha.; and to ratoons, in furrows 5–10 cm deep, close to the row at 15 kg (15 per cent a.i.)/ha, and then covered with soil (Carnegie 1974, 1976).

The West Indian cane fly

The West Indian cane fly, *Saccharosydne saccharivora* (Westwood) (Homoptera, Delphacidae), occurs in Florida, Louisiana, the West Indies and Venezuela, but is a pest of sugar-cane only in Jamaica and in parts of Venezuela.

Its life cycle has been described by Fennah (1969). The adults are pale green with transparent wings. Eggs are laid in batches of four to twelve in incisions in the midrib of the leaf on the underside, and have an incubation of roughly seven days. The nymphs which emerge, yellow at first, later become green and do not disperse. They pass through five instars in about thirty-six days. Many of the adults also feed locally, but some fly to other leaves or to the central whorl. Red rot develops in the oviposition slits and a sooty mould on the leaves, which wither and die if the infestation is heavy. The mould is a fungus which grows on the honeydew excreta of feeding adults. Few assessments of damage have been made but Ingram *et al.* (1939) estimated that 'a loss of at least 10 per cent would result' from a localized attack in Florida in 1936.

In Puerto Rico *S. saccharivora* is kept under control by indigenous parasites; *Stenocranophilus quadratus* Pierce kills the nymphs and adults; and the wasp, *Anagrus armatus* (Ashmead), destroys the eggs. Martorell *et al.* (1973) suggested that the same parasites were also effective in the Dominican Republic. In continental USA, however, control of nymphs and adults is achieved by the small wasp parasite *Gonatropa* sp., the larva of which forms 'a prominent, seed-like sac on the side of the fulgorid's abdomen' (Ingram *et al.* 1951a).

In Jamaica, where the use of an insecticide against this pest is often necessary, Metcalfe (1964) found that, with block treatment and careful timing in regard to the life cycle, a single application of malathion gave complete control for one year. The insecticide was applied from the air at a rate of 1 kg a.i. in 3.3 litres of water/ha. Bees are attracted to the honeydew of *S. saccharivora* and as many as 17,800/ha have been recorded in heavily infested fields. As bees are also susceptible to malathion, Metcalfe (1966) recommended that, before application, all bee colonies in the neighbourhood should be confined to their hives, which previously had been covered with wet sacking, until spraying ceased.

The sugar-cane leafhopper

The leafhopper, *Perkinsiella saccharicida* Kirkaldy (Homoptera, Delphacidae) is now important only as a vector of Fiji disease (Ch. 7). At the turn of the century, however, its other activities, similar to those of *S. saccharivora*, were of much greater conse-

Fig. 6.11 *Perkinsiella saccharicida* Kirkaldy (Sugar Industry Research Institute, Mauritius)

quence and caused severe damage in Hawaii. Accidentally introduced either from Fiji or Australia, it was first recognized on the
island of Oahu in 1900. In the absence of parasites it spread rapidly,
especially in the cooler and wetter parts of Hawaii Island; and its
effects were so disastrous that sugar production in the Kau district
fell from 17,080 tonnes in 1903 to 750 tonnes in 1906 (Baver 1960).

Its life history and biology have been described in detail by
Fennah (1969). The damage caused is to the cane leaves and comprises
laceration of tissue by the ovipositor, the subsequent development
of red rot in the oviposition slits, desiccation and the formation of
sooty mould as honeydew excreta spreads over the laminae. Control
was achieved by biological methods. Egg parasites were brought to
Hawaii from Australia, Fiji and Taiwan; and after initial disappoint-
ments proved to be effective. The wasp, *Paranagrus optabilis*
Perkins, was particularly useful. Nevertheless, severe local
outbreaks still occurred until 1920 when the egg predator, *Tytthus
(Cyrtorhinus) mundulus* (Breddin), was imported from Australia
and Fiji. It was soon established, and with the other natural enemies
has achieved almost complete control since 1923.

Perkinsiella saccharicida still occurs in many countries; but
nowhere is it a serious pest. In describing its significance in Australia
under the general heading 'Minor pests of sugar-cane', Mungomery
(1965) stated that the appearance of sooty mould was often more
spectacular than destructive; that leafhoppers increased rapidly after
unseasonable weather; but also that natural enemies quickly
regained control when growing conditions became favourable. The
only action recommended was good husbandry: the provision of
adequate irrigation and improved drainage, and the application of
appropriate fertilizers.

Defoliating insects

Outbreaks of leaf-eating caterpillars, usually called army worms,
occur in most cane-growing countries which have a pronounced dry
season. The parasites which normally hold these pests in check are
disproportionately reduced in number during a long drought but,
shortly after the onset of the rains, quickly reassert control. The
foliage of cane shoots, having been reduced to midribs without
laminae by first-brood caterpillars, is then restored. A list of some
of the insects which behave in this manner is given in Table 6.4.

In general the damage sustained is not regarded as serious; on the
contrary, Trinidadian folklore holds that an outbreak of army worm
is a good omen. This is not without reason: the drought which
affected the parasites also allowed the previous crop to be taken off
expeditiously and without the damage to stools which is caused by
reaping in wet weather. Experience in Queensland appears to have

Table 6.4　*Defoliating insects*

Species	Country	Reference
Cirphis latiuscula	Florida	Ingram *et al.* (1939)
Herrich–Schaeffer	Mexico	Van Zwualuwenburg (1951)
Cirphis cholica Dyar	Mexico	Van Zwualuwenburg (1951)
Pseudaletia unipuncta	Hawaii	Bianchi (1960)
(Haworth)	Java, Taiwan, Philippines, Queensland, Fiji	Pemberton (1963)
	Mexico	Van Zwualuwenburg (1951)
Cirphis loreyi Dup.	Taiwan, Philippines, Queensland, Fiji, Java,	Pemberton (1963)
	Mauritius	Williams (1963)
Laphygma exempta	Hawaii	Bianchi (1963)
(Walker)	Philippines, Queensland	Pemberton (1963)
	South Africa	Dick (1951)
Spodoptera mauritia	Hawaii	Bianchi (1960)
(Boisduval)		
Laphygma frugiperda	Continental USA	Ingram *et al.* (1951)
(Smith and Abbot)	Cuba	Scaramuzza & Barry (1960)
	Venezuela	Guagliumi (1969)
	Trinidad	Box (1954)
Parasa bicolor Walker	Taiwan	Pemberton (1963)
Pelopidas mathias	Taiwan	Pemberton (1963)
(Fabricius)		
Mocis repanda	Colombia	Potes (1954)
(Fabricius)	Cuba	Scaramuzza & Barry (1960)
	Guyana	Bates (1967)
	Mexico	Van Zwualuwenburg (1951)
Mocis latipes Guenee	Venezuela	Guagliumi (1960)
	Trinidad	Box (1954)

Note: In English-speaking countries all of these pests, with few exceptions, are known locally as army worms. The exceptions are: *C. latiuscula*, the Florida cutworm; *M. repanda*, the semi-looper of Guyana; and *M. latipes*, the striped grass looper of Trinidad.

been much the same as in Trinidad (Mungomery 1965). In Mauritius, however, Williams (1963) warned that considerable loss in yield, 14 tonnes cane/ha, could be caused by a second defoliation. Control is achieved easily by spraying.

The future

New pests are being recorded where cane has been planted for the first time on a large scale. Tan & Johnson (1974) mentioned several

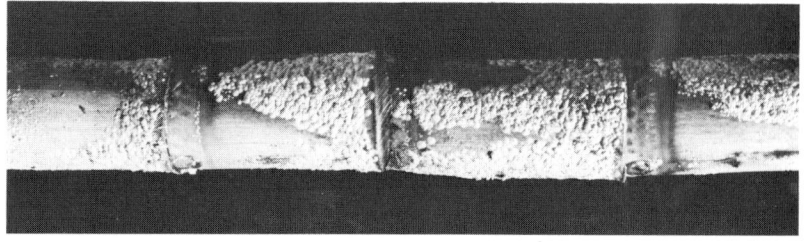

Fig. **6.12** *Aulacaspis tegalensis* (Zhnt.) (Sugar Industry Research Institute, Mauritius)

they met during the establishment of a 5,000 ha estate in north-western Malaysia, and in particular described an abnormal yellowing of the leaves caused by *Phaenacantha saccharicida* Karsch (Hemiptera, Lygaeidae).

Other pests, formerly of minor or local importance, are increasing in significance as cultivation expands. For example the white scale insect, *Aulacaspis tegalensis* (Zehnter) (Homoptera, Diaspididae), introduced into East Africa from south-eastern Asia at least fifty years ago, affected production only recently and caused a reduction of one-third at the TPC estates in Tanzania in 1969–70 (Fewkes 1972).

Eldana saccharina (Walker) (Lepidoptera, Pyralidae) also is causing greater damage than in past years. Dick (1951) first reported this stem borer in South Africa in 1939, but its distribution was restricted to the Umfolozi Flats. By January 1975 this pest had spread in South Africa and Swaziland and was increasing in intensity. Heavily infested cane was found not only at Umfolozi but also at Amatikulu, Empangeni, Felixton and Pongola (Carnegie 1974, 1976). Further afield *E. saccharina* caused heavy damage in 1956 at a jaggery estate in Tanzania, reached pest status on plantations in that country in 1966 and caused considerable losses thereafter (Waiyaki 1974). Girling (1972) confirmed that serious damage was suffered in northern Tanzania, noted that *E. saccharina* had been found on cane in Kenya in 1967, and stated that it was also the most serious borer in the new plantations at Kinyala in Uganda 'where the joint attack was up to 44.5 per cent in 1970'. With the continuing expansion of the industry in Africa, *Eldana* might be to sugar-cane in Africa what *Chilo* is in the rest of the Old World and *Diatraea* in the New World.

Chapter 7

Diseases

Many serious outbreaks of disease have occurred since the Bourbon or Otaheite cane, widely planted throughout the world, succumbed to a complex of diseases in Mauritius *ca*. 1840, to gumming disease in Brazil in 1869, to root rot in Hawaii and to that and other diseases in the West Indies and elsewhere towards the end of the nineteenth century. Most were overcome by planting resistant varieties.

In the major cane-growing areas measures are now taken either to prevent future outbreaks or, if they should occur, to minimize their effect. Seedlings grown at breeding stations are inoculated – manually or by insect vectors – with the causal agents of diseases against which protection is specially required in the countries which they serve, and reactors are rejected; strict quarantine measures are enforced; and the danger inherent in planting a disproportionately large area with one variety is recognized. When symptoms of a serious disease are noticed, usually the first action is to rogue and destroy infected plants, to select 'clean' seed-cane and, according to the disease, treat it with hot water, hot air or a fungicide immediately before planting. Meanwhile resistant varieties are sought.

Stevenson and Rands (1938) prepared the first comprehensive, annotated list of fungi and bacteria associated with sugar-cane and its products. In compiling a check list of sugar-cane diseases of the world, Martin (1951) assigned the causal organism to each pathological disease and included a number of physiological diseases. Martin's paper was brought up to date in the *Proceedings* of the ninth congress of the ISSCT.

The next major revision was presented at the fourteenth ISSCT congress (Egan *et al*. 1972). The nomenclature of Egan *et al*. (1972, 1974), with minor amendments, has been used.

The lists should be read with caution but without dismay. For many years most of the diseases have been and are of little or no economic importance in the countries in which their presence has been recorded. For example, in Part II, under the heading 'The sugar cane countries and their diseases', Egan *et al*. (1972) reported the following for Trinidad and Tobago: basal stem, root and sheath rot, chlorotic streak, dry rot, eye spot, leaf scald (a new listing), mosaic, pineapple disease, pokkah boeng, ratoon stunting, red leaf spot, red rot, red rot of leaf sheath, red spot of leaf sheath (not observed in recent years), red stripe, rind disease, ring spot, sooty mould, stellate-crystal fungus and wilt. Despite this daunting catalogue, for the past thirty years the cane growers of Trinidad (sugar is no longer produced in Tobago) were unaware that their crops suffered from any disease, and the appointment of a pathologist was not even considered. Nevertheless the sudden outbreak of smut disease in Guyana and Martinique (Clarke *et al.* 1975), and its recognition in Jamaica (Barbados SIR 1976) and Trinidad (Clarke *et al.* 1976) one year later, will dispel any complacency which might have developed. The price to be paid for freedom from disease is constant vigilance.

The authoritative work, *Sugar-Cane Diseases of the World*, is published in two volumes under the auspices of the ISSCT (vol. 1, Martin, Abbott *et al.* 1961; vol. 2, Hughes *et al.* 1964). Some of the more useful or interesting regional descriptions are given in Table 7.1.

Table 7.1 *Publications with Regional description of sugar-cane disease*

Region	Reference	Title
Continental USA	Rands & Abbott (1939)	Sugar-cane diseases in the United States
Louisiana	Abbott (1963)	Problems in sugar-cane disease control in Louisiana
Caribbean	Baker *et al.* (1954)	Sugar-cane diseases in the Caribbean
Puerto Rico	Johnston & Stevenson (1917)	Sugar-cane fungi and diseases of Puerto Rico
	Liu *et al.* (1967a)	Diseases of sugar-cane in Puerto Rico
Dominican Republic	Liu *et al.* (1967b)	Diseases of sugar-cane and their control at Central Romana
Lesser Antilles	Nowell (1923)	Diseases of Crop Plants in the Lesser Antilles
Brazil	Bitancourt (1939)	Diseases of the sugar-cane in Brazil
Peru	Abbott & Martin (1952)	The sugar-cane situation in Peru
Paraguay	Alvarez (1954)	Sugar-cane diseases in Paraguay
Argentina	Fawcett (1924)	Las enfermedales de la caña de azúcar en Tucuman
The Americas	Johnston (1918)	Diseases of sugar-cane in tropical and subtropical America, especially the West Indies

(Table 7.1 continued)

Region	Reference	Title
Hawaii	Cobb (1906, 1909)	Fungus maladies of the sugar cane
	Martin (1938)	Sugar-cane diseases in Hawaii
Java	Wakker & Went (1898)	Ze ziekten van het suikerriet op Java
Philippines	Ocfemia (1939)	A review of sugar-cane diseases in the Philippines
	Reyes (1954)	Postwar observations on sugar-cane diseases in the Philippines
Taiwan	Lo (1951)	A report on sugar-cane diseases in Taiwan
	Matsumoto (1952)	Monograph of sugar-cane Diseases in Taiwan
Fiji	Daniels *et al.* (1972)	The control of sugar-cane disease in Fiji
Australia	Hughes (1951)	The control of cane diseases in Queensland
India	Butler (1918)	Fungi and diseases in Plants
	Subramaniam (1936)	Diseases of sugar-cane and methods for their control
	Chona (1956)	Chairman's addresses, pathology section, ISSCT
Sri Lanka	Egan (1963)	The diseases of sugar-cane in Ceylon
Mauritius	Shepherd (1926, 1931)	Diseases of sugar-cane in Mauritius
	Wiehe (1963)	The control of sugar-cane diseases in Mauritius
Réunion	Horau (1967)	Sugar-cane diseases in Réunion island
Mozambique	Noronha (1972)	Sugar-cane diseases in Mozambique
South Africa	King (1956)	The major sugar-cane diseases of Natal
Egypt	Rosenfeld (1939)	Minor sugar-cane diseases in Egypt

In this chapter the more important diseases have been grouped according to their causal agents as fungal, bacterial and viral. Within each group they have been arranged in alphabetical order.

Fungal diseases

Eye spot

Drechslera sacchari (Butler) Subramaniam and Jain formerly *Helminthosporium sacchari* Butler

An account of this disease has been given by Martin *et al.* (1961a). It caused serious damage, by destroying leaf tissue, in limited areas of Hawaii (Martin 1938) and Puerto Rico (Johnston & Stevenson 1917); and in more widespread outbreaks in Cuba, Taiwan (Matsumoto 1952) and Florida (Edgerton 1955). In a description of the diseases of Australia, however, it was not mentioned (Hughes 1951); nor was its presence in India regarded as serious (Subramaniam 1936).

In general, eye spot is a widespread disease but causes little loss in yield.

Symptoms

Elongated water-coloured lesions, roughly 1–2 mm in length × 0.5–1 mm in width, develop on the youngest leaves of affected plants, their long axes parallel with the leaf veins. Within a few days they increase five fold in size, their centres become chocolate brown in colour and occupy most of the areas of infection. Runners, which destroy much more tissue than the original lesions, extend from the points of primary infection towards the leaf tips. They pass through colour changes similar to those of the original lesions and turn from pale yellow to dark brown.

If the infection is heavy, growth is retarded and short internodes are formed. In its most acute form, however, the disease can induce top rot and the death of cane stools.

Eye spot is most serious during the cooler winter months, especially if dew or light rain causes high humidity and allows water to remain on the cane leaves. Heavy rainfall washes the fungal spores from the leaves and infection thereby is greatly reduced. The climatic conditions which affect the incidence of eye spot in Mexico were described by Flores & Ramirez (1963), and the influence of temperature in Puerto Rico by Liu (1969).

Control

Control is achieved by replacing susceptible with resistant varieties. The resistance of new varieties to the disease may be determined by carrying out trials in known 'eye-spot localities'. Details of the method used in Mexico, based on Hawaiian experience, were given by Osada & Flores (1969).

Fusarium sett or stem rot

Perfect state: *Gibberella fujikuroi* (Sawada) Ito
                      formerly *Gibberella moniliformis* Wineland
Imperfect state: *Fusarium moniliforme* Sheldon

Fusarium sett or stem rot has been described by Bourne (1961). It is a disease of the basal parts of unwounded stems, and of setts, and causes their internal tissue to become dark red and to rot. The economic importance of stem rot in past years is difficult to assess with accuracy: several pathologists believe that much of the damage caused by it was assigned erroneously to the red rot fungus, *Glomerella tucumanensis* (Speg.) Arx and Muller, with which it is often associated. Bourne (1961) asserted that mycologists were unable to distinguish between the causal agent of stem rot and that of another cane disease, pokkah boeng, both caused by *Gibberella fujikuroi*, though Martin *et al.* (1961) suggested that the strains involved might be different. Be that as it may, by planting varieties resistant to disease, *Fusarium* stem rot now is of little importance in countries where once it is thought to have caused considerable damage. The most recent outbreaks occurred in Florida (Bourne 1954) and India (Khanna & Rafay 1953).

Pineapple disease

Perfect state: *Ceratocystis paradoxa* (Dade) C. Moreau
Imperfect state: *Thielaviopsis paradoxa* (de Seynes) Dade
An account of pineapple disease has been given by Wismer (1961). The causal agent, the fungus *Ceratocystis paradoxa*, attacks the setts shortly after they have been planted. It enters through the cut end surfaces and spreads rapidly. The tissue, which at first turns red, remains firm; and at this stage the smell of pineapple fruit might be noticed. Later, after the parenchyma has been destroyed, the interiors of the setts become black and hollow. Liu *et al.* (1967b) isolated a dark and a light strain of the fungus in Puerto Rico.

The presence of *C. paradoxa* has been reported in most sugar-producing countries of the world, and it has caused serious losses by the failure of infected setts to germinate. Control has been achieved by impregnating the setts with a fungicide at the time of planting. Organomercury compounds have proved effective for this purpose. Hughes (1951) recommended that, in Australia, setts should be treated with an aqueous solution of methoxy ethyl mercuric chloride, containing 0.015 per cent mercury, either by immersion in a dipping bath or by a series of sprays attached to the cutter-planter machine. In Hawaii, Wismer (1951) found that a solution of 0.025 per cent phenyl mercuric acetate, used as a cold dip or spray, was as effective as any other fungicide tested, and the least expensive. Satisfactory results with these and similar compounds were obtained in Brazil (Dantas & Da Silva 1956), Mauritius (Evans and Wiehe 1947) South Africa (McMartin 1949) and Taiwan (Chu & Wang 1949).

Liu *et al.* (1967b) pointed out, however, that although organo-mercurials successfully controlled pineapple disease, generally speaking they were phytotoxic and often left undesirable residues.

Seeking an alternative they noted that the systemic fungicide benomyl (Benlate) was four times more effective than several organo-mercury compounds in inhibiting the mycelial growth of *C. paradoxa in vitro*. The use of benomyl as a replacement for mercurials in the standard cold-water dips was recommended by Ricaud (1974a) on a large scale in Mauritius on the grounds of economy, and was to be applied at the rate of 200 g in 320 litres/ha. The addition of an indicator dye, such as rhodamine at 5 g/100 litres, was advised. Ricaud also gave directions for the incorporation of benomyl into the hot-water treatment of setts. Non-mercurials are also used in other cane-growing areas including Queensland.

Pokkah boeng

Perfect state: *Gibberella fujikuroi* (Sawada) Ito
                        formerly *Gibberella moniliformis* Wineland
Imperfect state: *Fusarium moniliforme* Sheldon

Descriptions of the disease called pokkah boeng, the Javanese term for a malformed or twisted top, have been given by Van Dillewijn (1951) and Martin *et al.* (1961b). It is induced by the fungus *Gibberella fujikuroi*, and has been recorded in almost all countries in which cane is grown commercially, but has caused severe damage only in Java where the widely grown variety, POJ 2878, was particularly susceptible to the disease.

Symptoms

As the name implies, its effect is to distort the cane tops. The earliest symptoms are seen on the young leaves, which become chlorotic towards their bases, twisted and wrinkled, and are narrower and shorter than normal leaves. Later, irregular reddish stripes develop within the chlorotic parts. If infection is limited to the leaves, the plants usually recover; if not, internal and external ladder-like lesions develop in the stems. The most serious injury is sustained when the fungus penetrated the growing-points, which die and rot.

Infection is by the entry of airborne conidia into the openings between partially unfolded leaves, which are formed during periods of hot, dry weather. Such conditions occur in Java immediately before the onset of the rains. When the rains start, the conidia are washed down to the susceptible parts of the spindles, where they germinate and pass through the soft cuticle to the inner tissues. This activity was reported by Van Dillewijn (1951).

Control

Control has been achieved by planting resistant varieties. At the Sugar Experiment Station, Pasuruan, Java, seedlings are tested for resistance to pokkah boeng by injecting a suspension of *G. fujikuroi* conidia into the leaf spindles 10 cm below the highest visible leaf joint. In Brazil, however, da Eire *et al.* (1974) found an interaction between the concentration of the inoculum and the variability of *G. fujikuroi*. To assess the resistance of varieties to the disease it was therefore essential to know the aggressiveness of the isolate being used. In their experiments the plants were inoculated 5 cm above and 25 cm below the apical meristems.

Red rot

Perfect state: *Glomerella tucumanensis* (Speg.) Arx and Muller
formerly *Physalospora tucumanensis* Spegazzini

Imperfect state: *Colletotrichum falcatum* Went

Red rot is one of the oldest and most serious diseases of sugar-cane and its causal agent has been described in much that has been written on *Colletotrichum falcatum*, the name of the imperfect stage of the fungus. The disease has been described by Hughes (1954), Edgerton (1955) and Abbott & Hughes (1961); and Chona (1956) gave its history, with special reference to India.

Red rot occurs in most countries in which cane is grown and was first identified by Went, in Java, in 1893. It may attack any part of the plant, but is specially important as a disease of the stems and of setts. By and large the injury caused to the leaves is unimportant, except that growths on the midribs are sources of infection for other parts.

Symptoms
Damage to the stems

The first symptoms of injury depend on the method of infection. If the fungus has gained entry into the stems through the nodes, wounds or pest injury, red rot will begin at those points and will extend slowly or rapidly according to the resistance of the variety. If the infection is rising up the stems from underground parts, the vascular bundles will first turn red before the other tissue also becomes red, sometimes interrupted by white blotches. The sucrose content of the affected parts is greatly reduced. External symptoms appear only in the later stages of the disease when ill-defined red patches may appear on the rind before the stems dry out and shrink.

Damage of this type in tropical countries is confined to the secondary infection of cavities made by moth borers and rodents, and its severity depends on the incidence of these pests, as well as on varietal susceptibility. In subtropical countries, on the other hand, heavy losses unconnected with injury have been suffered: in

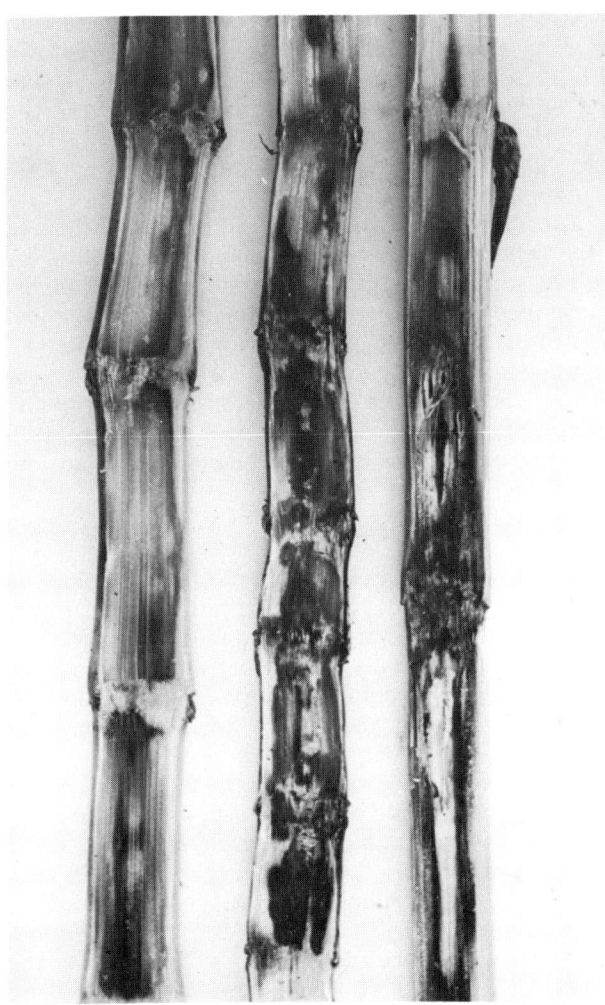

Fig. 7.1 Red rot symptoms (Photograph: South African Sugar Association, Experiment Station, Mount Edgecombe)

Australia (Hughes 1954), Mauritius (Wiehe 1944) and South Africa (Dodds 1943). A measure of the severity of such an attack at Bundaberg in Queensland in 1940 has been given by Hughes (1954). In fields of Co 290, 40 per cent of the cane was left on the ground, and much of that sent to the factory had been affected, with consequent deterioration in juice quality. The epidemic of 1938–40 in India was even more disastrous: the most widely grown variety, Co 213, was wiped out in the states of Uttar Pradesh and Bihar, which contained two-thirds of that country's cane area (Kar *et al.* 1974). The outbreak in Hawaii in the 1950s was equally severe in fields planted with H 38–2915, but other varieties grown at that time, H 32–8560, 37–1933 and 39–3633, were highly resistant to the disease.

Damage to setts

Red rot can also affect the establishment of young plant cane in subtropical countries where the temperature might be too low for quick germination but suitable for the development of the fungus. Edgerton (1951) described how this type of damage occurs. The leaf scars and bud scales of the seed-cane are already infected at the time of planting and, in favourable conditions, mycelia grow inwards from these foci. The buds on the setts are killed, or may germinate and then die, causing the stands of cane to be poor and gappy. Red rot is an important disease of seed-cane in Australia, Taiwan and especially in Louisiana.

Control

The most effective means of reducing loss from red rot is by planting resistant varieties. Chemical control is not practised. The treatment of setts with fungicide at the time of planting was unsuccessful because the causal organism already had penetrated beyond the reach of externally applied disinfectants. However, some reduction of infestation in standing cane by the use of fungicides has been reported.

Two types of varietal resistance are recognized: physiological, associated with qualities of the living cells which suppress or prevent the development of the fungus; and morphological, caused by structures in the tissues which physically retard its development. Physiological resistance is the more important. Most clones of *S. officinarum* are very susceptible to the disease; those of *S. barberi* and *S. sinense*, in general, also are susceptible; while those of the relatively few forms of *S. robustum* which have been tested are intermediate in their reaction to the disease. Physiological resistance appears to be restricted to *S. spontaneum* and is governed by one or more genes. Azab & Chilton (1952) studied its inheritance in the progeny of fourteen crosses and suggested that in some hybrids a dominant gene from *S. officinarum* might mask the effect of the resistance genes from *S. spontaneum*. In cane-breeding stocks, therefore, there is not that high degree of resistance to red rot, approaching immunity, which exists in respect of other diseases. Moreover, there is a far greater variability in the virulence of isolates of the fungus than occurs in most other pathogens of sugarcane. The methods used to select resistant varieties in Louisiana were described by Abbott (1956); in the Punjab, India, by Singh *et al.* (1956); and in Uttar Pradesh, India by Kar & Singh (1956).

In Hawaii, India and Taiwan, standing cane is inoculated to determine varietal resistance but in Louisiana (Abbott *et al.* 1967), and also in Taiwan, cultures or spore suspensions are introduced into setts. Whichever technique is used, after a period of incubation the varieties under test are classified according to their degrees of rotting in relation to standard varieties with known reaction to the disease.

In Uttar Pradesh, India, a preliminary selection is made before varieties are subjected to inoculation. A thick suspension of red rot spores is sprayed on to the test population. Resistance is then judged by the spread of lesions in nodal and internodal tissues; and only those which show little or no reaction are subjected to more rigorous examination. Great care is taken to ensure that the inoculum used is that of the prevailing red rot flora (Kar *et al.* 1974).

Root rot, *Pythium arrhenomanes* Drechsler

Root rot, caused by the fungus *Pythium arrhenomanes*, was responsible, at least in part, for the replacement of several widely grown noble canes. The dramatic failure of the Otaheite or Bourbon cane in the latter half of the last century, and thereafter, was attributed to root rot, though Nowell (1923) felt that red rot was the main cause of its breakdown in the West Indies. The decline of the same variety in Hawaii, there called Lahaina, was more gradual but was ascribed wholly to root rot: while in Louisiana, the Ribbon and Purple Cheribon, and later D-74, which had replaced the Creole and Otaheite, collapsed from a combined outbreak of mosaic, red rot and root rot diseases (Rands & Dopp 1938). Root rot also caused serious damage in Java and indeed in most countries in which sugarcane was grown, but before hybrid varieties were introduced.

A description of the disease has been given by Rands (1961).

Pythium arrhenomanes is spread by two means: by swimming zoospores after heavy rainfall and, at other times, by surface mycelia which grow from rootlet to rootlet. As the name implies, infected plants have deficient root systems. Young cane is gappy and has a typical unthrifty look; older cane is easily lodged and uprooted, and later becomes rotten. It is difficult to assess with accuracy the losses caused by root rot: it was so often associated with other diseases and, as noted earlier, the reasons for some crop failures ascribed to it were perhaps wrongly diagnosed. Rands & Dopp (1938) stated that root rot and red rot caused a reduction of 23 per cent in the average yield in Louisiana during the period 1910–20. Subsequently, mosaic disease caused a further reduction of 30 per cent. Conservative estimates of the final loss incurred because of these diseases from 1920 to 1927 were US $150 m. In the earlier epidemic in the Caribbean, between 1892 and 1894, losses variously estimated between 25 and 50 per cent were reported in Antigua, Barbados, Grenada, St Kitts, St Vincent and Trinidad (Nowell 1923).

All varieties being grown when and where disastrous outbreaks of root rot occurred were clones of *S. officinarum*. Later, *S. spontaneum* and *S. sinense* were found to be highly resistant, and *S. robustum* probably susceptible. Therefore control was achieved by planting resistant varieties, especially trispecific hybrids of *S. officinarum*, *S. barberi* and *S. spontaneum* (Abbott & Summers 1951). Sartoris (1947) stated that, by carrying out a series of complicated crosses to form such hybrids (Fig. 7.2), most of the varieties released from Canal Point, Florida, for general culture since 1941 were resistant to root rot, red rot and mosaic diseases. Similar breeding policies have been pursued elsewhere and now root rot is of little consequence.

Culmicolous smut, *Ustilago scitaminea* Sydow

Four smut diseases of sugar-cane are recognized: floral smut (*Sphacelotheca cruenta* (Kuehn) Potter), covered smut (*Sphacelotheca macrospora* Yen and Wang), false floral smut (*Claviceps* sp. plus *Epicoccum* sp.) and culmicolous smut. Of these, culmicolous smut, caused by the fungus *Ustilago scitaminea*, is the most widespread and has been of importance at one time or another in many cane-growing areas.

Because *S. barberi* and the Indian form of *S. spontaneum* are highly susceptible to the disease, and *S. officinarum* immune or highly resistant, for many years it was confined to the Old World. In various years, often widely spaced, attacks occurred in South Africa (McMartin 1945), Rhodesia (McMartin 1948) and Kenya (Robinson 1959); and it was described as the most important sugar-cane disease in the Malagasy Republic by Noronha (1972). In India it was considered by Chona (1956) to be second in importance only to red rot; in the Philippines it was present but caused little damage (Reyes 1954); and although not mentioned in a description of the cane diseases of Taiwan given by Lo (1951), recurred after an absence of thirty-four years in 1964 (Leu & Teng 1974). It was once of major significance in the subhumid irrigated areas of Mauritius, but now is of little consequence there (Antoine 1955); and is equally unimportant in Réunion (Horau 1967). Recently its presence was reported in Hawaii (Byther *et al.* 1971).

The first authentic report of smut in the New World came from Argentina (Fawcett 1924), since when outbreaks have occurred in Paraguay in 1944, (Alvarez, 1954) in Brazil in 1948 (RAM, 1948), in Bolivia in 1957 and, as mentioned on page 72, in Guyana and Martinique in 1975. Smut was recognized in Jamaica and Trinidad in 1976.

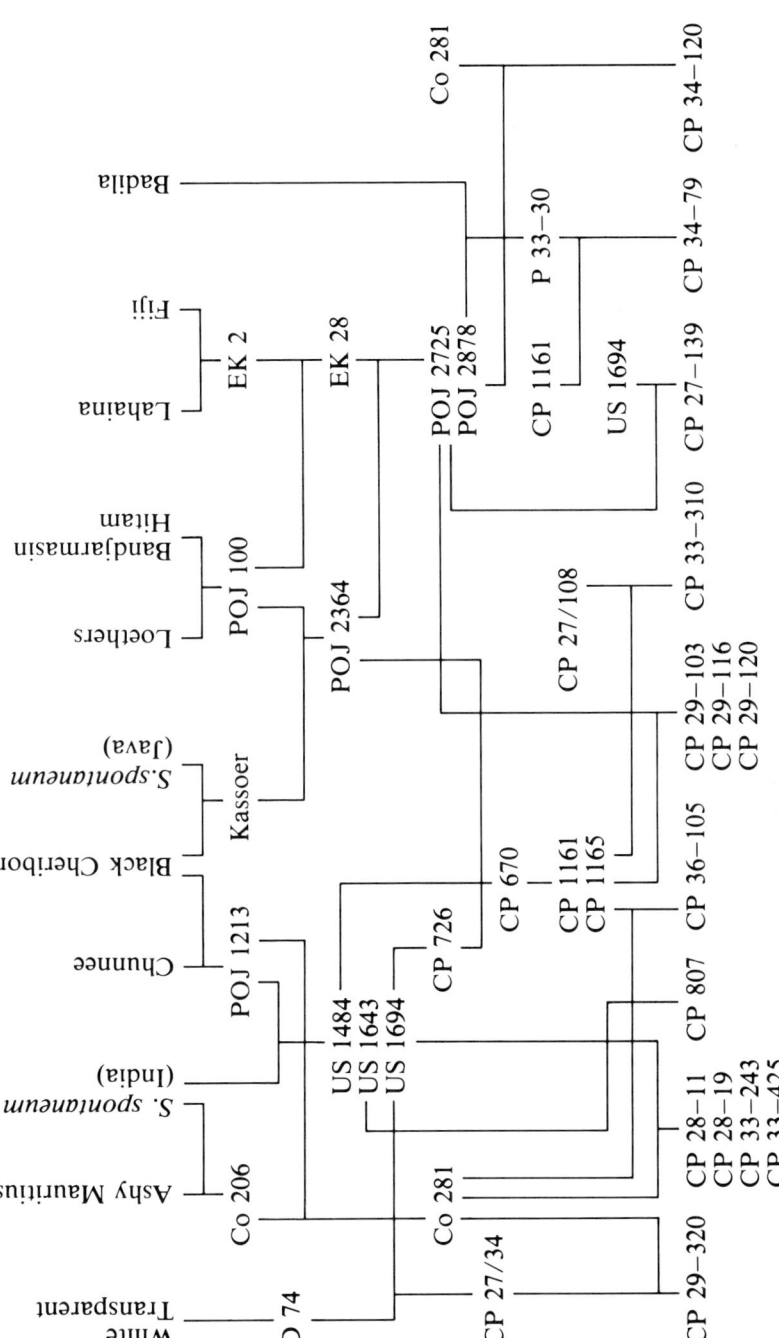

Fig. 7.2 Breeding for resistance to disease (From Sartoris 1947)

Symptoms

A description of the disease has been given by Antoine (1961). The symptoms are so distinctive that smut is one of the most easily diagnosed diseases of sugar-cane. They are the formation of characteristic whip-like, pencil-thick unbranched structures from the apices of affected stem (Fig. 7.3). Each structure comprises a core of parenchyma and fibrovascular elements surrounded by vast numbers of chlamydospores, enclosed at first in a thin, silvery sheath. Later, when the membranous covering splits, the exposed chlamydospores resemble a thick layer of soot. They are then dispersed, mainly by wind.

Smut is transmitted in two ways: by windborne spores gaining entry into standing cane through the buds; and by spores in the soil, or in irrigation water, entering planted setts. Infected buds may develop quickly into whips, or the mycelia may remain dormant within the buds to form a latent source of disease if the stems are used as a seed-cane. Whips appear in two distinct seasons: the first, early flush is the result of primary infection; the second, which occurs much later, is caused by secondary infection in the field.

Damage

It is not easy to assess accurately the damage caused by smut: by and large reports have been more in terms of the percentage of shoots and stools affected than of loss in yield. Moreover, severe outbreaks have often been followed rapidly by the complete eradication of the disease: for example those which occurred in Argentina between 1943 and 1946, and in Rhodesia between 1945 and 1948. Chona (1956) reported that, in experiments carried out in Delhi, there was a reduction in yield by weight of millable stems of 29 per cent (Co 312) and 23 per cent (Co 313) in diseased plant cane, rising to roughly 70 per cent in ratoons. Mohan Rao and Prakasam (1956) confirmed that plant cane (39–56% loss) was less severely affected than ratoon cane (52–73% loss). In addition, cane from healthy clumps gave juice of significantly higher quantity than that from 'smutted' clumps.

Control

Many methods of control have been suggested: hot-water treatment, the use of fungicides, roguing diseased stools – critically assessed, under Zimbabwe conditions, by James (1974) – and growing only plant cane; but the replacement of susceptible with resistant varieties is by far the most economic and satisfactory means of achieving this end. As indicated earlier, clones of *S. officinarum*, with few exceptions, are immune or highly resistant to smut. McMartin (1948) found that canes derived from crosses between *S. officinarum* and *S. spontaneum* were more resistant than those from crosses between *S. officinarum* and *S. barberi*; and that trispecific hybrids were inter-

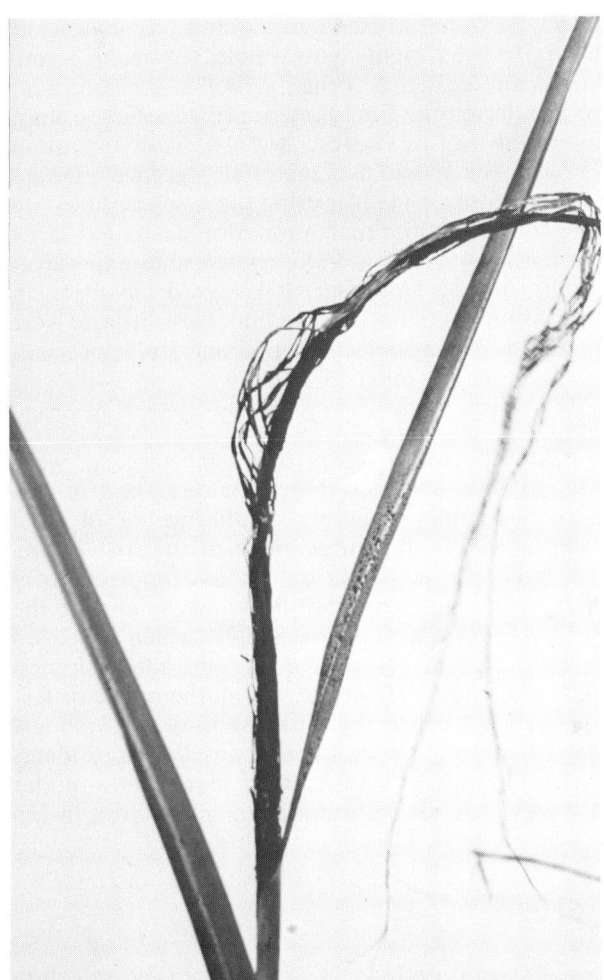

Fig. 7.3 A smut whip, the characteristic symptom of the disease (Photograph: South African Sugar Association, Experiment Station, Mount Edgecombe)

mediate in their reaction to the disease. Some of the bud characteristics which might cause such differences were examined by Muthusamy (1974) at Tamilnadu, India. He reported that, in standing cane, there was a positive correlation between smut incidence, bud size and bud sprouting. The sprouting was due to attack by the stem borer, *Chilo indicus*. The position of the nodal bud was subapical in most resistant varieties and apical in susceptible ones. Selection is complicated, as it is in determining resistance to red rot, by the existence of different strains of the fungus: although two isolates present in Taiwan were morphologically and physiologically indistinguisable, NCo 310 was highly susceptible to strain 1 and probably immune to strain 2 (Leu & Teng 1974).

The methods used to determine the resistance of varieties to smut in Taiwan were described by Leu & Teng (1974). After overnight incubation in vinyl bags, the inoculated material was then planted out. In Hawaii, Byther & Steiner (1974), using three susceptible and six resistant varieties, found that the reaction to dip, paste and spray inoculations of stem cuttings were generally comparable and correlated with susceptibility in the field; but that wound paste inoculations were less consistent. In the dip method the cuttings were submerged for ten minutes in a suspension containing 5 m spores/ml.

Miscellaneous fungal diseases

Roughly 100 fungi, each the causal agent of a cane disease, were listed by Egan *et al.* (1972). Most of those not described in this chapter are either of negligible economic significance or of local occurrence; and some are of both. An example of the first is ring spot (*Leptosphaeria sacchari* van Breda de Haan), reported from seventy-five countries, but which causes little damage; and of the last, veneer blotch (*Deightonellia papuana* Shaw), which has been identified only in New Britain, New Guinea and the Solomon Islands, and is equally innocuous. However, not all the minor or less well-known diseases can so lightly be dismissed: downy mildew (*Sclerospora sacchari* Miyake) at various times caused serious losses in Australia, Fiji, the Philippines and Taiwan, but is now under control in those countries. The measures taken to achieve this in Fiji were described by Daniels *et al.* (1972).

Bacterial diseases

Bacterial diseases are weak pathogens, and generally infection follows damage such as driving rain in hurricanes or from harvesting knives.

Gumming disease, *Xanthomonas vasculorum* (Cobb) Dowson

Gumming disease, caused by the bacterium *Xanthomonas vasculorum*, is one of the oldest recorded diseases of sugar-cane and one of the most serious. At various times it has caused heavy losses in many countries. The first outbreaks occurred in Brazil, Madeira, Mauritius, Réunion, Australia and Fiji in the nineteenth century; and in the West Indies, Colombia, S. Rhodesia (now Zimbabwe), South Africa and Madagascar in the twentieth century. Earle (1928) suggested that gumming disease was brought into Mauritius in cane imported from Brazil in 1869, and gained entry into Australia from Mauritius in similar fashion. Orian (1954), however, postulated that the much earlier (*ca.* 1840) collapse of the Otaheite cane in Mauritius and Réunion probably had been caused by gumming disease; that this disease was primarily one of palms, especially the white palm (*Dictyosperma album* Wendl and Drude), a plant native to Mauritius and the other Mascarene islands; and that the white palm, the royal palm (*Roystonea regia* O. F. Cook), maize and India tiger grass (*Thysanolaena maxima* (Roxburgh)) formed a reservoir of infection responsible for the continuing widespread occurrence of gumming disease in sugar-cane in Mauritius. Orian argued his case with compelling logic: for example by pointing to the absence of gumming disease from areas where sugar-cane originated and by casting doubt on the diagnosis made in Brazil by Dränert in 1869. If his theory is correct, Mauritius has been the most important source of the disease.

An account of gumming disease has been given by Hughes (1961); of its probable origin by Orian (1954); and of the evolution of two strains of the pathogen in Mauritius and Réunion by Antoine (1969) and Ricaud & Sullivan (1974).

Symptoms

The disease might be detected in its early stages by the formation of characteristic longitudinal streaks on the older leaves. Pale at first, later they turn yellow, sometimes flecked with red, and finally become greyish brown. They are 3–6 mm wide, with well-defined edges, though older streaks may become diffuse. Following the course of vascular bundles, they run straight but at a slight angle to the midribs. Diagnosis of gumming disease in its later stages is by the symptom which gave rise to its name: the exudation of yellow gum from the cut ends of infected stems (Fig. 7.4). Some of the vascular bundles become red, but this symptom is neither distinctive nor specific.

Gumming disease is spread locally by infected agricultural implements, especially cutting blades, but is transmitted further afield, and from one planting to the next, in diseased setts. Infested stools are usually stunted and produce weak, chlorotic shoots. If the growing-points of stems are killed, lateral buds might sprout. By and large ratoons are not as badly affected as plant cane. An unusual feature is the wide variation in the losses caused from one crop to the next: when growing conditions are favourable there might be little reduction in yield, even in susceptible varieties. Nevertheless, as already noted, devastating epidemics have caused serious damage in past years: for example the crop at Harwood in New South Wales, Australia, was reduced by 40 per cent in weight and by 17 per cent in sugar content during the years 1893–5 (North 1935). Moreover,

Fig. 7.4 Gumming disease *X. campestris* pv. *vasculorum* (Sugar Industry Research Institute, Mauritius)

the gummy juice from diseased cane caused difficulties in the factory and was not easily processed. The severity of similar outbreaks in South Queensland in the 1890s was largely responsible for the formation of the Queensland Bureau of Sugar Experiment Stations. In more recent years, however, the disease has been checked before such losses could be incurred. These bacterial diseases in Mauritius are spread from field to field in the hurricanes. The heavy rain and strong wind assist the bacteria to infect the cane leaves.

Control

Hughes (1961) described the action to be taken when the disease is recognized:

1. A quarantine area must be established, with its boundary at least 3.2 km from the diseased fields.
2. Within this area the planting of susceptible varieties must be prohibited.
3. Diseased fields, as they are ploughed out in the normal cycle, or earlier by compulsion, must be closely watched and volunteer stools destroyed.
4. Planting material of resistant or semi-resistant varieties must then be supplied from a clean locality outside the quarantine area.

Eventually it should be possible to reintroduce the old susceptible varieties, but much better to continue to grow resistant ones.

There is great variation in susceptibility to gumming disease in the clones of *S. officinarum*, *S. robustum*, *S. sinense* and *S. spontaneum*. Although some of each are susceptible, others are not; and a large reserve of genetic material resistant to the disease and satisfactory in other commercial qualities has been established. Therefore, breeders serving countries where gumming disease is a threat are able to offer a wide range of resistant varieties for cultivation.

Hughes also described how simple resistance trials might be carried out.

As with other diseases, a close watch must be kept for the appearance of new forms of the pathogen. An epidemic named 'gummosis I' was controlled in Réunion and Mauritius by 1948; but another outbreak, caused by a new strain of the bacterium and named 'gummosis II', occurred in Réunion in 1958 and in Mauritius in 1964. The remedial action taken was so prompt that, by 1968, 51 per cent of the cane area of Mauritius had been planted with varieties resistant both to 'gummosis I' and to 'gummosis II' (Antoine 1969).

Leaf scald, *Xanthomonas albilineans* (Ashby) Dowson

An account of leaf scald, caused by the bacterium *Xanthomonas albilineans*, has been given by Martin *et al.* (1960) and a symposium on the disease at the fourteenth congress of the ISSCT comprised papers by Egan (1972a, b) Koike (1972) and Steindl (1972).

Leaf scald is thought to have originated in the Old World. Serious outbreaks occurred in Australia in the 1920s and 1930s (North 1935; Hughes 1954) and at one time or another in Fiji, Hawaii, Java, the Malagasy Republic, Mauritius, the Philippines, Réunion, Taiwan and Thailand. Its presence in the New World was first reported by Arruda & Do Amaral (1945) in Brazil and was confirmed by Dale (1950) in Guyana. Later, in the 1960s, leaf scald was identified in continental USA, in most of the countries of Southern Africa, and elsewhere. Not that the disease had previously been absent from these areas: for example Baker *et al.* (1954) surmised that originally it was brought to Guyana and Surinam from Indonesia in illicit importations of seed-cane. Thereafter it escaped detection until the susceptible variety, B 34104, was planted on a large scale. One of the most dangerous features of leaf scald is that it might be carried in resistant varieties without the appearance of visible symptoms. It was for this reason that Egan (1972a) stated that, of all the sugar-cane diseases in the world, leaf scald: '. . . currently is the major potential troublemaker . . .'. Effective quarantine is difficult, the results of resistance trials suspect and there are many unidentified strains of the pathogen.

Symptoms

Martin and Robinson recognized two phases of the disease: chronic and acute. The external symptoms of the chronic phase are narrow white stripes on the leaves and leaf sheaths, stunted stems, etiolated leaves and the profuse development of side-shoots (called 'lalas' in Hawaii). As the leaves mature, the stripes might broaden and become diffuse. Whether or not this happens, the leaves wither downward from the tips along the lines of the stripes and thereby assume the scalded appearance from which the name of the disease is derived (Fig. 7.5). Internal examination of the stem shows fine red streaks, especially pronounced in nodal areas and at the junctions with side-shoots. In the acute phase large areas of cane rapidly wilt and die, without showing symptoms of disease, as if affected by drought. This phase is rare and is confined to extremely susceptible varieties. In it the symptoms of the chronic phase which, because of premature wilting, were not shown on the affected plants, eventually appear on ratoon shoots.

Leaf scald is transmitted from place to place in infected cuttings and is then spread by cane knives and by reaping machines.

Control

Many means of control have been attempted: regular inspection of seed-cane areas, selection of healthy setts, roguing diseased stools and disinfecting cane knives at frequent intervals. Most of them have been successful, at least in part, but effective control has been achieved only by planting highly resistant varieties.

The methods used in leaf-scald trials throughout the world, by and large, have been developed during the past fifty years in Australia, Hawaii and Mauritius. They were described by Koike (1972), who concluded that the quickest and most reliable results were achieved by cutting young shoots above the growing-point, inoculating the cut surfaces (Antoine & Ricaud 1961) and, except on cloudy days, covering them with aluminium caps (Koike 1965). Koike also recommended acceptance of the numerical system of rating for reaction to this, and other, diseases, proposed by Hutchinson (1969) and already adopted in Australia, Hawaii and else-

Fig. 7.5 Leaf scald *X. albilineans* (Sugar industry Research Institute, Mauritius)

where. In comparing results from different countries, however, he warned that differences in the strains of the causal organism, and the environment, must be taken into consideration.

An account of breeding for resistance to leaf scald has been given by Egan (1972b). Resistance is most readily found in clones of *S. spontaneum*. Clones of *S. robustum*, on the whole, are susceptible and those of *S. officinarum* vary in their reaction to the strain(s) of *X. albilineans* in Guyana (Stevenson 1957), Queensland (Hughes *et al.* 1968) and Hawaii (Egan 1972b). Relatively few forms of *S. sinense* have been tested, but in Guyana Stevenson placed them in an intermediate group, classified as 'tolerant' or 'susceptible-tolerant'.

Ratoon stunting disease: a bacterium?

Ratoon stunting disease (RSD) is an enigma. First recognized in Queensland in the summer of 1944–45, by 1974 its presence had been reported in forty countries. Some of these reports, however, should be treated with caution: the symptoms of the disease, though carefully described and well known, are difficult to detect. Moreover. there is still some doubt about the identity of the causal agent. For many years it was thought to be a virus, but Gillaspie *et al.* (1974) examined cane from Louisiana and found that an extremely small (5–10 µm × 0.3–0.5 µm), non-motile, rod-shaped bacterium was associated with RSD. Teakle (1974), working in Queensland, reported similar findings and concluded that a bacterium was probably the causal agent of RSD. Liu *et al.* (1974) went further and identified isolates from the vascular bundles of affected cane, grown in Puerto Rico, as *X. vasculorum*. Recently the theory has been advanced that a *Rickettsia* might be the cause.

The symptoms of the disease, its early history and the first control measures taken, which are virtually unchanged, have been described by Steindl (1961).

Symptoms

The external symptoms, uneven growth and an unthrifty general appearance, have no diagnostic value; and for several years RSD was thought to be without recognizable characteristics. Hughes, Steindl, and Egan (1968), however, eventually found two internal symptoms sometimes associated with disease:
1. In mature stems the leaf trace vascular bundles might become orange-red, but the discoloration does not extend into the internodes.
2. In immature stems there might be a diffuse salmon-pink discoloration in the younger nodes, spreading into the parenchyma of the upper internodes.

Some varieties show neither symptom, some show both and some show one but not the other. Moreover, the intensity of their expression is influenced by physiological and climatic conditions. Problems in the diagnosis of RSD were described by Ricaud (1974b) who felt that detection, especially in a new area, was a matter for the specialist.

Losses

Steindl (1961) asserted that in recent years RSD has probably caused greater losses to cane growers throughout the world than any other disease. It has been responsible also for the 'running out' of many varieties: for example Co 281 in South Africa (King 1956).

Such losses, however, vary with climatic conditions: in particular, diseased plants are much more sensitive than healthy plants to stress caused by drought. Estimates of loss are rare and most of the few that are on record are not agreed by all concerned. Indeed, in assessing the economic importance of RSD, Hughes (1974) referred to the problem of trying to convince farmers, plantation managers and even extension officers, that the disease was present at all. Nevertheless he estimated that where control measures were not being carried out in Queensland, the loss in weight was 10 per cent under normal conditions, and could be as high as 30 per cent in a dry year.

The quality of the juice is not affected by RSD.

Ratoon stunting disease is transmitted in setts taken from diseased plants and thereafter can be spread by cane knives and other cutting blades.

Fig. 7.6 Effect of RSD on cane growth. From left, N:C$_0$ 376 healthy and diseased; N 53/216 healthy and diseased; N 55/805 healthy and diseased (Photograph: South African Sugar Association, Experiment Station, Mount Edgecombe)

Control

The most acceptable method of control, the growing of resistant varieties, is not yet practicable. The existence of immune clones has been reported, but they are unsuitable for commercial use (Wismer 1971). Therefore, great care must be taken that planting material is free from the disease. To achieve this in Australia, stems are immersed in hot water (50 °C) for three hours before being used as seed-cane. Various combinations of time and temperature, usually within the range of two to three hours and 50–51 °C are used with success in other countries (Steindl 1974). In Louisiana, however, preference is given to treatment with hot air at 58 °C for eight hours (Schenxnayder 1956; Tantera & Steib 1972). Menon *et al.* (1972), working at Lucknow in India, also favour hot-air treatment but recommend that the temperature should be reduced to 54 °C in order that damage to the buds might be prevented.

Viral diseases

Fiji disease

Fiji disease, described by Hughes & Robinson (1961), is confined to the islands of the South Pacific, Australia, the Malagasy Republic and Thailand. It was first studied in Fiji, whence its name, was introduced into New South Wales, either from Fiji or New Guinea, but did not penetrate deeply into Queensland; it then went to the Philippines in cane cuttings from Australia. Its discovery in the Malagasy Republic, for the first time in the Indian Ocean area, caused concern in Mauritius: one known vector of the disease occurs there and susceptible varieties are grown (Barat 1954).

Symptoms

Plants suffering from Fiji disease are stunted and their leaves are shortened, distorted and discoloured. The symptom by which the disease is positively identified, however, is the occurrence of galls on the undersides of the leaves. The galls, long and narrow, vary in size: the smallest are barely visible but the largest may measure 5 cm × 2–3 mm with a height of 1–2 mm. Light green at first, with age they first become yellow and then brown. They are caused by the proliferation of the phloem, always occur on the undersides of the leaves and are parallel with the venation.

The disease, thought to be caused by a virus, is transmitted by leafhoppers: by *Perkinsiella vitiensis* in Fiji, *P. saccharicida* Kirkaldy in Australia and by *P. vastatrix* Breddin in the Philippines.

Control

Control has been achieved by planting resistant varieties and by insisting that only seed-cane from regularly inspected nurseries be used. The effectiveness of the campaign in Fiji has been described by Daniels *et al.* (1972).

From 1946 to 1952 the reaction to Fiji disease of varieties bred in Hawaii was determined in American Samoa; thereafter they were tested, with Australian and locally bred varieties, in Fiji. Robinson & Martin (1956) have given an account of the methods used in both places. Since it is not possible to transmit the disease by artificial inoculation, rows of the varieties under test were interspersed with rows of diseased cane. The most suitable variety for this purpose was Kassoer, a derivative of the Javanese form of *S. spontaneum*. The rows of Kassoer were planted six weeks before the other varieties in order that they might attract colonies of the vector, *P. vitiensis*. A faster, less expensive and more reliable method of testing, using an insectory, has been developed in recent years (Daniels *et al.* 1969; Husain & Hutchinson 1972).

Mosaic disease

Mosaic disease, in one form or another, has been identified in all the major sugar-producing countries except Guyana and Mauritius. It caused serious losses, especially in the New World, in the 1920s, but has been controlled almost everywhere by the replacement of susceptible by resistant varieties.

A description of the disease has been given by Abbott (1961).

There are many strains of the causal agent, a virus once thought to consist of rod-shaped particles 15 μm × 630 μm in size, but more recently described as threadlike and 750 nm in length (Handojo & Noordam 1972). Summer *et al.* (1948) recognized seven strains in Louisiana since when the list has grown. Strain L, was found in Georgia, USA, where it was affecting the growth of cane being grown for the manufacture of syrup (Zummo 1974). The strains are not confined to one area.

Mosaic is spread by insects, the most important vector in the Americas being the corn leaf aphid, *Rhopalosiphum* (formerly *Aphis*) *maidis* (Fitch).

Symptoms

The effect of the virus is to destroy the chlorophyll of the leaves, which develop a typical pattern of elongated yellow chlorotic areas interspersed with similarly shaped patches of light and dark green (Fig. 7.7). The loss of effective leaf area causes stunted growth, but the damage differs greatly according to the variety being grown and the strain of the virus.

Control

In general, clones of *S. officinarum* are highly susceptible to mosaic: hence the crop failures of the 1920s, when noble canes were widely grown. Clones of *S. barberi* are equally susceptible but suffer less severe damage. Forms of *S. spontaneum* and *S. sinense*, on the other hand, are highly resistant, but most hybrids of *S. sinense* have undesirable characteristics. *S. acclarum spontaneum*, therefore, has for long been the source of the genes which confer resistance to the disease in commercial varieties. The genealogical diagram (see Fig. 7.2) shows that, thirty years ago, the varieties bred at Canal Point, Florida, all resistant to mosaic, included either the Indian or the Javanese form of *S. spontaneum* in their ancestry, and some included both.

Fig. 7.7 Mosaic disease symptoms (Photograph: South African Sugar Association, Experiment Station, Mount Edgecombe)

It is standard practice at most breeding stations to determine the reaction of seedlings to the disease by inoculation with the strain(s) which exist in the areas which they serve, and to reject reactors. For example in the West Indies varieties to be grown in the Lesser Antilles are tested in Barbados; but those chosen for the Greater Antilles, where there is at least one different strain of the virus, are tested in Jamaica.

Disease resistance and the selection of varieties

The reaction of new varieties to the more important diseases is determined at the breeding stations at various stages in the selection programme. The scheme followed in Mauritius is given as an example (Fig. 7.8).

Principles of disease control in sugarcane*

Varietal resistance

Disease control in sugar-cane is achieved mainly by means of resistant varieties. Frequent inspections of new varieties at the various selection stages in the breeding programme, together with screening trials against some of the most important diseases, are intended to eliminate susceptible varieties. This ensures that the varieties eventually released to growers have a high measure of general resistance to disease. However, the resistance of new varieties may not be permanent. New problems or the reappearance of diseases that were previously important can occur, and varieties once adequately resistant may not remain so under changing circumstances.

The breeding and selection of new varieties to meet the industry's requirements is a lengthy process. It also takes a long time for growers to replace existing varieties with varieties that are more resistant to disease, particularly if the susceptible variety is widely planted. For both these reasons a variety cannot always be rapidly withdrawn from production. Disease problems, therefore, must often be contained by other means, pending the eventual planting of new resistance varieties. In the case of RSD, virtually no variety possesses adequate resistant or tolerance, and control of this disease depends entirely on methods other than varietal resistance.

The incidence of many diseases is related to specific environmental conditions. For example, smut is most prevalent in the warmer, northern areas where susceptible varieties, although suitable elsewhere, are not recommended. Mosaic is most likely to occur in the cooler, southern areas. Here the planting of resistant varieties should be considered if this will not adversely affect productivity.

Growers should try all new, resistant varieties that become available to see if they will be useful under the growing conditions on

---

* Reprinted from 'Sugarcane diseases in South Africa', *Bulletin No 1*, (revised), April 1980, Experimental Station of the SASA.

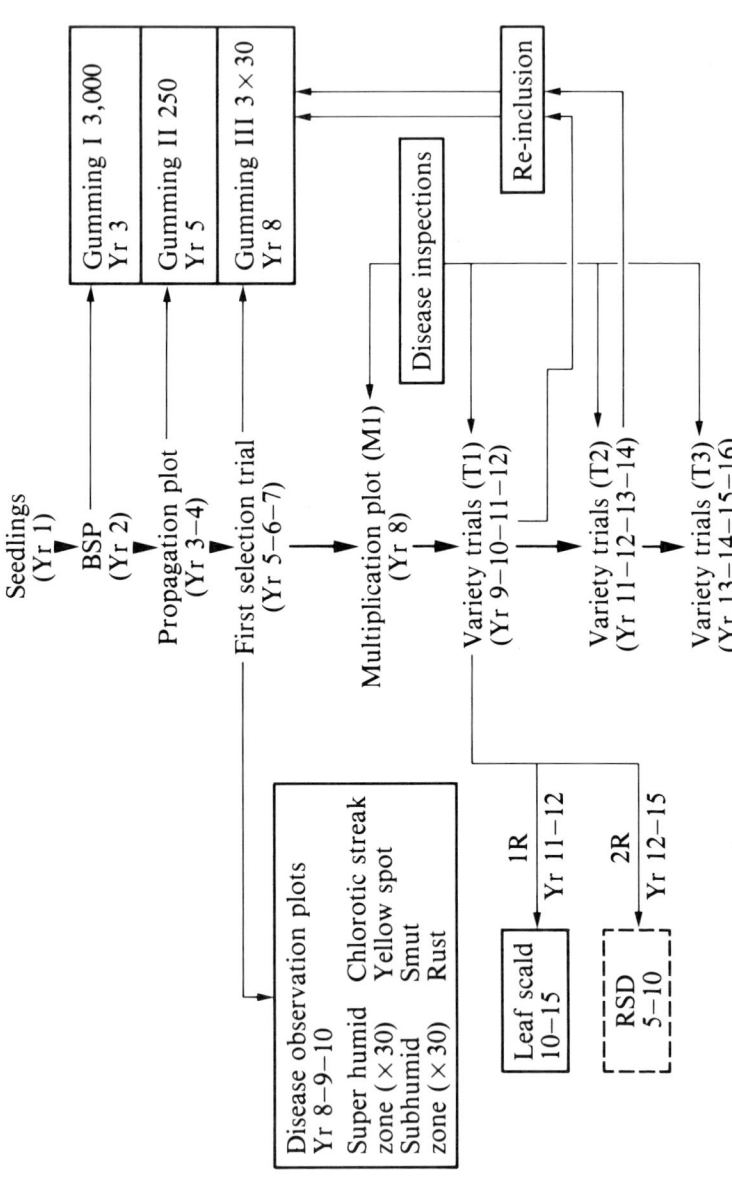

Fig. 7.8 The testing of new varieties for disease resistance in Mauritius (From Ricaud 1974)

their farms. Dependence on one dominant variety should be avoided where this is economically possible.

Seed-cane quality

Most important sugar-cane diseases, including RSD, smut, mosaic, leaf scald and, to some degree, red rot, are systemic, that is they are present within the cane stalk. These diseases, therefore, can be spread by planting infected seed-cane, they can persist in the stubble to recur after cutting and they also survive in volunteer regrowth to contaminate newly planted fields. Similar control measures are used to combat all of these systemic diseases.

The planting of healthy seed-cane is essential for general disease control. Growers should establish 'nurseries' with heat-treated stock to provide healthy, high-quality seed-cane to meet their annual planting requirements. Hot-water treatment, at 50 °C for two hours, is essential for the control of RSD and eliminates several other diseases, including smut and chlorotic streak. Seed-cane requirements should be estimated well in advance, so that adequate stocks can be produced. Seed-cane fields must be inspected regularly to ensure that they remain free of disease and only the plant and first ratoon crops should be used as seed-cane.

Field control practices

Healthy seed-cane must be planted into fields that are free from volunteer regrowth: if any volunteers present are diseased, much of the benefit of planting good seed-cane will be lost. It is essential to destroy the old crop effectively and to prepare the land thoroughly so that volunteers are eliminated before replanting.

The inspection and roguing of cane fields to remove diseased plants can do much to contain diseases at a low level and these are recommended practices where smut and mosaic are problems. The periodic inspection of fields also gives early warning of new problems as they develop and enables action to be taken at the most appropriate time. The ploughing out of severely diseased fields also contributes greatly to reducing the amount of infective material.

The incidence and effects of some important diseases, notably RSD and smut, are greatest when the cane crop suffers stress. Good crop management, including optimal nutrition, good weed control and adequate moisture, can contribute usefully to disease control.

Chapter 8

Ripening and reaping strategy

Ripening

Ripening and harvest are closely connected, and the ability to control the one and to carry out the other have strongly influenced the development of the many different patterns of agriculture and concepts of reaping strategy that exist in the industry.

After the boom period of growth, some of the stems change from the vegetative to the reproductive state and produce inflorescences called 'arrows' or 'tassels'. The conditions most conducive to this change, are cooler and longer nights, which usually occur in the Northern hemisphere from April to June. At such times appearances can be deceptive: when counts are made in fields in which each cane seems to have flowered, the proportion of stems without arrows is seldom less than 70 per cent. Those which have flowered ripen: sucrose is accumulated in their tissues (Fig. 8.1). The remainder continue to grow until low temperatures and/or an inadequate water supply cause them also to ripen.

At sea-level seasonal changes of temperature are greater and more important in the subtropics than in the tropics. This was recognized by Agee and Dass (1933) in Hawaii. They introduced the concept of 'day degree': the number of degrees Farenheit by which the maximum temperature each day exceeds 70 °F; and showed that there was a significant negative correlation between this index and ripening. Le Grand & Martin (1966) reported a similar connection in Florida.

Where irrigation is practised, ripening is induced by withholding water prior to harvest. An important advance in the scientific control of ripening – and of nutrition – under such conditions was made in Hawaii by Clements (1948, 1980), who developed the concept of charting a log for each field. The data for the crop logs are derived from the measurement of temperature, sunlight and growth; and from the analysis of young leaf sheaths for moisture, N, P and K. Clements's methods, modified to suit local conditions, have been

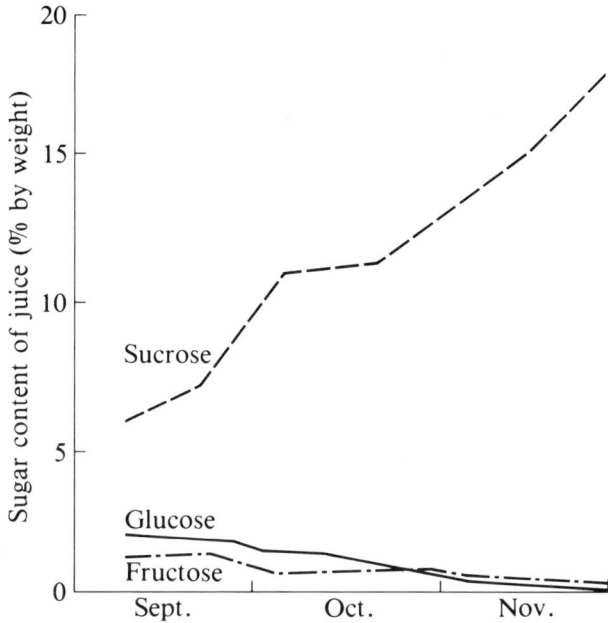

Fig. 8.1 The content of sucrose, glucose and fructose in cane juice during ripening in Louisiana (After Spencer & Meade 1945)

used with success in several countries: for example Humbert *et al.* (1967) claimed that, at one estate (Los Mochis) in Mexico, effective water control was largely responsible for an increase in yield of 24 kg sugar/tonne cane ground. At Haft Tappeh in Iran, close to the northern limit of cane cultivation, ripening is also brought about by withholding irrigation water and by falling temperature during the period April – October. Recently, however, Gowing and Baniaboassi (1978) reported that this process could be reduced or even halted if temperatures approached freezing-point, and was not resumed until rain fell. They concluded that ripening is promoted by moderately cool weather and by moderate drying off, but is interrupted by extremely cold weather and by too severe drying off. Chemical ripeners are being widely tested.

   Where irrigation is not practised, reaping is confined to the drier seasons(s) of the year, and ripening is dependent largely on the weather.

   Ripening is affected by several factors other than moisture supply and temperature: it is a function of age and is modified by nutrition and by soil conditions. The maximum sucrose content which might be achieved differs greatly between varieties. By and large the oldest cane is the first to ripen and, whatever its age, will do so more quickly if grown on well-drained light loams rather than on low-lying heavy clays. The influence of variety is twofold: some varieties, especially the noble canes, are inherently sweeter than others; and some, for example B 4362, reach maturity more rapidly. Early ripening varieties, by definition, will enhance juice quality and improve *rendement* if reaped at the beginning of the crop season; if their harvest is delayed, however, they deteriorate quickly and stems which have flowered are transformed into cork-like 'dry' canes.

Reaping strategy

The length of the harvest season

The course of ripening during the dry season at a typical non-irrigated estate (10°N), measured in terms of tonnes cane required to make 1 tonne of sugar, is illustrated in Fig. 8.2. The ratio is roughly 12 : 1 at the beginning of January but falls gradually, as ripening proceeds, to 9 : 1 by the end of March. Thereafter there is little variation until the onset of the rains, which frequently occurs in the third week of May. The quality of the juice then deteriorates, and does so more rapidly because pre-harvest burning is practised. The data recorded in Fig. 8.2 refer to a year of normal weather (1961); to one in which the rainy season started early (1968); and to one in which the rains were late (1960).

To achieve the highest possible output of sugar by confining the crop season to the period between the middle of March and the end of May, when the sucrose content of the cane is expected to be at its highest, is not practical. It would be uneconomic to invest in manufacturing plant and rolling-stock to be used for only nine or ten weeks in each year; and socially unacceptable if there were no employment for a stable workforce from mid-December to mid-March. Therefore, in the example cited, reaping proceeds throughout the dry season from the beginning of January to the end of May.

Maximum efficiency is achieved if all the cane grown for harvest is reaped and the mills grind at full capacity at all scheduled times. These targets are seldom reached. Many variables, especially weather and the incidence of pests, affect cane growth and usually there is an imbalance between field production and factory capacity. If the crop is poor, the start of reaping might be delayed for one or two weeks so that a larger proportion of the cane might be milled when its sucrose content is high. If, on the other hand, growth has been exceptionally good, and the market favourable, there are two possible courses of action: either the mill settings – the gaps between the rollers in the factories – might be widened to allow increased

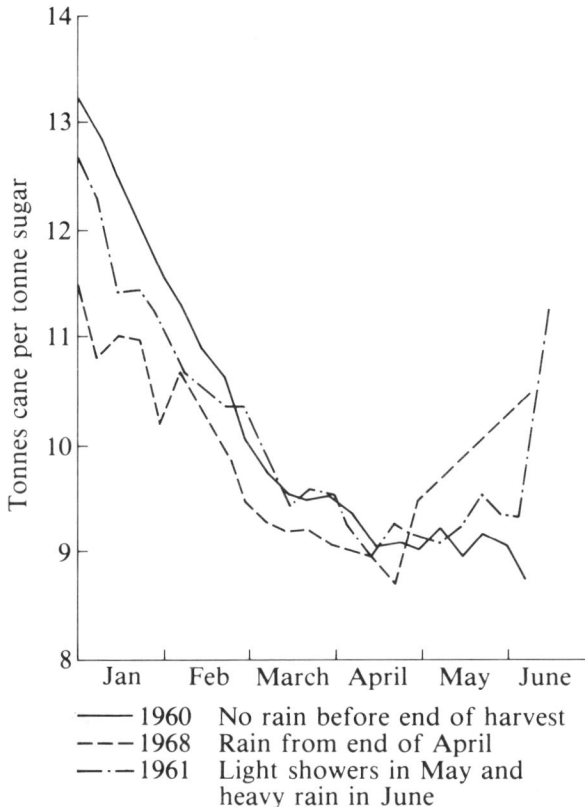

Fig. 8.2 The course of ripening at Brechin Castle, Trinidad (10°N.)

throughput, but with reduced extraction; or harvest might be extended into the wet season. The former is more attractive to the manufacturer, and less to the grower if payment for cane is based not on its sugar content but on factory *rendement*. If the latter is chosen, it is better to start early than to finish late. Fields damaged by reaping in wet conditions at the beginning of 'crop' can be prepared for replanting during the forthcoming dry weather, but fields damaged at the end of 'crop' cannot be ploughed as the soil is much too wet. They might be retained in cultivation for another year, but with poor results; or they might be abandoned until they can be renovated during the next dry season. Moreover, when reaping is protracted normal wet-season operations are delayed, with serious effect on the cost of establishment and subsequent yield of fields waiting to be planted. In short, sugar made after the onset of the rains is at the expense of future crops.

Under non-irrigated conditions, therefore, the more severe the

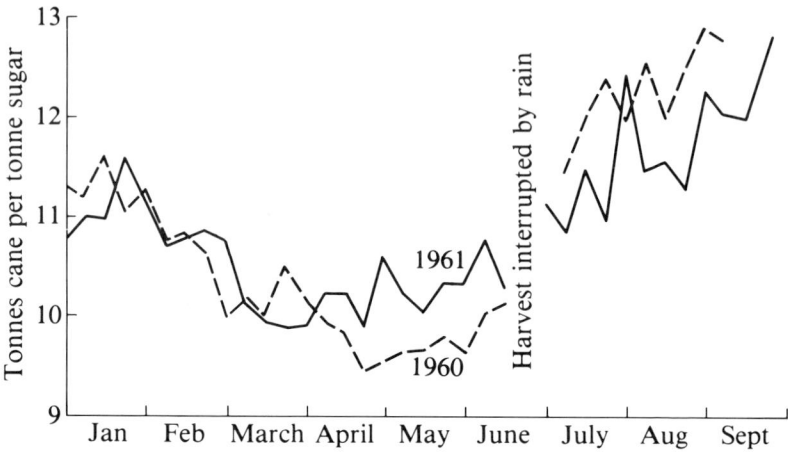

Fig. 8.3 The course of ripening at Monymusk, Jamaica (18°N.)

wet season, the more important is the timing of the harvest. Where little rain falls and growth is largely dependent on irrigation, reaping can take place during perhaps nine months of each year, for example at Ewa in Hawaii and on the plains of South Clarendon (Vere) in Jamaica (Fig. 8.3). Timing is then of less consequence; and the problem does not even arise in Peru, where there is complete dependence on irrigation and reaping takes place throughout the year (p. 69).

Other considerations

Whatever the rainfall pattern, reaping and cultivation should be closely co-ordinated. Fields to be replanted should be cleared at such a rate that the machines used in ploughing, harrowing, furrowing and other preparatory operations are fully employed; and the sequence should allow the equipment to spend as little time as possible in unproductive travel.

The menace of fire is described later in this chapter. It is diminished if reaping takes place along axes at right angles to the prevailing wind, and if large blocks of cane are quickly divided.

There is another consideration: areas in which it would be difficult, if not impossible, to operate machinery in bad weather – for example steeply sloping and low-lying fields – should be reaped when it is highly improbable that rain will fall (see Fig. 8.2). This proposition, though simple, is easily and frequently forgotten.

Decisions concerning the order in which fields are to be reaped are more easily made and accepted if there is common ownership of field and factory. Where some or perhaps all of the cane is produced by independent growers, however, problems can arise. Each grower would prefer to reap only when the weather is good and, depending on the basis of payment, when the sucrose content of the cane is high. In these circumstances the controlling authority – cane farmers' superintendent (West Indies), cane inspector (Australia), field superintendent and traffic manager (Fiji) – must ensure that the allocation of delivery quotas is seen to be fair. Good and bad reaping conditions should be shared proportionately by all.

An acceptable estimate of cane quality would be a useful guide to the preparation of reaping schedules especially where, as in Trinidad, harvest starts shortly after the end of a period of heavy rainfall. At that time, however, ripening is so uneven that it is impossible to take a sample of cane representative of a field: statistical analysis has shown repeatedly that within-field variation is as large as between-field variation. Later, as the crop matures, there is a greater uniformity; but in the absence of an overriding factor, such as variety or soil type, the older the cane the nearer it is to being ripe.

Testing for ripeness

In commercial practice reaping should be so arranged, within an agreed policy, that the more mature fields are harvested before the less mature.

As already mentioned, considerable progress has been made in the control of ripening, under irrigated conditions, by using the crop log. Where there is complete dependence on rainfall, control is more difficult. A field of cane comprises stems in different stages of development. They vary in age from young suckers, a few weeks old, to those which have flowered and ceased to grow. The difficulty of choosing a sample representative of such a heterogeneous population has been described by Chinloy & Innes (1954). In the discussion which followed the presentation of their paper, reference was made to the damage caused when samples for chemical analysis were being taken from well-grown fields in which some stems were lodged; to the considerable quantity of cane used; and especially to doubts concerning the validity of the sampling methods used. For example, whereas Chinloy and Innes (Jamaica) recommended the choice of individual stems, Evans (Guyana) preferred whole stools and stated that six stools from each field, roughly 4 ha in area, provided a satisfactory sample.

Little has been written about this difficult aspect of production on non-irrigated estates, though recently Hoekstra (1978) in South Africa found that nothing was to be gained by carrying out maturity tests. He concluded that '... the gains of maturity testing were not of sufficient magnitude or certainty to make it a worthwhile proposition'.

Nevertheless, pre-harvest testing for ripeness is still carried out in many countries. Samples of stools, canes or core punches, are taken at intervals of perhaps two weeks, crushed in laboratory mills, the juice analysed and estimates of the yield of sugar calculated. This procedure makes at least one worthwhile contribution to the success of the industry: it is a constant reminder, particularly to growers, of the importance of cane quality.

The hand refractometer

The density of cane juice, and by inference its sucrose content, can be determined readily in the field by using a hand refractometer (Fig. 8.4). This consists of a straight tube, with an adjustable tele-scope at one end and a fixed prism with a hinged cover at the other. A sample of juice, taken by inserting a grooved tool into the middle of an internode, is dropped on to the face of the prism. The cover is then closed, the eyepiece adjusted and a reading taken from a scale marked in degrees Brix.

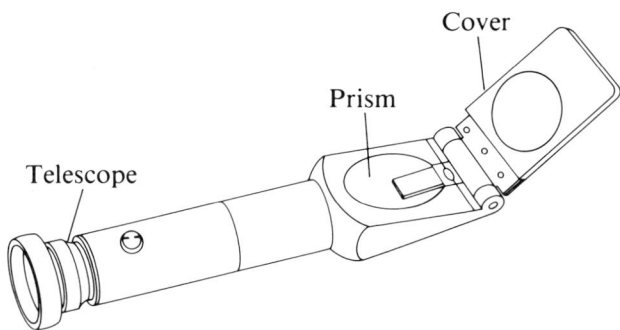

Fig. 8.4 The hand refractometer (After Browne & Zerban, 1941)

During periods of rapid vegetative growth the juice in the lower part of the stem is of much higher density than that in the upper part. As ripening proceeds the difference gradually disappears, a characteristic which has given rise to the top : bottom test for maturity. To carry out this test, samples of juice are taken from the central internodes of the upper and lower thirds of stems at intervals of perhaps three or four weeks, and are examined by hand refrac-tometer. When they are much the same, that is, when the top : bottom ratio approaches unity, the cane under examination is ripe.

Although the hand refractometer is not widely used by the

commercial grower, it is of considerable value to the cane breeder. For example, at the West Indies Central Sugar Cane Breeding Station in Barbados, selection in the first year (single stool) and second year (22 m row) seedling trials is based on weight, proportion of rotten cane, tendency to flower and density of the juice measured by hand refractometer. There has been little change in the procedure described by Inniss (1954). If the density of the juice falls appreciably below that of the standard commercial variety, planted for comparison, the seedling is discarded ' . . . unless it is obvious from the readings that the canes are very unripe, in which case the bottom readings will be quite high, and the top very low . . . '.

Many F_1 varieties used in cane-breeding programmes are so thin that it is difficult to take juice samples from them for examination by hand refractometer. Whole cane must be analysed, and a simple method of doing this has been developed by Trott (1969).

Pre-harvest burning

When the older leaves die, they either cling to the cane stems or gradually fall away; this is a varietal characteristic. The resultant mass of vegetation, called trash, is an impediment to reaping, whether by hand or machine; and when dry becomes a serious fire hazard.

During the years of slavery, when labour was cheap and plentiful and the price of sugar high, everywhere cane was cut without pre-harvest treatment.

Following the abolition of slavery, attitudes changed. It was noticed that if the trash had been removed by burning before reaping took place, the cane cutter's output was increased threefold: from roughly 2 to 6 tonnes/man-day. For example, in 1911 it was reported from Demerara that when fields were fired immediately before harvest, the cost of cane cutting was reduced by 5–6s. per ha, i.e. £0.7 per ha (Deerr 1911). The shortage of manpower during the war years of 1939–45 caused this custom to spread. It was accepted in Queensland and extended to all varieties grown in Trinidad, where previously it had been restricted to the thin-barrelled canes Uba and Co 213. In Hawaii the labour supply was reduced so drastically that a completely new reaping technique was developed. In more recent years economic pressure has caused the cane cutter, with his bill or machete, to be replaced in several countries by mechanical harvesters, and some machines work less effectively, if at all, in unburnt fields. Consequently, pre-harvest burning has gradually become standard practice in many cane-growing countries.

Great care must be exercised in its use and fires confined to the quantity of cane expected to be reaped within the next twenty-four hours. Provided that this is done, little or no loss is incurred.

After cane has been burnt the quality of its juice deteriorates. Under dry conditions there might at first be an apparent increase in the sucrose content caused by loss of moisture. Thereafter deterioration takes place at an ever-increasing rate, especially if rain falls. The sucrose in the juice is progressively broken down by hydrolysis into glucose and fructose, both of which fail to crystallize during the subsequent factory process and are discharged in the final molasses. This causes imbalance between the milling plant and the boiling house. The capacity of the latter is too small to accommodate perhaps two or three times the normal output of molasses. In most countries equipment is installed to handle 23–28 litres of molasses/tonne cane ground; but when only stale burnt cane is being processed the yield of molasses has been known to increase to 55 and even 80 litres/tonne. Moreover, for reasons not yet fully understood, though the formation of dextrans is to some extent responsible, in such conditions sucrose extraction becomes increasingly difficult.

In short, the losses sustained when burnt cane is allowed to become stale are:

1. Sucrose in the cane juice is transformed into molasses-forming substances.
2. More sucrose, otherwise recoverable, is retained in the vastly increased quantity of molasses produced.
3. The boiling house becomes congested. A decision must be taken whether to abandon the burnt cane; to discharge final molasses containing more sucrose than is usual; or delay reaping operations.

Many cane-growing countries lie in the trade wind belt and pre-harvest burning takes place in the evening, when the wind has fallen. Fire is set along the leeward side of the area to be burnt and proceeds slowly against a gentle breeze. Sufficient personnel and equipment must always be on hand to maintain control and to prevent the fire from spreading to neighbouring fields.

In practice there is need to impress on all concerned that their purpose is to produce sugar, not cane. Consequently:

1. The time interval between burning, and reaping and gleaning must be minimal.
2. Excuses made for day-time burning, when the wind is high, the fire difficult to control and so fierce that the cane might be damaged, should be carefully examined and accepted with reluctance. The opinions of amateur meteorologists who 'thought that it was going to rain' should be treated with scepticism.
3. In order to remove the danger of fires of 'unknown origin' being so extensive that the cane cannot be harvested in good time, the

planting programme must be so designed that reaping can proceed along axes at right angles to the prevailing wind, thereby breaking up large blocks of fields. This requires personnel engaged in these operations to move frequently from place to place, when it is much easier for them to remain for long periods in one area.

If there is doubt whether rigorous control of pre-harvest burning can be exercised, the practice should be discouraged. An example of this is embodied in the *Production of Cane Ordinance of Trinidad and Tobago*, 1947 (Ch. 23, No. 12), in force for more than twenty years, which was enacted to give legal authority to the relationship between the small growers, called 'farmers', and the manufacturers. Paragraph 6 of the Second Schedule reads as follows: 'References in this contract to canes of good quality shall be deemed to include burnt canes fit for grinding which have been burnt not more than three days before delivery: provided that in the case of burnt canes as aforesaid the sum of sixty cents per ton shall be deducted from the price determined by law'.

Careful watch must be kept for the possible disturbance by pre-harvest burning of satisfactory host : parasite relationships of sugar-cane pests. As yet none has been recorded. On the contrary, this practice has caused the eradication of at least one major pest: the weevil or beetle borer, *Rhabdoscelus obscurus* (Boisduval), which had been imported into Queensland from New Guinea and had so risen in importance that in the 1930s it was rated as second only to the notorious 'grey back' white grub *Dermolepida albohirtum* (Waterhouse). Parasites and predators exercised little check on its spread; nor were the use of clean planting material and resistant varieties of much help. However pre-harvest burning, adopted during the early 1940s as already noted, caused the pest to be reduced to one of minor importance (Mungomery 1954), a status confirmed eleven years later (Mungomery 1965). Fire has also been used to destroy rats.

Cane fires and their control

The removal of trash by controlled pre-harvest burning, in order to facilitate reaping, has been discussed above. Other fires, described collectively if sometimes inaccurately as of 'unknown origin', can and frequently do occur. They constitute one of the most serious hazards of cane-sugar production. The risk increases as the cane approaches maturity and is greatest during the early weeks of the harvest season, before fire-breaks have been formed by carefully planned reaping. So little has been written about this important aspect of production that much of this is based on the author's personal experience.

The causes of cane fires

Fields adjoining public roads are particularly at risk from accidental fires caused by the careless disposal of burning cigarette butts. Proximity to railways is almost equally dangerous: the sparks emitted in the exhausts of diesel-powered locomotives and the flames from inadequately lubricated wagon-wheel axles cause many fires. Others arise from negligent supervision when bush or trash is being burnt, or when blow-torches are being used to carry out repairs in the field. Lightning also can set fire to cane. Nor should the existence of pyromaniacs be ignored.

Nevertheless, the greatest damage is caused by irresponsible or dissatisfied people, not necessarily engaged in the industry, especially during times of political and industrial unrest. Fire may be used to ensure that the field will be the next to be harvested. Four methods are known to have been used:
1. Matches struck by children either mischievously or, because of their relative freedom from prosecution if caught, at older people's instigation.
2. Lighted candles with manifestations of obeah (voodoo), left alight in fields in the West Indies.
3. Rags, impregnated with kerosene, tied to mongoose tails, set alight, and the animals released to scurry through the fields.
4. Lighted cigarettes, with matches attached by rubber bands, flipped away at strategic places where the crop is most vulnerable. Acting like slow fuses, they burst into flames when the smouldering tobacco reaches the matchheads.

Prevention and control

Legislation has been enacted for the control of agricultural fires in many countries where burning is the standard method of clearing land for cultivation. It is usually required that application be made to a fire warden, who will then satisfy himself that adequate precautions have been taken before issuing a licence to burn during stated hours. Notwithstanding these and other measures, unexpected and unattended fires occur frequently in and around most cane-growing areas. With constant vigilance they can be detected early, confined to small areas, and at most times of the year prevented from causing serious damage. Control is achieved by cutting the surrounding cane to isolate the fire, by burning the affected fields from the leeward side, i.e. back-firing, to protect fields further downwind, or by using tractor-drawn fire-engines accompanied by mobile water tanks.

When the cane is most inflammable, however, these conventional methods can be wholly inadequate. On one memorable occasion in 1952, the Trinidad Government Railway line, with borrow pits (made by the removal of earth to form the railway embankment) and field roads on each side, 50 m in total width, was insufficient to prevent the spread of fire. Burning trash, in a high wind, fell well inside fields to leeward of the line. The fire progressed, crossed a wide main road and eventually reached the sea after 4,000 ha of cane had been burnt. Similar and even greater disasters have occurred elsewhere, for example in South Africa (Barnes 1974). The continuing menace of fire is illustrated by the following reports from the West Indies.

In Barbados in 1973, 1,121 malicious fires destroyed 3,870 ha of standing cane, roughly 15 per cent of the crop, of which 365 ha were a complete loss (WISA 1973). There was considerable improvement in 1975 and 1976, but not in 1977 when the incidence approached that of 1973 (Clarke *et al.* 1975, 1976, 1977), nor in 1978 when 2,526 ha were burnt.

In Jamaica in 1972, 613 ha were burnt by unscheduled fires and 10,390 tonnes of cane lost. There was little improvement in 1974 (WISA 1972, 1974). In 1975, however, the loss was reduced to 5,000 tonnes (SAC 1975).

In St Kitts in 1974, there was a 'spate of unauthorized fires' involving 20,000 tonnes of cane, 10 per cent of the island's crop (WISA 1974). Only 40 per cent of the cane reaped in that year was unburnt but, after a vigorous public relations exercise, 80 per cent was harvested as 'green' cane in 1975 (SAC 1975).

In Trinidad in 1973 more than 1 m. tonnes, roughly 50 per cent of all the cane grown, were burnt in fires of 'unknown origin'. Thousands of tonnes were abandoned, the harvesting schedule was disrupted and much of the cane ground was stale (WISA 1973). The incidence of unauthorized cane fires was halved in 1974 and this better, though far from satisfactory performance, was maintained in 1975 (SAC, 1975), but not during the next three years. With production at a disastrously low level, two major causes being drought and fire.

There is a considerable element of personal risk when attempts are made to control fiercely burning cane by back-firing. The wind is capricious and might suddenly change direction. When this happened at Cedar Hill Estate in Trinidad in 1965, several people were burnt, the acting manager so severely that he died shortly afterwards. To avoid injury in such circumstances it is best to lie face downwards, with hair protected, in the lowest part of the immediate vicinity – for example a drain – until the fire has passed over.

Barnes (1974) suggested that all those engaged in fire-fighting should be provided with protective clothing. He referred to this being done in Australia, where cotton-drill trousers and shirts were made flameproof by absorbing 10 per cent by weight of a 1 : 1 mixture of borax and boric acid.

The only proven method by which fires can be extinguished, or confined to small areas, when the crop is as dry as tinder and the wind high, is by the use of aircraft. MacIntyre & Keir (1967), in Trinidad, developed a successful system for the protection of 21,000 ha of cane within a radius of 15 km from a central air base. Careful attention was paid to the following:

1. The vigilance of the watchman on duty at the top of an observation tower, 18 m high, during the period when fires were most likely to occur; especially his quickness to report the first sight of smoke, with its bearing, to the pilots.
2. The instant readiness of three small aircraft (Piper Pawnees), one of which would take off immediately and, if necessary, attempt to extinguish the fire by depositing across its line of advance a load of 550 litres of water, impregnated with a 'thickening' or 'sticking' agent, usually an alginate. The thickening agent caused roughly 75 per cent instead of 25 per cent of the water to adhere to the cane leaves.
3. The support of the first small aircraft by the second and third, as required.
4. If these measures failed, a much larger aircraft (B.25) was brought into action. Field personnel were warned of the danger of being hit by water: the impact could be sufficient to cause serious injury.

The technique was given a rigorous test in 1969 when the first part of the crop was unusually dry and, by the end of February, 1,038 fires of 'unknown origin' had occurred. Control was effective and there was no serious dislocation of the reaping schedule. What otherwise might have been described as a 'disastrous drought' became 'ideal weather for reaping'.

Insurance

By and large insurance companies are reluctant, with good reason, to give adequate cover to cane growers for losses sustained by fire. Therefore in highly vulnerable areas it is prudent either to establish a contingency fire fund, financed by a levy on each year's profits, or to gain the insurers' confidence by establishing a record of several years without claim. The latter can be, and was, achieved by using the control measures described earlier.

Of the many aspects of insurance in general, those which are of particular importance to cane growers are: third-party cover and excess damage clauses; and, if the grower is also a manufacturer, protection against consequential loss.

Chapter 9

Reaping and transport

The reaping of cane by hand is notoriously labour intensive. The need to provide people to do this work has been described in Chapter 1. It led in great measure to the development of the slave-trade between West Africa and the Americas and, after abolition to the immigration of indentured labour from China, Madeira and India into the West Indies and Guyana and East Indians to Mauritius. Similar movements took place elsewhere: the Australian industry was established with the help of Kanakas and, much later, was sustained by new immigrants, many of Italian origin. People from several countries – China, Japan and the Philippines – went to Hawaii for similar purposes.

Reaping by hand will continue for many years in countries where labour is relatively inexpensive, as in many parts of Asia, South America and Africa; where much of the work is done by the family, as in India; and where drains and canals obstruct the free passage of machines. Nevertheless, mechanical combine harvesters are being used in increasing numbers not only in such countries as the USA and Australia, but also in the Philippines and Malaysia; and in new projects in Gabon, Nigeria, Swaziland, Ivory Coast, Indonesia and Sudan. In Guyana the physical difficulties imposed by an intricate drainage and irrigation system are being overcome.

Cutting by hand

Where cane is cut by hand the cutlass or machete is favoured by some, including most of Asian origin, and the cane knife or 'bill' by many of those from Africa. These are not necessarily the best tools for the job. Baldwin & Fisher (1969) carried out trials in Guyana with knives from Louisiana, Mauritius and Queensland, seventeen specially designed tools and the traditional Guyana cutlass. They applied industrial engineering techniques to their enquiry and recommended the use of two knives, one for cutting and the other for topping, which reduced the work content by 10 per cent and the

stump length by 30 mm. In South Africa the long-handled Australian cutlass, which enables the cutter to work efficiently without
stooping, if the cane is upright, has been introduced with considerable success.

As noted in Chapter 8, output is increased if the old leaves (trash)
are removed by pre-harvest burning. This is of particular importance
if the varieties grown have thin stems; but when piece-rate workers
have become accustomed to cutting burnt cane they do not take
kindly to reaping unburnt (green) cane. Fires of 'unknown origin'
usually occur when they are asked to do so. Some of the important
controls which should be imposed by management are:

1. If pre-harvest burning takes place, only the quantity of cane required for the next day's reaping should be burnt.
2. Where gleaning is practised (the high cost of labour is causing it
   increasingly to be discarded) it must be done immediately after a
   field has been reaped, otherwise the quality of the 'pick up' cane
   will deteriorate. Stale cane is easily recognized: the cut ends dry
   out and their surfaces become concave. In extreme cases pink fungal spores (probably *Trichothecium roseum* Link.) appear on the
   stems of burnt cane, especially at the nodes (Fig. 9.1).
3. The stems should be cut as close to the ground as possible and
   the trash removed. The tops should be severed, but without loss
   of millable joints.

Fig. 9.1 Stale cane (Photograph: A. L. Down)

4. To achieve this the knife most suitable to the cutter should be used. It might be necessary for the employer to provide the knife; but in many countries the Truck Acts allow the cost to be deducted from wages.

5. A grindstone should be available to each group of cutters, and should be in motion when sharpening takes place in order to prevent uneven wear. Alternatively, knives and sharpening stones are often issued to the cutters.

Loading and transport

Having been cut, the cane is then transported to the factory. There are many seventeenth- and eighteenth-century prints showing bundles of canes, bound together by bands of leaves, being carried on the heads of workers to the primitive mills operated in those days. As the factories increased in capacity so did the distances over which cane had to be transported. For many years it was hauled – and in some countries still is – by animal-drawn carts; but not always to the factory. With the development of large centrals, permanent rail systems were used. Advantage was also taken of water to float canes in flumes on Hawaii Island and to carry punts in Australia and Guyana. In northern Brazil cane grown on steeply sloping land is still formed into small bundles (*feixas*), roughly 1 m in length and wrapped in trash, and carried on the backs of donkeys to transfer points.

By and large, however, road haulage by trucks and tractor-drawn trailers has replaced other forms of transport. The vehicles may be loaded by hand or by machine. Many of the machines developed solely for loading require that the cane be placed by the cutter in such a position that it can easily be grasped, for example in heaps across the ridges on which the crop was grown.

Alternatively, the cane might be cut by machine and loaded by hand. In general none of these half-manual, half-mechanized systems of cutting and loading has proved to be entirely satisfactory. Usually they were introduced either because of labour shortage or of increasing costs. Thereafter the grower, more often than not, incurred unacceptably higher costs for the labour-intensive half of the operation which had not been mechanized, or was faced with industrial unrest. A fundamental principle of mechanization is that cane should not be touched by hand until it is delivered into the vehicle which will carry it either to a transfer station or to the factory. There is an exception to this general statement: self-loading trailers, which are used with success in several countries to transport cane which has been cut by hand.

Self-loading trailers

The Bell trailer

The Bell self-loading trailer, designed and developed in South Africa, has a low platform carried on two wheels at the rear of which is a loading ramp, fitted with skids which can be moved into an upright position when not in use. It is drawn by a wheeled tractor. A winch, driven by the power-take-off shaft of the tractor, is mounted either on the tractor itself or on the front of the trailer.

Cane is cut manually and formed into heaps or bundles. When loading takes place the trailer is manoeuvred into position and its skids are dropped. A wire or chain attached to the winch is then passed around the bundle and draws it up the ramp into trailer (Fig. 9.2). Bell trailers are also used, with local modifications, in Trinidad and, since 1968, in the Philippines. Donawa (1967) described the system in operation in Trinidad and made special mention of the ability of tractors and trailers to load cane in hilly areas where combine harvesters are unable to operate. On such land the bundles are built across the slope and are drawn uphill and loaded into the trailers. The cane is then carried to a mobile crane, stationed close to the field and placed either on to a stock pile or

Fig. 9.2 Modified Bell trailer

into road or rail trucks for transport to the factory.

In South Africa wire ropes are passed around the bundles after they have been formed; but in Trinidad the bundles are built on sling chains, with quick-release latches, which previously have been laid on the ground. The capacity of the trailers is much the same in both countries: from 2 to 6 tonnes in South Africa, according to type; and 5 tonnes in Trinidad. The trailer's load comprises one bundle in South Africa and two in Trinidad.

Thompson (1976) discussed the development of this and other self-loading trailers in Africa: the Perry which uses hydraulic arms to pull in the cane bundle, the side-loading Mascane and the versatile Jacobyl.

The Mascane system

In end-loading systems, such as the Bell, a single trailer of 5 tonnes capacity is drawn by each tractor and, because the load is so small and end loading precludes the use of more than one trailer per tractor, its economic distance of operation is limited to roughly 2.5 km. In the side-loading Mascane system, however, as many as four trailers may travel in tandem behind one tractor and therefore can be hauled economically over much longer distances. Even with two trailers the limit is extended to 12 km. Bundles of cane, each of 4–5 tonnes (one trailer's capacity) and made in parallel rows, are loaded by an auxiliary tractor which has a winch and a high frame. The winching tractor operates from the side of the trailer opposite the bundles. Portable ramps are placed against the empty trailer and a cable from the winch is passed over the frame, across the trailer, and attached to the chain and grip latch encircling the bundle. The bundle is then drawn up the ramps into the trailer. In the latest model the winch and frame are side-mounted on a tractor which allows the loading tractor to travel parallel with the line of trailers. The Mascane system is widely used in Kenya, Malawi, South Africa, Tanzania and Zambia.

The Jacobyl trailer

The Jacobyl trailer, like the Bell trailer, was contrived in South Africa. It is extremely versatile. With little change it can be rear-loaded or side-loaded, either from left or right, and therefore can be operated in tandem. Loading is by the standard drum winch and cable (see Fig. 9.3) but ramps are not required.

Mechanical harvesters

After the start of hostilities between. Japan and the USA in 1941 field employees in the sugar industry of Hawaii, most of whom were

Fig. 9.3 Jacobyl three-side loader with off-loader (Photograph: Jacobyl (Pty) Ltd, Natal)

of Japanese origin, were interned. To overcome the consequent labour shortage the push rake and grab system of harvesting was quickly devised and soon became standard practice throughout the territory. Elsewhere increasing interest in the mechanization of reaping occurred during the 1960s and 1970s, and now more attention is paid to this than to any other aspect of sugar-cane technology; so much so that descriptions of modern procedure may well be out of date before they are even published.

The design and development of cane harvesters has taken place chiefly in Hawaii, for the unique conditions of the industry in those islands; in Australia; in continental USA (Louisiana and Florida); and in Puerto Rico. The reader is referred to a paper by Clayton (1969) in which experience with the machines used in these areas is described. A parallel development of considerable significance took place in Belize, and especially in Trinidad. Cary harvesters from Louisiana were drastically modified to reap 'tropical cane' on an estate scale (Deacon 1969). Machines made subsequently to handle this type of cane have incorporated many of the features built into the Cary harvester.

In general, except in Hawaii, mechanical harvesters are of two types: those which deliver whole canes (whole-stick harvesters) and those which cut stems into billets roughly 0.35 m in length (cut–chop harvesters). Whole-stick harvesters have the advantage that they fit readily into existing systems of reaping, transport, storage and feeding to the mill. Cut-chop harvesters, on the other hand, require specially built ancillary transport and, in certain circumstances, also need expensive transload stations

Moreover, the vastly increased numbers of cut surfaces exposed to infection cause rapid deterioration in cane quality. This factor is of particular importance if the factories close for long periods at the end of each week; and when processing is delayed. For example Clayton (1969) noted that cane deterioration 'almost stymied the introduction of harvesters' in Australia in 1966. The problem was overcome by the 'earlier milling of chopped cane, especially before the weekend shutdown'. Similarly Wells & James (1976) reported that, following the unscheduled stop of twenty-nine hours at a factory in Queensland, the loss due to cane deterioration approached $1,800 per hour.

Various lactobacilli, mostly *Leuconostoc* spp., and especially *L. mesenteroides* and *L. dextranicum*, cause rapid loss of sucrose and increases in reducing sugars and polysaccharides (mainly the 1,6-dextrans). The stale cane is then usually processed as quickly as possible, with severe loss in recovery, in order that the boiling house should not be 'gummed up' by dextrans. Egan *et al.* (1978) reviewed the work done in Australia to promote the degradation of dextrans by adding dextranase to the mixed juice. In this new field there is, understandably, considerable divergence of opinion concerning the frequency with which the dextranase should be used, and at which station in the factory it should be introduced. The enzyme cannot overcome the basic problem of deterioration, the loss of sucrose, but at some cost it will allow stale, burnt cane to be processed without causing difficulties in the boiling house.

The problem is much less acute if the cane has not been burnt: with chopped green cane the time-span for significant changes to occur in the composition of the juice is measured in days rather than in hours.

Whole-stick harvesters

In Louisiana the whole-stick method has been pursued and the harvesters developed there – Cameco and J & L – are similar in design and operation (Figs. 9.4 and 9.5). The canes are gathered together, cut at the base with a rotary blade, topped, drawn into the machines butts first, cleaned and placed across adjacent ridges. From three to six rows are formed into a single line and, in a second operation, are loaded mechanically into trailers of 5 tonnes capacity, drawn by wheel tractors. If the distance is short the tractors travel directly to the factory; if it is long the cane may be transferred at a transload station into trucks of 15–20 tonnes capacity. In recent years the trend has been to use the trucks or much larger trailers in-field with the harvesters.

The principal limitation to the use of the existing Louisiana type of whole-stick harvesters has been their inability to deal with heavy crops of lodged and tangled cane. The machines deal very successfully with upright cane, which can readily be passed, as individual sticks, through the gathering and piling mechanisms but, at present, the dividing and gathering mechanisms cannot provide adequate separation and alignment of tangled sticks. This constraint has been one of the main factors leading to the development of the cut–chop machines, which handle the cane in bulk instead of individual sticks.

For many years, on the organic soils of Florida, hand-cut cane has been very successfully loaded by the J & L continuous loader. In this machine the cane is picked up by a conveyor-type elevator, with the 'points' of the elevator passing into the organic soil, ensuring that all the cane is picked up, without damage to the stand of cane or the pick-up mechanism. At the top of the elevator the cane is cut into billets, by a series of circular saws, and the cane can then be handled in a similar manner to cut–chop material. The continuous loader has a very high output, at over 160 tonnes/hour under good conditions.

A recent innovation to its use has been to load hand-cut cane in mineral soil areas, in a mixed hand-cut and cut–chop operation. The cane is grown on a bank and the hand-cut cane laid across the banks, which allows a suitably modified pick-up mechanism to pass underneath the cane. The chopped cane is then delivered into the same transport as used in the cut–chop operation. The continuous loader has been operated successfully at night, in the Swaziland situation, allowing twenty-four-hour delivery of cane and full use of the cane

Fig. 9.4 The J & L R-6 continuous loader (Photograph: J & L/Honiron Engineering Co. Inc., Louisiana)

(a)

(b)

Fig. 9.5 (a) The Cameco 1000-B combine harvester (b) The J & L S-30 single-row wholestalk harvester (Photographs: J & L/Honiron Engineering Co. Inc., Louisiana)

haulage equipment, with a reduced requirement for storage of cut–chop cane, in the situation where the cut–chop machines are only used during daylight hours.

Cut–chop harvesters

There has been a fundamental difference in cane-growing techniques in Hawaii, Louisiana and Australia, which led to the early and successful development of the whole-stick harvester in Louisiana and the rapid development of cut–chop harvesters in Australia. In Hawaii, in surface-irrigated areas, the cane was planted in a rather deep furrow and maintained in the furrow, with a well-defined bank in between. This situation presented a major difficulty to the development of equipment which could cut and lift high tonnages of

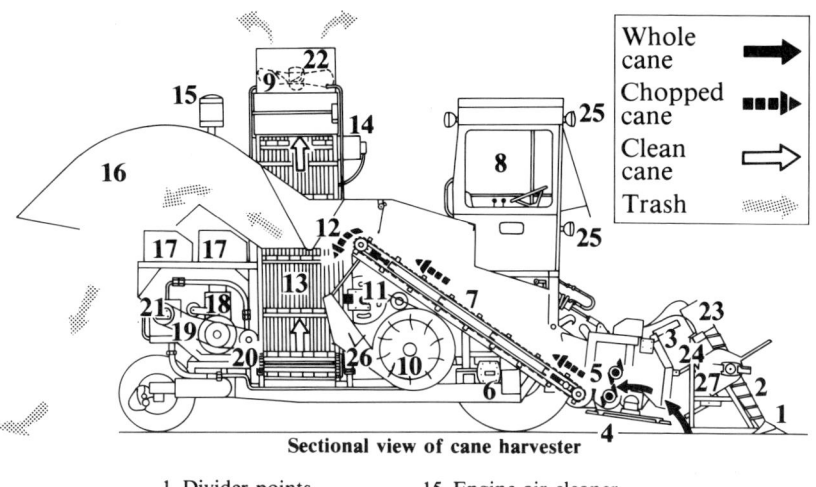

Sectional view of cane harvester

1 Divider points
2 Rotary dividers
3 Guide bar
4 Cutting discs
5 Chopper drums
6 Ground drive hydraulic motors
7 Main elevator
8 Driver's cab
9 Axial-type cleaning fan
10 Cleaning fan
11 Main drive
12 Rear deflector
13 Cross elevator
14 Cross elevator hydraulic motor
15 Engine air cleaner
16 Harvester rear hood
17 Hydraulic oil tanks
18 Engine
19 Ground drive hydraulic pump
20 Hydraulic lift pump
21 Hydraulic pump for rotating dividers, side cutter
22 Oil motor for axial blower
23 Oil motors for rotating dividers
24 Oil motor for side cutter
25 Working (night) lights
26 Hydraulic pump for cross elevator axial blower (secondary cleaning)
27 Side cutters

Fig. 9.6 The Claas 1400 Harvester (Diagram: Claas OHG, Harsewinkel, Germany)

heavily lodged cane without causing damage to the cane stools. In Louisiana cane is grown on a substantial bank and in Australia the final soil surface is generally flat. In both instances the restriction of cutting and lifting the cane out of a furrow has been removed, giving much greater flexibility in developing the base cutter and cane-gathering mechanisms. Now, in Hawaii, with the almost universal use of drip irrigation, the restriction imposed by the use of furrow irrigation has been removed.

Most of the harvesters developed in Australia, on the other hand, are of the cut–chop rather than the whole-stick type. An exception is the Creighton which used to be made by Massey Ferguson. As with whole-stick harvesters, the stems are gathered, severed at the base, topped and drawn into the machines butts first; but are then cut into pieces either by meshing rollers (Massey Ferguson, Claas and Cameco) or by a rotating knife (Toft and Don Mizzi).The cane is then cleaned and delivered by conveyers into side supporting vehicles. At present the Claas 1400 (Figs. 9.6 and 9.7) made in Germany, is one of the most popular cut–chop harvesters. A similar

Fig. 9.7 The Claas 1400 harvester at work (Photograph: Claas OHG, Harsewinkel, Germany)

machine is being made by Cameco. Other machines in production are the Don Mizzi in Australia, the Santal in Brazil and several home-made units in Cuba.

If showers might fall it is important that the auxiliary transport and harvesters should have similar flotation characteristics. Otherwise soil conditions could allow one to operate, but not the other.

Of the many harvesters operated in Australia, the Massey Ferguson (Fig. 9.8) and the Toft are perhaps the most widely used and machines have been designed to suit the small grower and groups of small growers. It has been shown in Australia that mechanical harvesters are not the prerogative of large plantations.

Fig. 9.8 The Massey Ferguson harvester

Extraneous matter

Whether the harvesters are of the whole-stick or cut–chop type, the adjustable topping devices inevitably must be set too high for some stems and too low for others because not all are of the same length: the optimum position is decided by the operator. Moreover, there is the problem of extraneous matter. In manual operations it has been shown that the trash content of cane at the factory seldom exceeds 5 per cent, whereas in mechanically harvested cane the proportion for many years rarely fell below 10 per cent. Consequently, increasing attention has been paid to the development of de-trashing devices and their incorporation into the harvesters. For example Cil (1974) described the advantage achieved in Cuba when two blowers were fitted to the KTP-1 harvester. All cut–chop harvesters now include cleaning devices, and suction fans at the discharge point of the elevators. The Claas and Massey Ferguson machines have delivered cane containing less than 1.5 per cent extraneous matter when reaping burnt cane, and less than 10 per cent when the cane was unburnt. Similar claims are made for the new Cameco harvester.

Humbert (1974), on the other hand, examined the effect of applying leaf desiccants prior to burning, especially at the beginning and towards the end of the harvest season, when rain might be expected. The experiments were carried out in many countries. He reported that the most effective desiccant, Gramoxone, when applied at the rate of 1.5–3 litres in 70–80 litres of water/ha, caused trash in cane to be reduced to 3–5 per cent. It was important that the droplets were large enough to penetrate to the lower green leaves: low-volume sprays were much less effective. Both Cil and Humbert reported higher output, when their innovations were used.

Also important is the increased quantity of soil which, because of the fixed position of the base cutting blades, accompanies mechanically harvested cane to the factory. This material, especially if it contains a high proportion of sand, causes abrasion of the mill rollers and damage to pump impellers and pipes. Clay and other soil constituents can also interfere with clarification and other processes in the factory.

The Barbados Sugar Producers' Association (BSPA)–McConnell harvesting system

The machines developed at the main centres of activity have been modified and adapted to meet different conditions in other sugar-producing areas; but only in some degree. Hudson (1978) felt, quite rightly, that in many countries, and particularly in Barbados, agricultural practices had been altered to accommodate the machines rather than *vice versa*; hence the development of the BSPA–McConnell harvesting system.

The BSPA–McConnell harvesters exploit two principles:
1. The weakest points in a cane stem are at ground level and at the junction between mature and immature joints; it is at these places that hand cutting should take place, and at which the stem can be broken.
2. Unburnt cane can be cleaned, without being chopped, if it is drawn **tops first** into a simple machine.

Reaping takes place in three stages. In the first the stems are

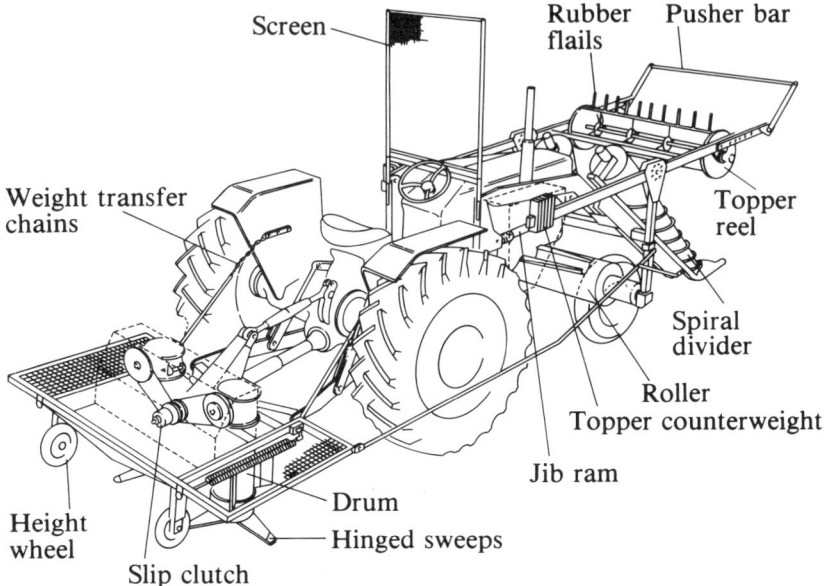

Fig. 9.9 The BSPA–McConnell cutter (From Hudson 1974)

topped, broken off at ground level, partially cleaned and left in a swath along the cane rows. Hudson (1974) described this operation and the machine which performs it (Fig. 9.9). In the second stage a tractor drawing a cleaning machine travels along the rows in the opposite direction to that taken in the first stage. The stems are drawn, tops first, by ground-following hinged sweeps into the cleaner, the leaves are removed and regular heaps of clean cane are formed (Fig. 9.10). Finally, the cane is transferred into conventional trucks by a simple slew loader (Fig. 9.11).

In describing this complete system for whole-stick harvesting, Hudson (1978) stressed the ease with which it could be introduced gradually to replace manual reaping; in every sense it conformed to Schumaker's *Appropriate Technology*. It has other advantages:

1. The various devices are fitted on to ordinary wheel tractors which, when not required for reaping, are free to be used for other purposes. Consequently the capital cost is less than that of many alternative systems.

2. Stones and rocks do not affect the machines' performance: there are no cutting blades.

3. Unlike other harvesters they can operate on any slope which can be negotiated by a wheel tractor.

4. Extraneous matter is usually less than 10 per cent, and often less than 5 per cent, depending on variety and yield.

Fig. 9.10 The BSPA–McConnell cleaner–bundler (From Hudson, 1978)

Fig. 9.11 The BSPA–McConnell loader (From Hudson 1978)

The output of the machines, under Barbadian conditions, is:
   Stage 1. Green cane, up to 40 tonnes/hour; burnt cane, up to 60 tonnes/hour.
   Stage 2. Green cane, up to 25 tonnes/hour.
Much interest throughout the world, especially in South Africa and Puerto Rico, has been shown in this system of reaping (Allison 1974) but the extent to which it will be used in commercial production is a matter for speculation. In South Africa significant progress has been made in combining a tractor and stages 1 and 2 into a single machine, by removing the front wheels of the tractor and using an articulated power steering mechanism to attach the combined stage 1 and 2 units to the tractor, giving a compact and manoeuvrable unit.

Reaping in Hawaii

The whole-stick harvesters used in Louisiana, and their cut–chop counterparts in Australia, were developed to reap upright or slightly recumbent cane seldom yielding more than 120 tonnes/ha. Most of these machines incorporate a spiral-shaped 'gatherer' or 'divider' which raises the recumbent stems so that they may be cut and topped together with upright stems. In Hawaii, however, where there is no climatic restraint to growth, fields are reaped when the cane is two and even three years old. Consequently, yields of 300 tonnes/ha are not unknown.

It is difficult to reap such a tangled mass of vegetation. As mentioned earlier, in the 1940s the push-rake system of harvesting was developed to overcome a sudden and acute labour shortage. In this method the cane is bulldozed into windrows by powerful crawler tractors equipped with bulldozing rakes and shearing blades. The push rake moves along the rows, cutting the stems at their bases and rolling them into windrows. Large crawler-type mobile cranes, equipped with grabs, then load the cane from the windrows into vehicles of 15–40 tonnes capacity, in which it is transported to the factory.

In high-rainfall areas an intermediate operation might be necessary: the use of buggies for in-field transport to a point at which the cane is transferred into road vehicles.

The push-rake system has two important defects: in wet weather stools are uprooted and at all times large quantities of extraneous matter accompany the cane. The museums of unusual pieces received for processing at the factories in the late 1940s included rocks, trees and even more remarkable objects. Consequently the cane has to be cleaned before being processed. First it floats across a water gap, in which soil, rocks and heavy objects are removed; and then is elevated on a carrier and passed between Olsen rolls, which strip off some of the leaves and tops, before being fed into the first mill. Considerable quantities of sugar are lost in these laundries, which also are expensive to build: the cost of a cleaning plant for a small (80 tonnes cane/hour) factory in 1948 was US $1 m.

To overcome this defect the Hawaiian Sugar Planters' Association (HSPA) developed a harvester with a rotating cutting blade, front-end pick-up roll and a scissors-like shearing device. The cane is elevated through the harvester on to a cross conveyor, from which it is discharged to form continuous heap rows. These are similar to those produced by the push rake system, but the cane in them is much cleaner and free from stones and other debris. Thus, as in Barbados, the new harvester fits neatly into existing loading and transport systems.

The HSPA also designed and tested a small cut–load harvester primarily for use in rain-fed areas where, in the absence of furrow irrigation, the soil surface can be uniformly smooth. It operates with an in-field buggy which carries the cane either to a truck at the side of the field or to a transload station. The harvester comprises a base cutter, with pick-up chains on each side, a rotating divider knife and a cross-flow cutter.

The push-rake system is still the most widely used method of reaping on irrigated plantations, but the newly developed machines are receiving increasing attention elsewhere in Hawaii.

The co-ordination and control of reaping

It is important that all the cane scheduled for harvest should be reaped. It is equally important that the factory should grind at full capacity from the first day of 'crop' to the last. To achieve this ideal state of affairs is not easy. Before reaping starts estimates are made of the yield of each field, or of each small grower, or both, according to the method of production. Delivery quotas are then allocated to each estate, farm or other unit, *pro rata* to the estimated daily grinding capacity of the factory(-ies). On occasion some will not be met: rain may have fallen in one area, harvesting machinery may have broken down in another or, if manual reaping is practised, there may be a dispute concerning the size of task to be performed or the rate to be paid. Should any of these occur, a central authority is notified. Increased production by more favourably placed groups is then arranged and appropriate changes in the assignment of vehicles made by the traffic controller. Regular communication by radio telephone is helpful if the operation is small, and essential if it is large.

Reaping statements are prepared at the end of each week. They

Table 9.1 *A typical crop progress report*

Caroni Limited	Week ended 23 May 1969					21st week's operations	
	1969 Cane tonnage estimate				Cane deliveries		
			Difference		1969 crop		1968 crop
	Original	Pro-gressive	Above estimate	Under estimate	% of pro-gressive estimate delivered	Tons cane delivered	Tons cane delivered
Curepe	68,633	74,428	5,795	—	90.74	67,540	60,449
Wilderness	74,698	79,834	5,136	—	91.36	72,855	71,369
J. Junction	46,173	44,795	—	1,378	95.52	42,790	44,270
Todd's Road	40,828	34,758	—	6,070	98.71	34,310	37,607
Caroni Area	230,332	233,815	3,483	—	93.02	217,495	213,695
Woodford Lodge	69,628	69,451	—	177	92.46	64,215	64,747
Edinburgh	71,727	71,850	123	—	95.85	68,869	65,417
Felicite	82,497	83,569	1,072	—	94.98	79,377	76,992
Waterloo	88,793	88,077	—	716	98.67	86,910	85,638
McBean	37,208	34,666	—	2,542	95.53	33,118	34,045
Woodford Lodge Area	349,853	347,613	—	2,240	95.65	332,489	326,839
Esperanza	54,474	58,488	4,014	—	90.54	52,954	52,573
Brechin Castle	61,063	68,284	7,221	—	94.10	64,254	58,252
Montserrat	62,786	63,366	580	—	93.74	59,397	59,667
Phoenix Park	49,885	53,298	3,413	—	88.35	47,091	45,136
Exchange	69,194	73,578	4,384	—	92.74	68,232	72,227
Perseverance	42,702	44,343	1,641	—	95.80	42,482	40,160
Gordon Area	340,104	361,357	21,253	—	92.54	334,410	328,015
Cedar Hill	72,609	70,681	—	1,928	84.93	60,033	63,301
Reform	62,798	62,197	—	601	91.36	56,822	56,273
La Gloria	100,000	92,126	—	7,874	94.12	86,710	95,315
Williamsville	45,972	44,225	—	1,747	81.94	36,236	40,900
A/M East Area	281,379	269,229	—	12,150	89.07	239,801	255,789
La Fortunee	62,434	61,349	—	1,085	91.62	56,210	59,396
Picton	71,048	60,578	—	10,470	92.91	56,282	59,967
Bronte	60,258	53,125	—	7,133	90.18	47,906	48,866
Petit Morne	79,652	76,956	—	2,696	89.08	68,553	70,481
S/M West Area	273,392	252,008	—	21,384	90.85	228,951	238,710
Company cane Total	1,475,060	1,464,022	—	11,038	92.43	1,353,146	1,363,048
Farmers' cane Brechin Castle	221,000	241,000	20,000	—	95.79	230,858	218,125
Ste Madeleine	519,000	519,000	—	—	96.93	503,056	470,817
Total	740,000	760,000	20,000	—	96.57	733,914	688,942
All cane total	2,215,060	2,224,022	8,962	—	93.84	2,087,060	2,051,990

Table 9.1 (Continued)

			1969 crop		1968 crop
		This week	To date	To date
Tons cane ground*	— Brechin Castle factory	38,404.01	863,481.74	845,618.80
	— Woodford (Lodge) factory	11,998.14	259,627.81	251,993.40
	— Ste Madeleine factory	34,236.60	867,983.62	863,759.22
	— Reform factory	5,290.00	108,180.00	108,165.00
	Total	89,928.78	2,099,273.17	2,069,536.42
Tons sugar made* (TQ)				
	— Brechin Castle factory	3,864.00	83,743.00	84,624.00
	— Woodford (Lodge) factory	1,123.63	24,940.63	23,887.12
	— Ste Madeleine factory	3,114.96	85,241.03	84,563.50
	— Reform factory	565.00	11,557.00	11,946.00
	Total	8,667.59	205,020.62	205,481.62
TC/TS (TQ)	— Brechin Castle factory	9.94	10.31	9.99
	— Woodford (Lodge) factory	10.68	10.41	10.55
	— Ste Madeleine factory	10.99	10.18	10.21
	— Reform factory	9.36	9.36	9.05
		10.37	10.22	10.09

* Tons cane ground and sugar made as at 24 May 1969.

show amended estimates of the size of the crop and the progress made in reaping it, section by section. The example given (Table 9.1) shows the position at the end of the twenty-first week of harvest at a company operating four factories in Trinidad. Two-thirds of the cane is grown on the company's plantations (estate cane) and one-third by independent growers (farmers' cane). The two estate areas supplying the Ste Madeleine and Reform factories (S/M East and

S/M West) had reaped 89.07 and 90.85 per cent of their crops respectively; the corresponding figures for the areas supplying Brechin Castle factory were: Caroni 93.02 per cent and Gordon 92.54 per cent; and 95.65 per cent of the crop in the Woodford Lodge area, supplying the factory of the same name, had been reaped. Operations are more susceptible to dislocation by wet weather at Woodford Lodge and less at Ste Madeleine; therefore it was satisfactory that the harvest was more advanced at Woodford Lodge. The allocation and fulfilment of reaping quotas between the several areas had been well co-ordinated and the grinding season soon ended in an orderly manner. The figures illustrate another important principle: to sustain good community relations it must be seen that farmers' cane is treated not less favourably than estate cane. Consequently, it was also satisfactory that 96.57 per cent of the farmers' crop and only 92.43 per cent of the estate crop had been reaped.

A measure of operational efficiency often used in this highly integrated industry is 'time lost per cent gross available grinding time' at the factory. The loss of time has many causes: holidays, predetermined weekend maintenance and cleaning periods, weather, industrial disputes, machinery failures and 'out of cane'. To make this index more effective as a tool of management holidays and weekend closures might well be excluded: it is not the fault of operational staff that, in some countries, factories cease to operate for several days at Easter; or because Carnival is being celebrated; or that grinding is restricted to five days per week. 'Time lost per cent available grinding time' is the critical figure. Too many mechanical failures might be a reflection of the standard of preventive maintenance; while increases in the hours 'out of cane' might suggest that the co-ordination of reaping could be improved. Some examples of 'time lost per cent gross available grinding time' are given in Fig. 9.12. They show that, during the years 1968 to 1973, there were considerable fluctuations in Guyana and Jamaica. For these there were well-known explanations: weather and industrial disputes. In Trinidad, however, the index rose steadily from less than 24 per cent in 1968 and 1969 to 34 per cent in 1973 and must have been the cause of considerable anxiety.

While it is important that the quantity of cane reaped by each production unit should be carefully controlled, it is perhaps of even greater consequence that the cost of doing this should be under intense scrutiny. In many countries expenditure on reaping is the largest single item in the cost of production, greater even than the cost of operating the factories. Weekly statements are therefore produced and explanations are required from the supervisor of any section or unit whose expenditure is greater than that demanded by the wages schedule. An example of such a statement, concerning the

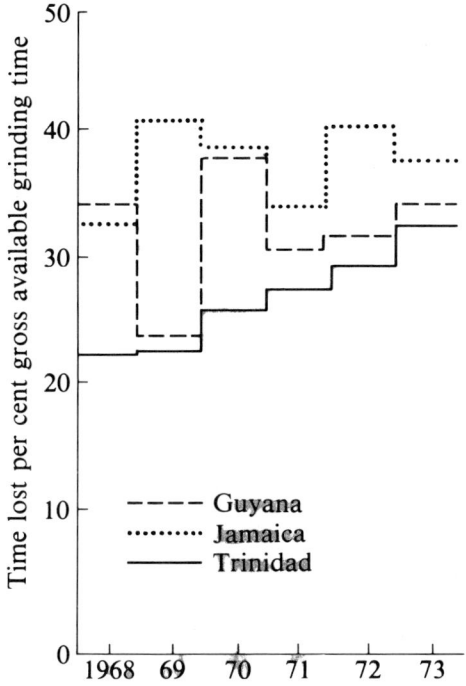

Fig. 9.12 The loss of grinding time (From West Indies Sugar Association Inc. Annual Reports 1972 and 1973)

same estates mentioned in Table. 9.1, is given as Table 9.2. The high cost of cane cutting and carting at one section, brought out so clearly, did not escape the attention of management.

It is interesting to note that while the operational cost of conventional harvesting was TT$4.03 per ton, that of self-loading trailers, used in the most difficult terrain, was TT$3.43 per ton; and that of combine harvesters TT$0.90 per ton. It was because proportionately more cane had been reaped by mechanical means in 1969 than in 1968 that, despite a large increase in wages, the average cost of reaping, TT$3.46 per ton, was the same to date in both years. The capital cost and depreciation of the equipment used is not included in these figures.

Timing is important: figures must be available for immediate action by the appropriate controlling authority. In the example cited it was stipulated that both the cost and tonnage statements should be compiled and submitted not later than the Tuesday following the end of the previous week's reaping.

Table 9.2   *The cost of harvesting: a typical weekly report*
CARONI LIMITED

ESTATE CANE HARVESTING – WEEK ENDED 31 JANUARY 1969
TT $ COST PER TON CANE

	Conventional harvesting						Self-loading trailers				Mechanical	
	Cane cutting		Loading and carting estate carts		hired carts		Cutting cane and bundling		operating S/L trailers		(crane) loading	
	This week	To date	This week	To date	This week	To date	This week	To date	This week	To date	This week	To date
Curepe	1.20	1.43	1.21	1.23	2.21	2.19	—	—	—	—	0.63	0.40
J. Junction	1.46	1.49	1.28	1.41	1.80	1.79	—	—	—	—	0.26	0.27
Wilderness	2.22	2.29	2.37	1.82	1.81	1.80	—	—	—	—	0.28	0.35
Todd's Road	—	—	—	—	—	—	2.45	2.68	0.47	0.52	0.37	0.37
Caroni Area	1.53	1.55	1.43	1.46	1.90	1.91	2.45	2.68	0.44	0.49	0.38	0.33
Woodford Lodge	1.42	1.58	1.30	1.45	1.66	1.80	—	—	—	—	0.27	0.28
Edinburgh	1.68	1.74	1.22	1.45	1.80	1.81	—	—	—	—	0.33	0.33
Felicite	1.43	1.48	1.29	1.36	1.71	1.77	—	—	—	—	0.19	0.22
Waterloo	1.54	1.73	1.83	1.70	1.08	1.50	—	—	—	—	0.27	0.30
McBean	1.21	1.52	1.05	1.51	1.84	1.86	—	—	—	—	0.20	0.31
Woodford Lodge Area	1.48	1.62	1.37	1.50	1.57	1.74	—	—	—	—	0.25	0.28
Esperanza	1.55	1.49	1.42	1.41	—	2.03	3.55	2.80	0.50	0.29	0.45	0.46
Brechin Castle	1.64	1.41	2.07	1.43	1.66	1.91	—	2.28	—	0.53	0.51	0.39
Montserrat	1.15	1.57	2.76	1.90	—	1.99	2.39	2.82	0.33	0.33	0.47	0.39
Exchange	1.31	1.34	1.61	1.57	1.84	1.87	—	—	—	—	0.44	0.35
Perseverance	1.55	1.52	1.67	1.66	1.67	1.77	—	—	—	—	0.46	0.41
Phoenix Park	—	—	—	—	—	—	2.97	2.77	0.36	0.30	0.37	0.37
Gordon Area	1.45	1.43	1.72	1.54	1.74	1.53	2.71	2.80	0.37	0.31	0.45	0.39
Cedar Hill	1.41	1.51	1.81	1.95	—	—	2.25	2.66	0.32	0.42	0.33	0.34
Reform	—	—	—	—	—	—	2.55	2.65	0.42	0.43	0.33	0.32
La Gloria	—	—	—	—	—	—	2.30	2.61	0.31	0.34	0.28	0.33
Williamsville	—	—	—	—	—	—	2.78	2.77	0.38	0.43	0.42	0.38
S/M East Area	1.41	1.51	1.81	1.95	—	—	2.44	2.66	0.35	0.39	0.33	0.34
La Fortune	1.42	1.54	1.85	1.97	—	—	2.30	2.62	0.41	0.37	0.40	0.35
Picton	1.34	1.55	1.97	2.04	—	—	2.43	2.65	0.47	0.41	0.37	0.37
Bronte	—	—	—	—	—	—	2.41	2.69	0.28	0.32	0.33	0.36
Petit Morne	1.26	1.52	1.57	1.87	—	—	2.27	2.67	0.25	0.28	0.33	0.32

| Picking up cane | | | | Total harvesting costs | | | | | | | | Average to date | |
| Field cane | | Road cane | | Conventional harvesting | | Self-loading trailers | | Mechanical harvesters | | Harvester pick-up | | | |
This week	To date	This week	To date	This week	To date	This week	To date	This week	To date	This week	To date	This year	Last year
0.10	0.15	0.02	0.02	3.54	3.69	—	—	—	0.90	8.55	5.59	3.20	3.38
0.20	0.19	0.02	0.02	3.48	3.54	—	—	—	—	—	—	3.53	3.08
0.24	0.24	0.04	0.04	5.02	4.72	—	—	0.92	0.90	8.55	5.59	3.08	3.37
0.09	0.07	0.05	0.03	—	—	3.43	3.67	—	—	—	—	3.63	3.57
0.19	0.19	0.03	0.02	3.80	3.79	3.49	3.71	0.92	0.90	8.55	5.59	3.14	3.33
0.19	0.18	0.05	0.05	3.38	3.67	—	—	0.92	0.90	—	5.59	3.31	3.31
0.14	0.16	0.03	0.04	3.63	3.64	—	—	—	0.90	—	—	3.84	3.44
0.21	0.20	0.01	0.01	3.31	3.42	—	—	—	—	—	—	3.42	3.16
0.24	0.26	0.03	0.02	3.49	3.94	—	—	0.92	0.90	8.55	5.59	2.99	3.04
0.17	0.19	0.03	0.04	2.49	3.69	—	—	—	0.90	—	5.59	3.28	3.56
0.20	0.20	0.03	0.03	3.39	3.70	—	—	0.92	0.90	8.55	5.59	3.33	3.26
0.35	0.27	0.07	0.05	3.84	3.70	4.57	3.60	—	—	—	—	3.59	3.26
0.18	0.17	0.07	0.05	4.23	3.65	—	3.25	—	0.90	—	—	3.48	3.28
0.52	0.26	0.03	0.03	4.93	4.16	3.22	3.57	—	—	—	—	3.62	3.26
0.17	0.15	0.07	0.06	3.70	3.56	—	—	—	0.90	—	5.59	3.43	3.45
0.30	0.24	0.08	0.05	4.06	3.92	—	—	0.92	0.90	—	5.59	2.67	3.29
—	—	0.04	0.05	—	—	3.74	3.49	—	—	—	—	3.49	3.22
0.23	0.20	0.05	0.05	3.91	3.70	3.58	3.55	0.92	0.90	—	5.59	3.42	3.30
0.21	0.21	0.01	0.01	3.77	4.02	2.91	3.43	—	—	—	—	3.73	3.90
—	—	0.04	0.03	—	—	3.34	3.43	—	—	—	—	3.43	3.60
—	—	0.01	0.01	—	—	2.90	3.29	—	—	—	—	3.29	3.55
—	—	0.01	0.01	—	—	3.59	3.59	—	—	—	—	3.59	3.75
0.21	0.21	0.01	0.01	3.77	4.02	3.13	3.40	—	—	—	—	3.47	3.68
0.16	0.19	0.02	0.02	3.85	4.07	3.13	3.36	—	—	—	—	3.75	3.87
0.12	0.16	0.01	0.01	3.81	4.13	3.28	3.44	—	—	—	—	3.83	3.89
—	—	0.02	0.02	—	—	3.04	3.39	—	—	—	—	3.39	3.64
0.13	0.15	0.07	0.06	3.36	3.36	3.92	3.33	—	—	—	—	3.69	3.82

Table 9.2 *(continued)*

	Conventional harvesting						Self-loading trailers				Mechanical	
	Cane cutting		Loading and carting				Cutting cane and bundling		operating S/L trailers		*(crane)* loading	
			estate carts		hired carts							
	This week	To date	This week	To date	This week	To date	This week	To date	This week	To date	This week	To date
S/M West Area	1.33	1.54	1.78	1.95	—	—	2.36	2.66	0.33	0.34	0.36	0.35
Total – all areas	1.45	1.55	1.56	1.64	1.70	1.82	2.49	2.70	0.35	0.36	0.35	0.34
To date – last year		1.52		1.72		1.80		2.65		0.33		0.30

Factory supplies

Manual reaping is confined to the hours of daylight; but, in order to reduce capital investment and to avoid the technical difficulties associated with intermittent grinding, factories are expected to operate for twenty-four hours each day and, if possible, for seven days per week (except for weekend cleaning and maintenance). Therefore stockpiles must be made during the day, the cane in them to be transported and processed during the night. At all times there must be a reserve at the factory sufficient to allow grinding to continue if, for any reason, there is a temporary disruption of supplies. The reserve might comprise a heap in the factory yard, from which cane is transferred by crane to the mill carrier; or it might be loaded rail trucks in the marshalling yards.

Mechanical reaping, on the other hand, need not be interrupted by nightfall. To achieve maximum output per man and machine, the harvesters should cease to operate only when being adjusted, cleaned and refuelled: but inevitably breakdowns will occur. Therefore it is common practice to attach an extra machine to each group of harvesters. The standby and operating machines used in rotation are regularly serviced, and maximum efficiency is achieved. Rolling-stock (with the exception of factory reserves, which nevertheless are gradually but continuously replaced in order to prevent deterioration in cane quality) should also be in motion, except when being loaded and unloaded. A precept of wide currency within the industry is that 'rolling-stock should roll'.

When new machines and methods are being introduced it is advisable, even in the experimental stages, that their operators

Picking up cane				Total harvesting costs								Average	
Field cane		Road cane		Conventional harvesting		Self-loading trailers		Mechanical harvesters		Harvester pick-up		to date	
This week	To date	This week	To date	This week	To date	This week	To date	This week	To date	This week	To date	This year	Last year
0.14	0.16	0.03	0.03	3.64	4.03	3.08	3.38	—	—	—	—	3.68	3.81
0.19	0.19	0.03	0.03	3.63	3.80	3.22	3.43	0.92	0.90	8.55	5.59	3.44	3.46
	0.19		0.03		3.78		3.31		0.79		—		3.46

should be employed on the precise terms, especially with regard to payment and hours worked, as those envisaged when the operations become standard practice.

Loading and unloading

In former years the loading of secondary transport vehicles was a tedious business; an example of this is the mule walking around a 'stiff-leg' derrick, shown in Fig. 9.13. Cranes have now replaced such slow methods of transfer in many countries and only a few minutes are required for this operation to be carried out.

In several newly developed areas cane is loaded mechanically in the field into bins, which are then taken either directly to the factory or are transferred by fork-lift trucks on to road or rail vehicles. Systems of this type is now in use in Cameroun, Ivory Coast, Sudan, Swaziland and Texas, USA. In Fig. 9.14 a road haulage unit is being loaded with two bins.

At the factory cane is discharged on to feeder carriers aligned at right angles to the milling tandems. The most widely used method of emptying a railcar is to clamp it to a tipping platform, to unlatch the locking device at the bottom of the side next to the carrier and then to tip the platform The gate, attached to the top bar of the truck, swings open and the cane is released. Road vehicles may be emptied in similar fashion, but the 'Hilo net' – developed in Hawaii, as its name suggests – became more popular. Chains fastened to the top of one side of the container form a bed for the cane. They pass underneath and around the load and are attached to a bar at the top of the other side of the container which, during transit, is loosely

(a)

Fig. 9.13 The old and new methods of loading vehicles: (a) by mule-driven derrick; (b) by crane (Photographs: A. L. Down)

held in open brackets. When unloading takes place, the loose bar is raised by a specially designed hoist and the chains roll the cane over the side of the truck on to the feeder carrier (Fig. 9.15). While suitable for cane reaped by hand, or by whole-stick harvesters, Hilo nets require considerable modification if required to deal with the billets produced by cut–chop harvesters. For this type of cane road trucks or motor-drawn trailers, equipped with top-hinged sides, might be attached to the tipping platforms and off-loaded in the same way as railcars. Alternatively, the billets may be transported to the factory or to a transloading station in wire mesh containers fitted with bottom-opening doors. Each container is hoisted by a crane, the door released, and the cane falls either into another vehicle or on to the feeder carrier. The development of a small unit of this type, to replace carts and tractor-drawn trailers in St Kitts, is described by Cole (1973).

In new projects preference is being shown for side-discharging trailers and trucks in which the body of the vehicle pivots about its top outer edge. When the body cage is lifted at the factory cane is

(b)

Fig. 9.14 Road haulage unit being loaded with two bins (Photograph: H. F. E. Deacon)

Fig. 9.15 Off-loading by chain (Hilo) net (Photograph: A. L. Down)

tipped from the open top on to the carrier. Both cut–chop and whole-stick cane can be accommodated by this single transport and unloading system. An adaptation made in Australia and in the USA enables vehicles loaded in-field to be off-loaded into larger containers at transfer stations without the use of auxiliary machines.

Uniformity and versatility

Although the laws relating to traffic and transport vary from country to country, there are two principles which are of general application:
1. The fewer the types of prime movers used, the better. If several different makes and models of tractors and trucks are operated, they require a multiplicity of spare parts; and capital which could be put to more productive use is then represented by excessive quantities of ironmongery in the stores. Moreover, the operators and maintenance crews develop special skills concerning the machines with which they are familiar.
2. Versatility as well as standardization should be sought. For example it is possible to design a trailer which might be drawn either for short distances by a wheel tractor or for longer distances if attached to a truck (also laden). When the inevitable breakdowns occur there is then flexibility: the trailer normally drawn by a truck which is under repair can be attached to a tractor, or vice versa.

Chapter 10

Manufacture

The method of manufacture, whether it be in a cottage industry or in a large factory, is to extract juice containing sugar from the cane either by a crushing or washing process, to purify this juice and then to secure from it sugar free from unacceptable contamination and of the desired quality. In doing this considerable quantities of molasses are made as a by-product. Molasses is a highly viscous syrup which contains some sucrose and also other sugars, salts, colouring matter and organic compounds which have no place in the final product.

Economic operation usually requires that most of the steam and power needed for manufacture should be provided by the combustion of bagasse, the fibrous material remaining after juice has been extracted from the cane.

The manufacturer is faced with a choice of several extraction and purification procedures at almost every stage of the process. The designers of factories, for their part, must allow for regional differences in the composition of cane juice, and on seasonal variations likely to occur during each year's crushing campaign.

Marketing requirements, especially the quality or grade of sugar which must be produced, are also of great importance.

Types of sugar

There are two main types of sugar: centrifugal and so-called non-centrifugal sugar. In the former the molasses has been separated from the sugar crystals with the aid of centrifugals; in the latter it has not. There is also a third category: syrup. Some small factories produce syrup which is sent to larger factories for further processing; others make edible syrup. These various products may be grouped as follows:

Centrifugal sugars   Raw sugar – to be sent to re-
                                        fineries for final purification and

preparation in the customer's desired form.
Direct consumption sugars – golden
brown, yellow, washed grey, standard white, plantation white and
khandsari.
Refined sugar

Non-centrifugal sugars   *Gur, jaggery, panela* and *desi* –
usually produced in cottage indus-
tries for local consumption and, in
the case of jaggery, also for the
manufacture of alcoholic beverages
by fermentation.

Liquid sugars   Fancy molasses
Edible syrup

Analytical terms

In order to understand the basic language of the industry, and therefore to describe the manufacture of these various types of sugar, it
is essential to know the meaning of three widely used terms: polarization (pol), Brix and purity.

Polarization

Polarimetry is the science of measuring phenomena exhibited by
polarized light, and its use in the estimation of sucrose makes it the
most important analytical tool of the sugar industry. It exploits the
fact that certain crystals and molecules are able to influence beams
of light.

A beam of ordinary light vibrates in many planes. However,
when transmitted through a polished rhombohedral crystal of transparent Iceland spar, which has been cut along a specific axis and
cemented with Canada balsam (a Nicol prism), the emerging beam
vibrates only in one plane; it has become polarized due to the optical
activity of the crystal. If the beam of polarized light emerging from
a Nicol prism is subsequently passed through solutions or crystals
of other optically active substances, the plane of polarization is
twisted or rotated, and the degree of rotation is specific to the
substance concerned. In the case of solutions it is directly
proportional to their concentration and to the length of the light
path through the liquid. Sucrose is one of these optically active
substances and causes the beam to be rotated to the right; that is,
it is dextrorotatory. A polarimeter (sometimes called a polariscope)
is used to measure the degree of rotation. Basically the polarimeter

comprises a tube compartment with a Nicol prism (the polarizer) at one end, close to a source of light. At the other end of the tube there is a second prism, called the analyser, and this can be rotated. The analyser is connected to a telescope above which is a scale-reading eyepiece on which are marked the angles of rotation (Fig. 10.1). Therefore, if the material under examination is introduced into the tube, its optical activity can be measured in terms of its rotatory power by adjusting the analyser.

In the course of time scientists in the sugar industry found empirically a convenient range of concentrations and tube lengths for examining sucrose solutions, and took matters a step further by calibrating their polarimeters not in angular degrees but directly and more conveniently in sugar degrees, that is in per cent sucrose. These specialized instruments are called saccharimeters. Thus if a standard or 'normal' weight (26 g) of pure sucrose is dissolved in a 'standard' volume of distilled water (100 ml) and examined in the 'standard' length of tube (200 mm) the saccharimeter will indicate 100 per cent sucrose or 100° polarization (pol) when the analyser has been turned to compensate for the rotation of the light beam caused by the solution. If the sample is impure then a lower reading will be given, reflecting precisely the proportion of impurity present.

The last sentence should be accepted with caution. The impurities present in a sugar solution might themselves be optically active and so exert an influence on the net polarization of the material under examination. Two of the most important of these are glucose (dextrorotatory) and fructose (laevorotatory). Besides occurring naturally in sugar-cane juice, they also are the products of the hydrolysis of sucrose. Since an equimolecular solution of these two compounds is laevorotatory, collectively they are called invert sugar (because they change the direction of net polarization). The greater the proportion of optically active compounds in the sample the less precise is the measurement of sucrose by polarization.

To overcome these difficulties an expensive and time-consuming procedure, double polarization, must be carried out. In the Clerget method a sample is first measured for polarization. It is then treated with hydrochloric acid, in order to hydrolyse all the sucrose present, and is measured again. Since the change in polarization must be attributed to the formation of invert sugar from sucrose, the amount of the latter can be deduced.

The methods of conducting saccharimeter readings, the adjustment for temperature, the difficulties encountered with coloured, partly opaque, impure solutions, or the accounting for other optically active compounds which might be present, form a science themselves, and will not be discussed further. Suffice it to say that there is no other chemical or physico-chemical method available to estimate the sucrose content of materials which can rival the conven-

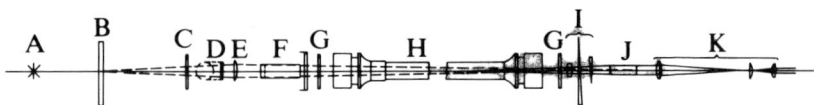

Plan view of path of light and observation
tube shown twice size of view below

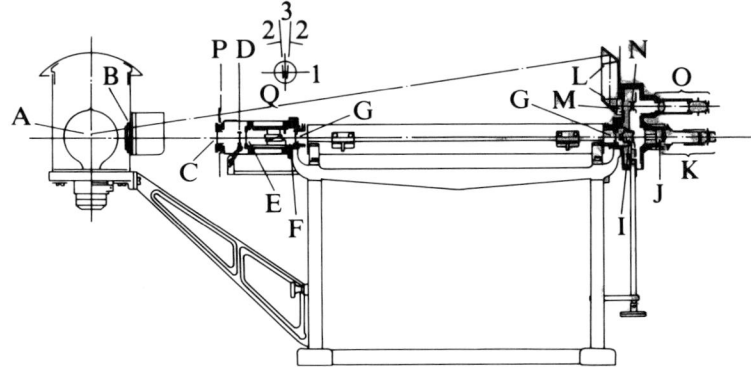

Diagram of Bausch and Lomb polariscope

A Light source;
B Ground glass lamp window
C Cover glass
D Glass filter replacing bichromate cell
E Condensing lens
F Polarizer
G Splash glass
H Observation tube
I Quartz compensation
J Analyser
K Observing telescope
L Scale illuminating prism
M Condensing lens
N Scale and vernier plates
O Scale-reading eyepiece
P Glass filter in scale-illuminating system
Q Photometric field as seen from eyepoint
1 Plane of polarization of analyser
2 Planes of polarization of semifields
3 Half-shade angle

Fig. 10.1 Diagram of a polarimeter (From Spencer & Meade 1945)

ience and accuracy of polarimetry, despite its acknowledged shortcomings. As an aid to the management of factories or as a measure,
in the commercial sense, of the value of sugar and molasses, or
indeed of sugar-cane itself, it is used and accepted throughout the
world.

The raw sugars of commerce are required to have a pol of 96°, the practical standard of a generation ago. International agreements usually include clauses which invoke a penalty for cargoes which do not reach this standard, and a premium for those of higher quality. Invariably the penalty is greater than the premium. Nevertheless the tonnages exported and imported under such agreements are *tel quel* and in former years, in times of plenty, some exporters deliberately produced high-quality raws in order to dispose of surplus sugar. Nowadays the pol of raw sugar is usually between 97° and 99°.

Direct-consumption centrifugal sugars made locally for the domestic market usually have a polarization between 99.0° and 99.5°.

Solids

The next most important analysis is the estimation of solids in solution. For generations this has been done by hydrometry, i.e. by measuring the density of the liquid intermediate and end-products of the industry. The density of a solution of pure sucrose in water is directly proportional to its concentration. Therefore if the changes in density of the solution are calibrated on a scale reflecting this, it is possible to determine quickly and easily its sucrose content. This scale is known as the Brix scale.

At one time hydrometers based on a different scale were used almost exclusively in the sugar industry. This, the Beaumé scale, has no convenient relationship to the density of sucrose solutions and was replaced by the Brix scale long ago. There is one exception to this generalization: the USA Bureau of Standards' Beaumé scale is still used in the edible molasses trade.

The traditional Brix hydrometer or spindle (Fig. 10.2) comprises a cylindrical floating tube, with a long thin extension at the top which is graduated in degrees Brix (and usually in specific gravity as well). A weight in the glass tube (sometimes metal) causes the hydrometer to float in a vertical position when placed in a solution. The extent to which it sinks reflects the density, and hence the sucrose content, of that solution. The degrees Brix is then read from the scale on the spindle. If the solution is impure, the reading will reflect the concentration of all dissolved substances in terms of sucrose. Other factors must also be considered, especially undissolved solids, the aeration of the solution and its temperature; therefore technologists have looked for a different method to estimate solids.

The most advanced, which has gained considerable acceptance despite the extra cost of the instruments used, is the refractive index. For pure sucrose solutions this, like density, is a property which bears a linear relation to concentration. Therefore a scale can be

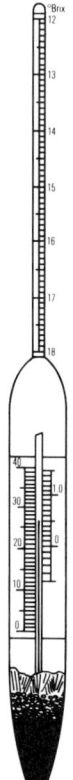

Fig. 10.2 The Brix hydrometer
(From Spencer & Meade 1945)

constructed to give the sucrose content as refractometer degrees Brix.

The refractometer comprises a prism on to which a small sample of the solution under examination is placed with the aid of an optical glass plate. A telescope is hinged around the prism and is used to determine the angle of internal reflection of a beam of light as indicated by a sharp shadow line. The angle is dependent on the change of refractive index at the surface between the liquid sample and the glass prism and hence gives an immediate measure of the refractive index of the solution. Like the hydrometer, the scale is calibrated to give directly the sucrose content as refractometer degrees Brix.

Since many of the disadvantages of hydrometry are eliminated in refractometry, its acceptance as a suitable method for estimating the solids content (or apparent solids content) is assured. It does not give precisely the same results as hydrometry, but this is due mainly to slight departures from linearity of both the density and the refractive index relative to the concentration of sucrose.

For the ultimate measurement of solids, i.e. true solids, some form of drying must be used. It is by no means a simple matter. Sucrose can form highly concentrated solutions of a syrupy, viscous nature; therefore it is difficult to ensure uniform heating throughout the sample. Inert sand is often used as an extender to overcome this difficulty. As the water molecules become partly associated in the chemical sense with the sucrose, especially in concentrated solutions, it becomes almost impossible to drive them off without using temperatures high enough to cause combustion of the sucrose itself. Vacuum oven techniques, allowing evaporation at low temperatures, offer some improvement but, since any method of drying to determine the moisture content is difficult and time consuming, investigators are constantly striving to improve reliability by constructing appropriate conversion scales to relate solids determined by refractometer with true solids determined by drying. Again, in practice the analyses made in the factory are influenced by other impurities present (for example inorganic salts or other organic soluble material) so that the results achieved are those essential to the successful technical and commercial control of the enterprise and are not absolute measurements.

Purity

The purity of a solution containing sucrose is the proportion by weight of sucrose to all dissolved solids, expressed as a percentage.

$$\text{True purity} \quad = \frac{\text{per cent sucrose}}{\text{per cent solids}} \times 100$$

However, in practice sucrose is estimated as pol and solids as Brix (hydrometer or refractometer), each of which is subject to the shortcomings already described. Therefore reference is usually made to the apparent purity of the solution.

$$\text{Apparent purity} \quad = \frac{\text{pol}}{\text{Brix}} \times 100$$

Since the purpose of sugar factory operations is to produce sugar containing nearly 100 per cent sucrose, and molasses with (ideally) all the impurities, everyday operations depend on the constant testing of juice, syrup and molasses for purity. Apparent purities are adequate in most circumstances. The one exception is molasses, in which the high proportion of invert sugar (20–25 per cent) would create difficulties in interpretation unless true purities were determined.

The manufacture of centrifugal sugars

At least two types of sugar are made in most modern factories: that for export, known as raw sugar, and that for local consumption. The processes used in their manufacture differ only in the later stages.

Raw sugar manufacture

Conventional raw sugar factories have two distinct though interdependent components: the mill house in which juice is extracted from the cane; and the boiling house in which it is clarified and concentrated, and in which sugar is crystallized and separated from its mother liquor, molasses. Traditionally, engineers are responsible for efficient milling and for providing steam power and other services, while chemists see to the production of sugar and maintain chemical and statistical control. However, their spheres of influence cannot be so clearly defined: engineers must ensure that machinery in the boiling house functions adequately, and chemists have concern for the quality of juice extraction. The flow chart of typical raw sugar factory is illustrated in Fig. 10.3.

Preparation

Cane is transferred on to a slowly moving conveyor, consisting of metal slats, secured across longitudinal chains, by the methods described in Chapter 9; and also from stockpiles in the yard by cranes, either wheeled or travelling on gantry bridges. From these carriers it is discharged into the milling train and passes under two sets of knives in order that the stems can be chopped into finely divided pieces and reduced to a bed of uniform thickness before entering the first mill. The knives consist of blades attached to rotating horizontal shafts placed across the carrier. The first knives level the tangled cane and the second, set much closer to the carrier, with perhaps 75 mm clearance, complete the preparation. The first are known as 'leveller' knives and the second as 'heavy duty' knives.

Juice is then extracted by crushing the cane in a series of three-roller mills, the cane proceeding from one mill to the next on slightly inclined, slatted intermediate carriers. Factories are therefore described as having a six-, nine- twelve-, fifteen- or eighteen-roller tandem(s). In some factories each three-roller mill may have additional feed rollers and pressure chutes, while in others cane is passed through a shredder. A few older plants also use a two-roller crusher. The shredder provides a more powerful and more intensive form of preparation (though it is greatly susceptible to damage by

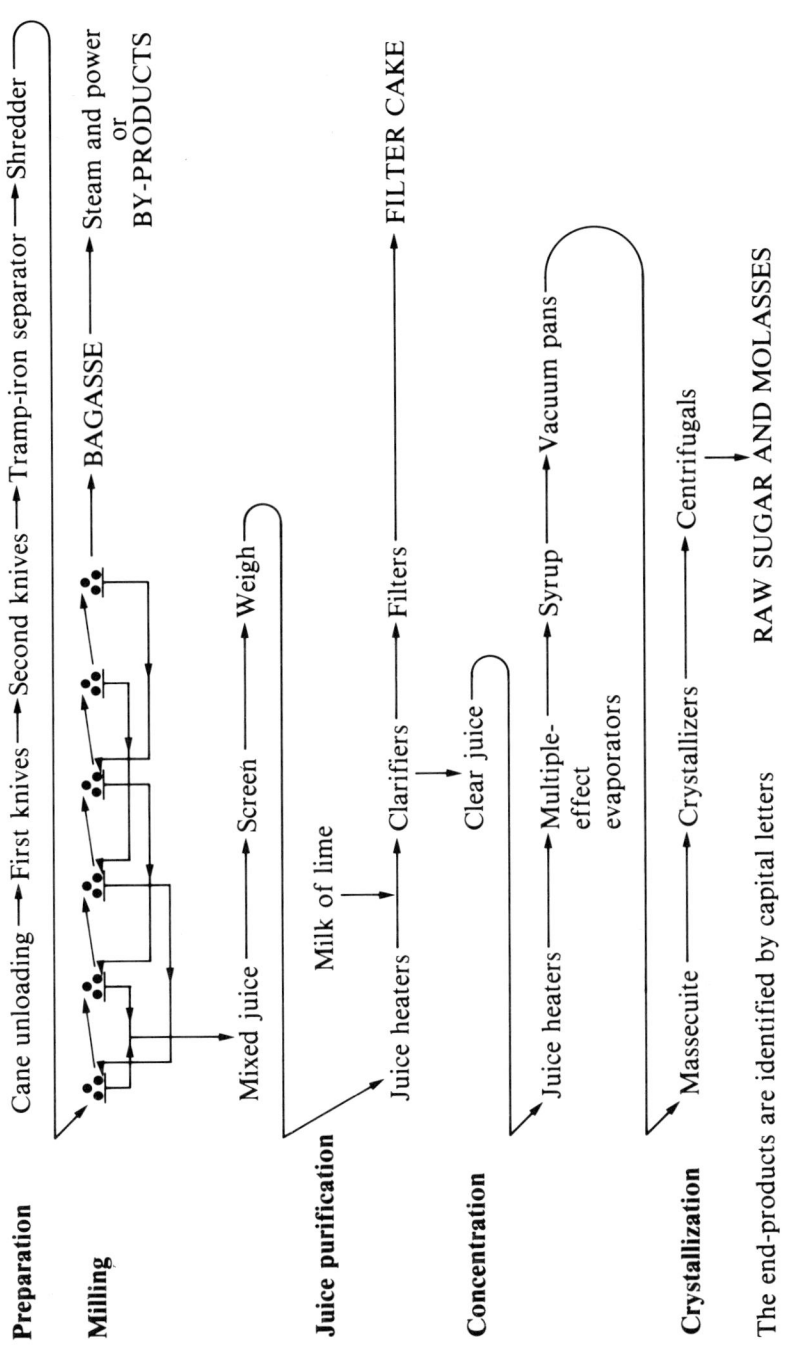

Fig. 10.3 The flow chart of a raw sugar factory

foreign bodies in the cane) while the two-roller crusher is now regarded as uneconomic in comparison with a three-roller mill. What is necessary is the installation before the first mill of a magnetic device to remove 'tramp iron': sling chains, railway fish-plates and any other ferrous material which, inserted into the cane accidentally or deliberately, could cause considerable damage if allowed to pass into the tandem.

The mill house

A three-roller mill comprises a bedplate, which also acts on a juice tray, and two vertical headstocks which support the rollers. Rollers have steel shafts, or gudgeons, on which cast-iron shells are pressed or shrunk. They are so arranged that cane is squeezed twice, under pressure, in each mill between the top and the two bottom rollers. The surfaces of the rollers are grooved to increase their gripping power and to facilitate juice drainage. They are set with great care so that the grooves of the top roller intermesh with those of the bottom rollers. The distance between the grooves, called their pitch, and also their depth, decreases gradually from the first to the last mill in the tandem. The rollers are provided with scrapers to keep the grooves clean. In the past various means of improving juice drainage were adopted, the most widely used being perhaps Messhaert grooves, spaced 63 mm apart, and almost as deep. Special scrapers were required to prevent the accumulation of bagasse in them.

As indicated in Fig. 10.3, juice extraction is not a simple operation. After having been crushed, the residue of the cane still contains sugar which might be recovered. Therefore water is added to the expanding mass of fibre as it leaves the penultimate mill. Juice from the last mill, too low in sugar content to be sent to the boiling house, is applied in similar fashion to the extrusion from the ante-penultimate mill; and so on. The process is known as compound maceration, and the water added is called 'maceration' or 'imbition' water. It is only from the first and second mills that the 'mixed juice' is pumped to the boiling house, weighed and processed. There is therefore a countercurrent extraction effect, with the cane yielding juice and maceration water gaining it.

The fibre leaving the last mill, called bagasse, contains roughly 50 per cent moisture and, according to the quality of the milling, from 1.5 to 4 per cent sucrose. It is passed through rotary sieves to remove fine particles (*bagacillo*) for use as a filter aid later in the process.

Much of a bagasse is then fed to the boiler furnaces where its heat of combustion provides the energy required for the operation of the

factory. The surplus is a valuable by-product on which many ancillary industries are based (Ch. 11).

It is advisable that, where such industries have been established
or are contemplated, the furnaces should be so designed that they
can operate efficiently on natural gas, oil or bagasse; thereby the
availability of bagasse can be increased at will. Where no such
industries exist, and where the disposal of surplus bagasse is
expensive, it might be attractive to induce incomplete combustion.
When this happens, the black particles of partly burnt fibre
discharged from the factory chimney cause a nuisance to those living
to leeward, and fly-ash separators might have to be installed by
Court Order, or in the interests of good public relations.

The boiling house

Clarification

The mixed juice pumped from the mill house to the boiling house
is opaque, dark green-brown in colour and acidic. Having been
screened to remove large suspended particles, it is heated just in
excess of boiling-point (approximately 105 °C) and milk of lime is
added in two stages, before and after heating. The purpose of
adding the milk of lime is twofold: to prevent the inversion of
sucrose which takes place in acidic conditions, and to coagulate and
precipitate impurities. The treated juice then enters the clarifiers in
which organic substances such as chlorophyll, anthocyanins, poly-
phenols, wax, gums albumin and pectins, as well as calcium phosphate (formed by the milk of lime) coagulate to form flocs and are
precipitated as mud. Clarified juice, straw coloured and of low
turbidity, is drawn from the top of the clarifier and sent to the evaporators. Mud is removed from the bottom. The mud still contains
sugar which can be recovered; it is therefore mixed with *bagacillo*
to make it cohesive, sucked on to the mesh of rotary drum vacuum
filters and sprinkled with water to elute more sucrose. The filtrate,
containing most of the mud's residual sugar, is returned to the clarifier; the residue, known as 'filter cake' or 'filter press mud' (plate
and frame presses were the predecessors of rotary filters) is useful
fertilizer. Cake from rotary filters should contain not much more
than 3 per cent sucrose, but higher losses must be accepted where
presses are still used or where *bagacillo* is not added to the mud.

Evaporation

An evaporation station consists of three, four or five connected vessels, known respectively as triple-, quadruple- or quintuple-effect
evaporators. Each vessel comprises a large cylindrical body sur-
mounting a calandria. In the calandria are tubes through which juice

flows, is heated and vapour is driven off. At the top of the evaporator is a baffle which recovers and returns to the juice any droplets entrained in the vapour. Steam is fed into the calandria of the first vessel, vapour from the first vessel into the calandria of the second, from the second into the third and so on. The decreasing temperature gradient from the first to the last vessel is created by a similar pressure gradient, the last vessel being under high vacuum. A typical triple-effect evaporator is shown in Fig. 10.4 and indicates how the pressure and temperature of vapour might alter from vessel to vessel.

Multiple-effect evaporation allows considerable economy in the use of steam and therefore is crucial to the energy requirements of the factory. Evaporation is considered complete when the syrup density reaches 65 °Brix, which is just below the point of saturation when crystallization will begin. The syrup is then transferred to the crystallization section of the boiling house where evaporation is continued under more closely controlled conditions so that sugar crystals of the desired size can be formed.

Crystallization

Although the instrumentation of crystallization is now accepted in many countries and is used with some degree of success to control the formation and development of sugar crystals in the syrup, pan boiling world-wide is still more an art than a science. In some places the art has been handed down from one generation to the next; indeed the operators in Trinidad, mostly of Guyanese stock, insist that they are so unique that they must be represented by their own union, the Pan Boilers' Association; and so they are.

A vacuum pan is related in design to the evaporator vessels. It, too, is a vertical cylindrical vessel equipped with a steam-heated calandria and a condenser to maintain vacuum while evaporation in single effect takes place. However, the tubes of the calandria are of much wider diameter because through them flows not juice but a viscous mixture of sugar crystals and syrup called a 'massecuite'. A graining charge of syrup, perhaps one-third of its capacity, is introduced into the pan. Evaporation continues, the solution becomes supersaturated and crystals form in due course, a process which can be promoted by 'seeding': the addition of small sugar crystals of powder size. More syrup is added as water is evaporated; the crystals grow in size and their development is carefully watched by the boiler who, at suitable intervals, examines samples withdrawn from the pan in a proof stick. His purpose is to produce crystals of uniform size and, above all, to prevent the formation of 'false grain', the tiny crystals which appear if the balance between the evaporation of water and the introduction of syrup is incorrect. When a pan is full, and the crystals have grown to the desired size, the massecuite is discharged into crystallizers.

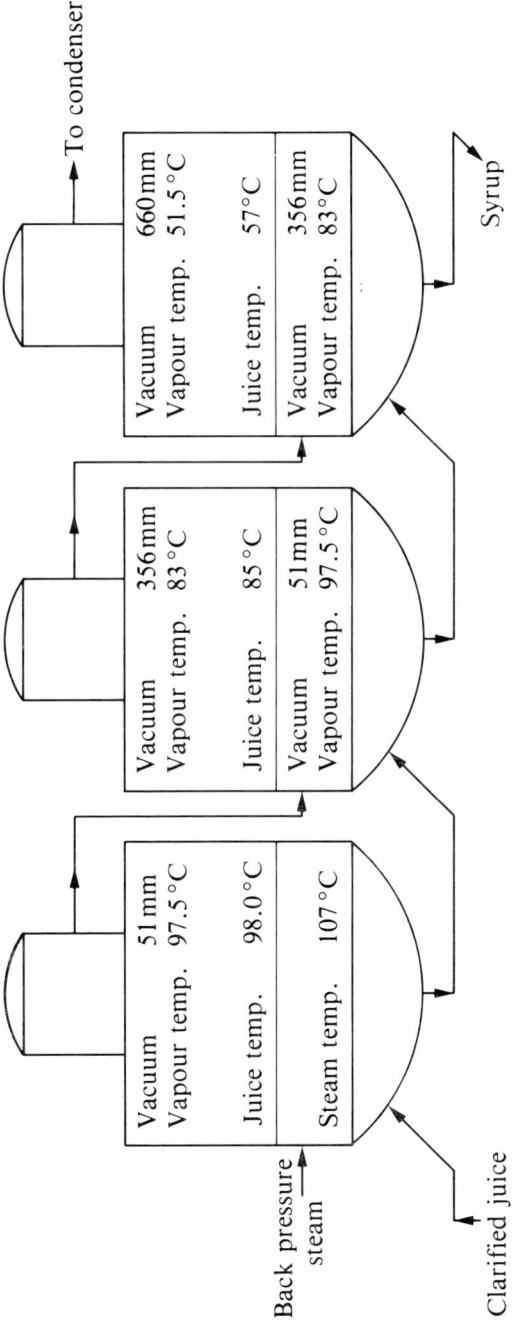

The vacuum figures are mercury gauge pressure

Fig. 10.4 A triple-effect evaporator

A crystallizer is a long, semi-cylindrical, open vessel, fitted with a horizontal helical stirrer which keeps the massecuite in motion while crystallization continues as the temperature falls (Fig. 10.5).

Separation

The massecuite is passed from the crystallizers to the centrifugals, in which sugar is separated from molasses. A centrifugal comprises a cylindrical casing or basket suspended on a vertical shaft. The

Fig. 10.5 A crystallizer (Photograph: A. L. Down)

basket is perforated and fitted internally with a metal screen made of copper, brass or stainless steel, which also is perforated.

The basket spins at high speed, the resultant centrifugal force causing the massecuite to be thrown outwards to form a layer of sugar 15–17 cm in thickness on its internal surface; the crystals are retained on the screen and molasses passes through the perforations both of the screen and the basket. When the charge has been 'cured' or dried the machine slows down and the sugar is removed by a blade, called a plough, and falls through an aperture in the bottom of the basket exposed by a movable cover. The molasses is pumped away and is either subjected to further crystallization or is stored as final molasses.

The recovery of as much sugar as possible from the syrup is usually achieved in three stages by boiling a series of graded massecuites: 'A', 'B' and 'C'. Syrup yields 'A' sugar and 'A' molasses. The 'A' molasses, however, still contains sugar which can be extracted. It is therefore mixed with more syrup to form a 'B' massecuite, which in turn is boiled and cured. The process is then repeated; a little syrup is added to the 'B' molasses and a 'C' strike is boiled. At each stage the process becomes more difficult and protracted, the increasing concentration of impurities in the molasses slowing up the subsequent stage.

Centrifugals in the 'A' and 'B' batteries usually operate on a batch system. They are automatically charged and discharged, automatically accelerated and braked, and recycling follows a set programme. Continuous centrifugals have been developed in recent years, but are used mainly for curing 'C' massecuites. They have cone-shaped baskets into which the massecuite is fed in a steady stream and from which sugar is thrown, uninterruptedly, over their lips.

'C' molasses is regarded as being exhausted of commercial sugar, has a true purity in the range of 36–40, and is known as 'final' or 'blackstrap' molasses. 'A' and 'B' sugars, which have a polarization between 96 and 98.5 are the 'raws' of commerce; 'C' or 'Java' sugar, on the other hand, is not of sufficiently high polarization to be acceptable for refining, and is used as footings to seed the 'A' and 'B' massecuites.

This is a simplified description of pan boiling; there are many variations depending on the purity of the syrup and of the 'A' and 'B' molasses. Nevertheless the flow sheet (Fig. 10.6) gives an outline of the general pattern of manufacture.

Process control

The weighing of juice is important because process control is based on the equation:

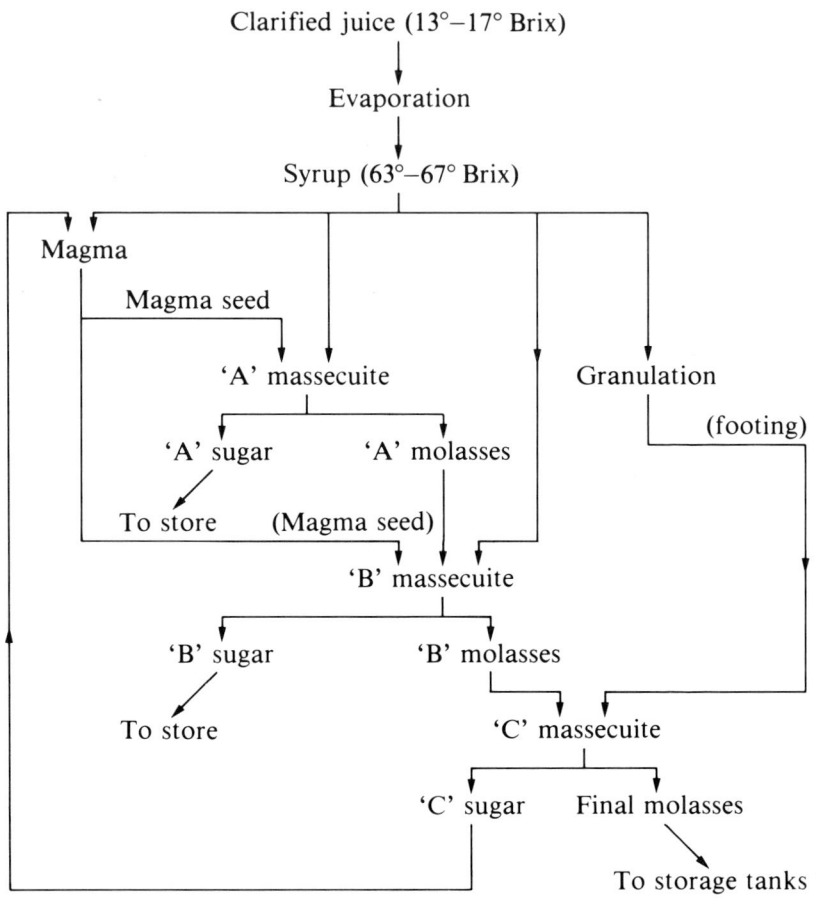

Fig. 10.6 The flow sheet of a boiling house making raw sugar

cane + maceration water = mixed juice + bagasse

The cane and the mixed juice are weighed and the maceration water is measured; but the weight of bagasse is determined by difference to balance the equation. This is a serious weakness. One factory known to the author was grinding cane which apparently had the highest fibre content in the world; that was until a leak in the bedplate under the first mill was discovered. For control purposes the juice lost through the hole had become bagasse; however, it was because the factory reports were being carefully examined that the hole was discovered. Bagasse can be weighed – and, for that matter, extremely accurate sampling and analysis of cane and the intermediate and final factory products can be carried out – but at considerable cost. By and large, therefore, methods of control are chosen to suit the economic conditions in which each factory operates.

Quite apart from this, the equation is not accurate. Davies (1942) reported that, even if all the items were weighed, the equation held good neither in the factory nor, without maceration water, in a small laboratory mill. There appears to be no evidence to contradict this disturbing observation.

In most countries factory managers and chief chemists are not charged with having received specific quantities of sugar as cane; they produce their own accounts, which are not subject to internal audit. The integrity of their reports is the measure of their value. There is at least one exception to this general comment: Australia. In that country cane is meticulously sampled and analysed before being processed and the grower is paid accordingly; but it is expensive to do this. Be that as it may, daily and weekly reports are provided in all modern factories to give information concerning the performance of both the mill house and the boiling house. The tonnage ground, the reasons for stoppages, the quality and quantity of intermediate and final products are all provided in detail. Examples are given as Tables 10.1 and 10.2.

The manufacture of direct-consumption sugars

Direct-consumption sugars are also made in raw sugar factories. They do not always conform to the standards of hygiene which are mandatory for refined sugar in many countries, but nevertheless provide a highly satisfactory product for much of the world's population.

White sugars (plantation whites)

White sugars or 'plantation whites', as they are sometimes called, form by far the highest proportion of this category of sugars. Two extra processes are used in their manufacture. First the clarified juice is treated with sulphur dioxide. This is usually done in spray towers in which the juice falls down a series of trays or baffles and meets an upward current of sulphur dioxide, blown in by ventilator fans. The sulphur dioxide causes an increase in acidity and more lime must be added to preserve the juice's neutrality. This causes the formation of a precipitate of calcium sulphite which improves clarification; meanwhile the juice is bleached by the sulphur dioxide. Consequently, when the syrup is boiled all the intermediate products are much lighter in colour. The second modification is that, despite the bleaching action of sulphur dioxide, the 'A' sugars must be

Table 10.1 *Specimen daily report*

WEEKLY PRODUCTION MEMO   Run No . . . . . .   Week Ended . . . .

Milling	No. 1 Tandem		No. 2 Tandem		Total	
	This week	To date	This week	To date	This week	Crop to date
Tons cane ground						
Grinding rate – tons cane/hour						
Pol % cane						
Fibre % cane						
Crusher juice – brix						
Crusher juice – purity						
Mixed juice – purity						

Production	This week	To date
Yellow crystals		
Granulated		
Washed greys		
Export – bulk		
96° Sugar in process		
Total produced		
Tons cane/ton comm. sugar		
Tons cane/ton 96° sugar		

Recoveries	This week	To date
Pol % in juice		
Pol % available pol in juice (Winter Carp)		
Pol % pol in cane		
Pol extraction		

Signed . . . . . *Chief Chemist*

Table 10.2 *Specimen weekly report*

MANUFACTURING & laboratory report no . . . . . WEEK ENDING										
				Cumulative figures to date						
		This week			*This crop*			*Last crop*		
Cane No. 1. Tandem – tons ground										
No. 2 Tandem – tons ground										
Total tons ground										
	No. 1	*No. 2*	*Over-all*	*No. 1*	*No. 2*	*Over-all*	*No. 1*	*No. 2*	*Over-all*	
Milling Tons cane/hour										
Tons fibre/hour										
Pol extraction %										
Reduced pol extraction (12.5%)										
Abs. juice lost % Fibre (pol)										
Mixed juice % Cane										
Imbibition % cane										
Imbibition % fibre										
Java ratio										
Cane pol %										
Fibre %										
Bagasse Pol %										
Moisture %										
Fibre %										
1st exd. juice ° Brix										
Purity %										
Abs juice ° Brix										
Purity %										
Mixed juice ° Brix										
Purity %										
R/S pol ratio										
Last exd. juice ° Brix										
Purity %										

Table 10.2 *Specimen weekly report*

Report No . . . . .
MANUFACTURING

| | This week | Cumulative figures to date | |
		This crop	Last crop
Clarified juice ° Brix			
Purity %			
Syrup ° Brix			
Purity %			
Filter cake Pol %			
% Cane			
Final molasses ° Brix			
Apparent purity %			
Sucrose % solids			
Invert % solids			
Ash % solids			
Solids %			
Ideal %			
Gals at 88° Brix/ton cane			
Sugar ° Pol – washed			
° Pol – yellow crystal			
° Pol – raw export			
° Pol – local			
° Pol – average all Sugars			
Moisture – refined			
Moisture – washed			
Moisture – yellow crystal			
Moisture – raw export			
Moisture – local			
Moisture – average all sugars			
BHE No. (SJM)			
BHE No. (Winter – Carp)			
Tons cane/ton sugar (Comm.) overall			
Tons cane/ton 96° sugar (Copp) overall			
Tons cane/ton 96° sugar (Copp) Farmers			
Tons cane/ton 96° sugar (Copp) Estate			
Steam (f and a 212 °F) % 96 ° sugar			
Steam (f and a 212 °F) lb/lb bagasse			
Lime (factory) % 96 ° sugar			
kwh used in factory/ton 96 ° sugar			
Maximum load on factory alternators			

Table 10.2 *Specimen weekly report*

Report No . . . .
TIME ANALYSIS AND DAILY CANE DELIVERIES

	No. 1 Tandem				No. 2 Tandem			
	Cumulative figures to date				Cumulative figures to date			
	This week	This	crop	Last crop	This week	This	crop	Last crop
	Hours	Hours	% TAT	% TAT	Hours	Hours	% TAT	% TAT
(1) Total available time								
(2) Weekend stoppage								
(3; Net available grinding time (1)–(2)								
Time lost Machinery repairs								
Chokes								
Steam								
Electrical								
*Other reasons								
Total from factory causes								
Out of cane								
Holidays								
(4) Total time lost								
All causes								
(5) Actual time grinding (3)–(4)								

	Mon.	Tues.	Wed.	Thu.	Fri.	Sat.	Sun.
Cane delivered to cane yard – tons							
Cane ground – tons							
Stock at a.m. tons							

*Details: Tandem No. 1   Tandem No. 2

Table 10.2

POL BALANCE BASED ON CANE GROUND FOR SUGAR

| | This week | | Cumulative figures to date | | | |
| | | | This crop | | Last crop | |
	% pol in cane	% pol in juice	% pol in cane	% pol in juice	% pol in cane	% pol in juice
Pol – Recovered in Sugar						
Pol – lost in final molasses						
Pol – lost in filter cake						
Pol – lost in ref. filter cake						
Pol – lost in ref. molasses						
Pol – ref. undet. loss						
Pol – unaccounted for						
Pol – total in juice						
Pol – lost in bagasse						
Pol – total in cane						

PRODUCTION

| | | This week | Cumulative figures to date | |
			This crop	Last crop
Cane ground				
Estate	Tons			
Farmers	Tons			
Total	Tons			
For sugar	Tons			
Other purposes	Tons			
Sugar				
Refined	Tons			
Washed	Tons			
Yellow crystals	Tons			
Raw Export	Tons			
Raw local	Tons			
In process	Tons			
Total made and in process	Tons			
Final molasses Made and in process at 88 °Brix galls.				

Signed:

. . . .

Factory Manager.   Chief Engineer.   Chief Chemist

                                              Superintendent of Manufacture

washed in the centrifugals to remove a large proportion of the molasses coating on the crystals' surfaces. The final product is white in colour and has a pol of 99.5°. It is customary to reprocess the 'B' sugars.

For convenience, plantation whites sometimes are produced as crystals smaller than those of raw sugar but, unlike raw sugar, they must be dried in a granulator before being packaged.

In recent years there has been increasing use of the polyacrylamide *Talodura* for clarifying syrup in the manufacture of this type of sugar. Some sulphitation is still necessary, but *Talodura* greatly facilitates the process.

Washed greys

Where the local market favours crystalline sugar, but not in large quantities, 'washed greys' are manufactured. Washed greys are simply 'A' sugars from which part of the molasses coating of the crystals has been removed by spraying water into the centrifugals.

Refined sugar

Much more sophisticated treatment is required if refined sugar is to be made. The designation 'refined' implies freedom from all forms of contamination and can be achieved only by melting raw sugar, filtration and treatment of the liquor and recrystallization. Nevertheless the production of refined sugar at a raw sugar factory provides the opportunity to form an economic, highly integrated operation provided, of course, that the centre of consumption is not far distant.

Several methods of purifying (defecating) the liquor are used: phosphatation followed by filtration through diatomaceous earth; treatment with lime followed by carbonation to create chalk as a filter aid; or by sulphur dioxide to make calcium sulphite, also for use as a filter aid. The liquid is then decolorized by powdered carbon (which is discarded), granulated carbon (regenerated in kilns) or animal charcoal (also regenerated), the traditional material used in offshore refineries.

In recent years a new system, the *Talofloc* system, has been devised for the simultaneous defecation and decolorization of liquor. It has been adopted throughout the world in refineries attached to raw sugar factories. A description of this new concept of refining was given by Tate and Lyle (1975). *Talofloc*, a high-molecular-weight polyacrylamide, is a cationic surfactant which precipitates colouring matter and other high-molecular-weight anionic impurities from the melt liquor. Consequently the filtration requirements are considerably reduced. In practice the installation of the *Talofloc* system has met with great success. Refined sugar is then crystallized from the purified liquor, recovered in the centrifugals and packaged. Its pol approached 100°, whereas that of washed greys is 98°.

It is essential that high standards of sanitation be maintained. At one refinery sugar was found to be infected with *B. coli*. Pond water, instead of distilled water (condensate), had been used in its manufacture.

Special sugars

There is also a limited though attractive market for 'yellow crystals', 'coffee' or 'Demerara' sugar in the confectionary trade. It is produced only from syrups of high purity, and for many years the glittering yellow cast of the crystals was achieved by the addition of stannine (stannous chloride). Recently, however, ordinances regarding the tin content of food have caused stannine to be replaced by titanous chloride.

The manufacture of raw sugar has been described in many books, that by Spencer and Meade in its several editions being the chemists' 'bible'. Yet apart from the papers by Davies (1938) and Yearwood (1940), little has been written about the production of yellow crystals. It might be useful, therefore, to record a short description of their manufacture.

Clarification:	Mixed juice of pH 5.0–5.5 is sulphited to pH 4.1–4.4. Phosphoric acid is added, followed by milk of lime, until the pH is 6.8–7.0. The juice is heated almost to 104 °C and then filtered.
Evaporation and Screening:	The same as for raw sugar.
Crystallization:	Magma footing is topped with 10 per cent yellow crystals 'A' molasses and sodium tetrathionite is added, but the massecuite is not allowed to cool.
Curing:	A dilute solution of titanous chloride is added to the massecuite in the basket, which is then purged at 1,100 – 1,200 r.p.m.

The analysis of a typical yellow crystal is as followed:

	per cent
Sucrose	98.08
Invert sugar	0.65
Ash	0.26
Moisture	0.35
Organic matter	0.66

The content of titanium is roughly 500 p.p.m. and the grist (size of the crystals) 1.76 mm.

To produce a really good sugar of this type there are two requisites:

1. The cane must be fresh; if stale, no chemical can produce the brilliance required of the crystals.
2. Automatic liming and temperature control instruments should be used; if the pH exceeds 7.0 at any stage in the process, the final product will be ruined.

Khandsari

The manufacture of *khandsari* is a cottage industry in India. Juice, extracted by primitive animal-driven mills, is subjected to simple clarification, boiled to the consistency of a thick syrup, and allowed to stand until sugar crystals are formed. The small, light yellow crystals are then separated in hand-operated centrifugals and dried in the sun.

Non-centrifugal sugars

It is in Asia, South America and Africa that non-centrifugal sugars are made for direct consumption: *gur* in India, *desi* in Pakistan, *panela* in South America and *jaggery* in Africa. Because these sugars do not form part of international trade their importance is often overlooked. Production statistics, for example those of the ISO, refer only to 'centrifugal sugars', but it could well be that most of the cane grown in the world is for the manufacture of low-grade non-centrifugal sugars. For example in Colombia in 1972, 80 per cent of the 462,000 ha planted with cane was for the production of *panela* in thousands of small factories. The difficulty of compiling reliable figures is illustrated by the following: in some South American countries the price of centrifugal sugar is controlled by law, but that of *panela* is not; consequently an undetermined quantity of refined sugar is reprocessed and converted into the more profitable *panela*. These sugars vary as much as their names.

Panela is produced in cakes, roughly 15 cm in diameter, which resemble rock buns. The factories in which it is produced have changed little since the eighteenth and early nineteenth centuries. They comprise open-hearth. *taiches* from which the increasingly concentrated liquor is transferred from one vessel to the next with intermittent skimming, until a glutinous mass is extracted from the last vessel, formed into cakes and left to cool.

Jaggery and *gur*, on the other hand, are the products of cottage industries. The final product contains roughly 10 per cent water and has the disadvantage of being hygroscopic. The term *desi* is more difficult to define. It is all-embracing and covers a range of products which vary from molasses-like substances to crystalline sugars.

Liquid sugars

For many years a syrup known as 'fancy molasses' has been made in Barbados for export to the New England states of the USA, the Maritime Provinces of Canada and Newfoundland. The name 'fancy molasses' was preferred because of the higher duty levied on imports described as 'syrup'. Similar products, known as 'edible molasses' are made in Louisiana and Georgia in the USA, in Cuba and also in South Africa.

Fancy molasses

Fancy molasses used to be exported from Barbados by schooner and for many years one of the sights of Bridgetown was that of 'spiders' carrying the puncheons to the harbour. The reciprocal cargoes were salt fish and wooden shingles. Shipment is now made in tankers.

A comprehensive history of this industry was written by Taussig (1940). Fancy molasses, to be confused neither with final (blackstrap) molasses nor choice molasses (a by-product of the manufacture of muscovado sugar), is made from clarified, concentrated juice from which no sugar has been extracted. To produce it sucrose is hydrolysed into the isomers glucose and fructose (invert sugar) by the addition of an acid:

$$C_{12}H_{22}O_{11} + H_2O \rightarrow {}_6H_{12}O_{6+} C_6H_{12}O_6$$

sucrose   water   glucose   fructose

For many years acetic acid (vinegar) was used for this purpose. Cane juice was set aside and the sucrose in it converted first by yeast into alcohol, and later by bacteria into acetic acid. It was a wasteful process and in the 1930s sulphuric acid gradually replaced fermented juice as the promoter of hydrolysis. The stronger sulphuric acid required more accurate process control. If too much was added inversion would proceed so far that glucose might crystallize in the puncheons. Similar disasters could occur if the syrup was too highly concentrated. Consequently limits of not less than 30°–35° pol (checked by chemical methods in the smaller plants, where a polarimeter was a luxury) and a density of not more than 41.5° Beaumé were set.

The price premium attracted by fancy molasses allowed several windmill and small steam plants in Barbados to continue to work for many years, and it was not until 1946 that the last windmill plant ceased to operate. The production of fancy molasses is now confined to one factory and much stricter quality control as being attempted.

The standards set are as follows:

1. Colour, clarity and flavour: as high as possible, with a light absorption value of not more than 60 per cent.
2. Temperature: not more than 43°C (110°F), measured on discharge.
3. Density: not less than 41.5° Beaumé (at 27.5°C).
4. Polarization: between 24° and 34°.
5. Total sugars: not less than 70 per cent (sucrose plus reducing sugars as determined by volumetric analysis).
6. Ash: not more than 2 per cent (ash after sulphitation, multiplied by 0.9).
7. Sludge: not more than 0.5 per cent by volume (determined by standard centrifuge procedure).
8. Not more than 94 per cent of the total soluble solids must be sugars: it is a requirement of the US Customs that the ratio

$$\frac{100 \times \text{soluble solids not sugar}}{\text{total soluble solids}}$$

must exceed six.

The considerable measure of success achieved can be judged from the data given in Table 10.3. In production statistics prior to 1954 fancy molasses was equated to raw sugar on the basis of 330 US gallons (known locally as wine gallons)/ton, but from that year the ratio was changed to 290 gallons/ton. An account of the vicissitudes of this unusual, though not unique industry in Barbados was described by Saint (1953).

In 1976 the recovery was 35.51 gallons/tonne cane (for comparison 8.46 tonnes of cane were required to make 1 tonne of

Table 10.3 *Analyses of crop composites of fancy molasses in Barbados, 1975–78*

Year	Light absorption (%)	Density (°Beaumé)	Pol	Total sugars (%)	Ash (%)	Sludge (%)	US Customs ratio
1975	63	41.5	28.2	72.5	1.48	0.5	6.8
1976	79	41.5	27.5	71.4	2.03	0.6	8.2
1977	72	41.6	27.9	72.5	2.01	0.6	7.6
1978	76	41.6	29.8	70.1	2.23	0.6	10.0

Note: In 1978 the daily average temperature on discharge never exceeded 43 °C (110 °F), and the crop average was 38 °C (100 °F).
Source: Adapted from Smith & Brooks (1978).

sugar). The equivalent of roughly 6,000–8,000 tonnes of raw sugar was manufactured as fancy molasses in 1976 and 1977.

Diffusion

Although the traditional method of extracting juice for the production of sugar is to crush cane in a series of mills in tandems, there is an alternative: extraction by diffusion.

For many years this process was thought to be misnamed: it was thought that extraction was by lixiviation (washing) and not by dialysis (diffusion). Graham *et al.* (1969), however, presented evidence that the mechanism is in fact molecular diffusion. Now it is generally accepted that both washing and a type of diffusion (not necessarily dialysis) are involved. Whichever theory is correct, the term 'diffusion' has currency within the industry and will be used here.

In the older diffusion plants the cane is crushed in one or more three-roller mills, in which roughly 70 per cent of the juice is expressed, and then the bagasse is subjected to continuous treatment with water or dilute juice. This takes place in a tank (a diffuser) divided into seven or eight compartments (eleven in the more modern De Smet type) by baffles, towards the bottom of which is a conveyor-belt. Either hot or cold water is sprayed on to the mat of bagasse in the last compartment, percolates through, and removes some of the residual sugar and other soluble substances. The dilute juice is then collected, sprayed on to the bagasse entering the previous compartment and becomes enriched as it filters through. It is again collected and the process repeated until the most concentrated solution is applied to the bagasse emerging from the mill(s). Thereafter the mixed juice is clarified and sugar extracted in the conventional manner. The countercurrent method of extraction used in diffusion is remarkably similar in concept to compound maceration.

The bagasse emerging from the diffuser contains too much water to be readily combustible. It is therefore passed through either one or two three-roller (dewatering) mills, its moisture content is reduced to approximately 50 per cent, and it is then fed to the boiler furnaces. The expressed liquid is sprayed on to the bagasse in the second and third compartments of the diffuser (Fig. 10.7).

Diffusion first found favour mainly in Egypt. Walter (1954) compared the cost and production efficiencies of two conventional mills with those of two diffusion plants (Abou Kourgas and Nag Hamadi), all owned and operated by the Société Générale des Sucreries et de la Raffinerie d'Egypte. He concluded that diffusion plants were more effective, provided that the cane was of good

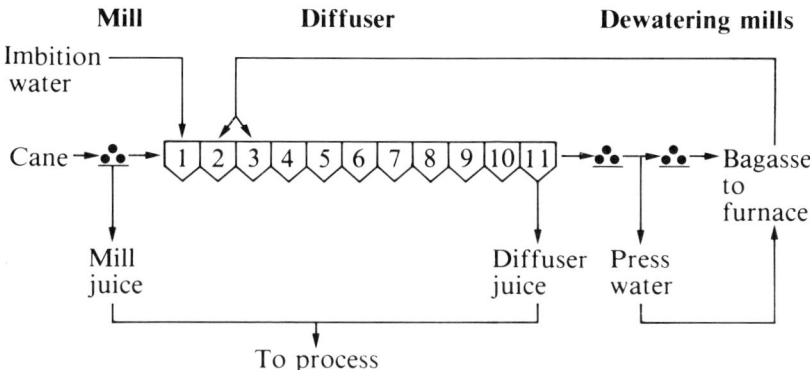

Fig. 10.7 The flow diagram of a simple diffusion plant

quality and that there was an abundant supply of cheap labour. However, in the discussion which followed the presentation of a paper on the same subject by Schmidt & Wise (1956), none could report on the operation in Egypt. Reference was made only to two plants, one in Trinidad and the other in India; both had been discarded long ago. Nevertheless some twenty years later Sayed *et al.* (1974) insisted that the diffusion system was superior to the mill train, and supported their contention with data from the Kom-Ombo factory in Egypt, in which juice is extracted by both methods. Moreover, they made no stipulation concerning the quality of the cane or the cost of labour; they based their arguments on lower capital and operating costs, and on the better extraction of higher quality juice.

By that time interest in diffusion had been shown in many other parts of the world. Idehara (1966) described the development of a ring diffuser at the Pioneer Mill in Hawaii and Payne (1969) reported that, in 1968, six factories in several different countries were using this newly designed system. The novel features of the ring diffuser are:

1. After preliminary light knifing the cane is passed not through a mill but under two sets of swing hammers: the Buster and the Fiberizer.
2. The prepared cane is delivered on to a rotating bed through which liquid is passed in eighteen stages. Again, starting with water, the countercurrent method allows it to become increasingly enriched with sugar as it moves from compartment to compartment.

Payne's paper was part of a symposium on diffusion presented at the thirteenth congress of the ISSCT. Several types of commercial diffusers were on the market and Wu (1969) described some of them, including the Silver Ring (Hawaii), De Smet (South Africa), DDS (Denmark), BMA (South Africa) and Fairymead (Australia). He placed them in two categories: those with and without pre-extraction; and subdivided the former into those in which the press water is returned to the diffuser, as described, and those in which it is heated, limed and passed to the clarifier. Typical flow diagrams are given in Fig. 10.8.

Of fourteen new factories built throughout the world during the period 1965–72, eight were equipped with diffusers and six with conventional milling tandems. By 1973 ten diffusion plants were in operation in the Philippines and six in South Africa, with a seventh on order. Diffusers were also being used in several other countries: Australia, Mauritius, Nicaragua, Tanzania and Venezuela; and, of course, in Egypt and Hawaii. However, this picture could be misleading: during recent years many conventional mills from declining areas of production have been relocated elsewhere, several with increased capacity. For example in 1978 Constancia factory from Puerto Rico started to operate in the Risaraldo Valley of Colombia.

Fitzgerald & Lamusse (1974) reviewed the development of diffusion in South Africa, discussed the results achieved and reported that the advantages were the same as those claimed by Sayed *et al.*:

1. Reduced capital costs.   R1,200,000
Three mills with drives (250 t.c.h.)
    One bagasse diffuser   R 500,000
                  Saving   R 700,000
2. Reduced maintenance costs
Two mills   R   67,000
One diffuser   R   29,000
                  Saving   R   38,000
3. Higher extraction of sugar.

Nevertheless not all technologists are convinced of the superiority of the diffusion system, or even of its viability.

The amount of sugar material in process at any given time is very much larger than in a conventional factory; therefore heavier losses are sustained if breakdowns, or stoppages for any reason, occur. A second factor which must be taken into consideration is the greater rigidity of a diffuser: it is not easy to adjust its throughput, whereas that of a conventional mill can be increased or decreased with relative ease. The advantages and disadvantages of diffusion will continue to be discussed with great interest.

Separation

An entirely new concept of manufacture, the separation process, is being developed in Barbados with the help of Canadian capital. In

With pre-extraction and treatment of press water

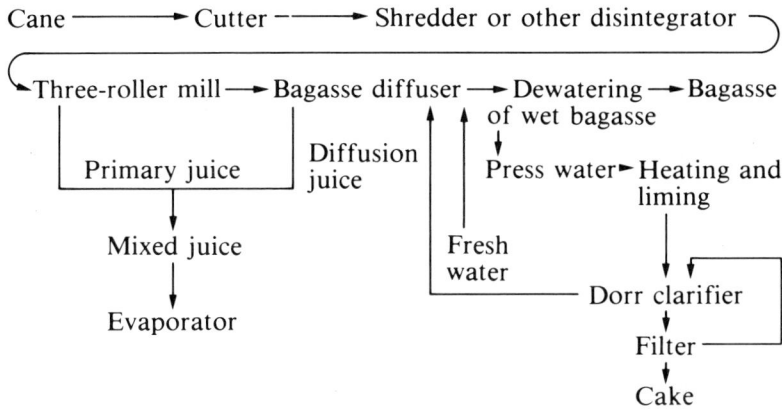

With pre-extraction and without treatment of press water

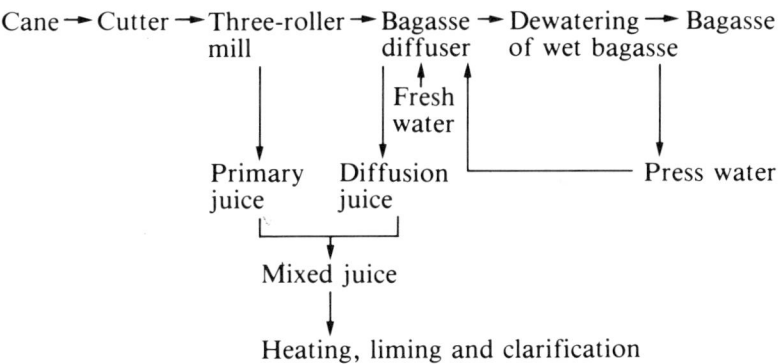

Without pre-extraction and with treatment of press water

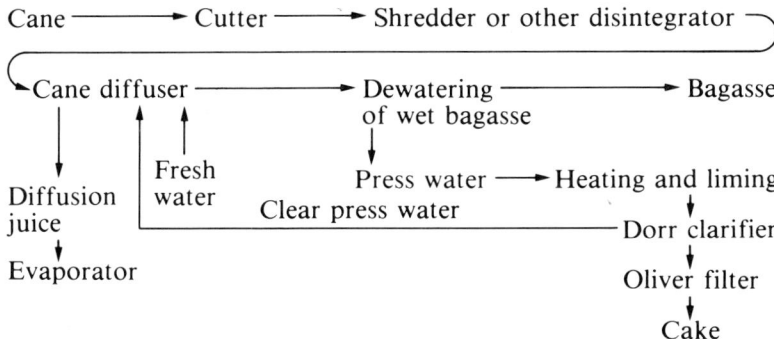

Fig. 10.8 The flow sheets of different types of diffusion plants (After Wu 1969)

it the cane stems, having been cleaned, are divided into three primary components before any juice is extracted. These components have been given the names dermax, comrind and comfith. Dermax is the cuticle, with adhering wax; comrind is the ring of dense, woody fibre (the rind) which lies immediately beneath the cuticle; and comfith is the inner matrix of parenchyma, with fine intravascular fibres (the pith), in which most of the sugar in the stems, 84 per cent or more, is stored. Each may be processed separately: dermax as a source of wax; comrind, having been purged of its sugar, is suitable material for the manufacture of particle board and/or paper; and comfith for sugar.

Fuller (1976) described the new process. First the stems are cleaned, aligned and chopped into billets roughly 30 cm in length. Then the billets are fed into separator machinery in which they are split in half, lengthways. The half-billets continue through the separator, comfith being removed from the inner side and dermax from the outer, leaving a strip of comrind. The three components are discharged independently.

The advantages claimed for the separation process are:

1. Dermax, only 3 per cent of the cane by weight, contains wax in a readily recoverable form, whereas its extraction from the filter mud produced by conventional milling is complex and costly.
2. Juice from comrind, containing roughly 14 per cent of the sugar in the cane, is recovered, clarified and combined with that from comfith. The bundles of long fibres of high mechanical strength which remain can be made into veneers and panel boards with little effort and without the use of large volumes of water.
3. The pol extraction is greater than that achieved by traditional methods.

Scale of factory operations
H. A. Thompson

A major problem in developing sugar production is the marked increase in the cost of factory buildings and machinery, even compared with only a few years ago. The minimum economic output size of a raw sugar factory is now in the region of 50,000 tonnes and, with a financing cost in the $1,500 per tonne of sugar output, the raising of finance for a project is a major undertaking (figures refer to end of 1977).

The cost of providing the local infrastructure in an undeveloped area can be high and can reach 15 per cent or more of the total development costs. As a result of the increasing cost and the difficulties in providing the infrastructure, in a reasonable period of time, alternate ways and means are being considered of arranging and planning developments.

As a result there has been an increasing interest in the possibility of establishing a number of mini factories – 'khandsari sugar mills' – with cane requirements in the region of 200 tonnes/day, compared with the 3,000–4,000 tonnes required for an economically sized full-scale factory, or the 700–1,000 tonnes/day capacity of the small factories which were common in past times. The main attraction is that the mini factories could be established without an extensive, and expensive, infrastructure and a gradual build-up of the cane area could be achieved, to provide, in time, the eventual cane supply for a major producing unit.

The mini-factory concept is rather different from the usual *jaggery* plant, in that the quality of sugar produced can be much higher. The operation of a mini factory is also a much more sophisticated process compared with the *jaggery* plant since it involves juice sulphitation, and the use of crystallizers and centrifuges, with the major difference, compared with a large-scale factory, being the use of open boiling pans instead of evaporators operating under reduced pressure. Undoubtedly a much higher level of technology is required to produce direct consumption sugar from *khandsari* plant, compared with that required to produce *jaggery* from a plant crushing about 1–2 tonnes/hour.

In actual terms of capital cost, related to the cost per ton of milling capacity, the mini factories may not be less expensive than a major sugar factory. In terms of sugar recovery, relative to the sugar content in the cane, the mini factory is significantly less efficient with a recovery of the order of 60 per cent or less, compared with the 78–84 per cent recovery in a conventional factory. In some circumstances, in terms of the economic use of natural resources, it could be difficult to justify the establishing of mini factories, but the practicalities of a particular situation may well justify their use.

It is clear that the assessment of the development potential and the form development should take cannot be judged solely, or mainly, in economic terms. The strategy of Government, relative to the needs to give priority to the development of specific areas, the overall planning of development of roads, irrigation water supplies and infrastructure in general, all may have a controlling influence on both the location and form appropriate for the development of a particular area.

Storage and Handling

Molasses

Molasses is stored at the factory or, if required for export, is transported by road or rail to tanks located close to deep water. No special precautions are required for its safe keeping. On first delivery to a tank its temperature should not exceed 43 °C (110 °F), otherwise excessive frothing might occur.

Direct-consumption sugar

Sugar for direct consumption may be delivered in bulk to those requiring large quantities, for example to manufacturers of confectionery or condensed milk; in packets or polythene lined bags to retail merchants; and in specially designed sachets to large hotels, airlines and the like. For these also no special precautions are required, except those dictated by common sense; but a high standard of hygiene must be maintained.

Raw sugar

The storage and handling of raw sugar is another matter. For centuries raw sugar was exported to the refineries in wooden containers variously called hogsheads, tierces or barrels, each of 701 or 812 kg (14 or 16 cwt) capacity. These were replaced by jute bags in the late nineteenth and early twentieth centuries; but in recent years, as with other commodities, the trend has been towards shipment in bulk rather than in bags.

Much experience has been gained in the design and operation of bulk stores. They must, above all, be waterproof. Attention must also be paid to reception, weighing and sampling; to the formation of heaps of sugar and the prevention of caking by frequent movement; to the conveyor systems: one to place sugar in the store and the other to discharge it evenly by 'throwers' into the holds of bulk carriers. Osborne (1967) described the store built at Point Lisas in Trinidad in 1965, with comments on the installations in Barbados and Guyana. It is illustrated in Figs. 10.9 and 10.10.

Raw sugar consists of sucrose crystals, each surrounded by a film of molasses. The degree of supersaturation of the molasses film is such that, if it could be maintained, microbiological action would not take place. This is difficult to achieve because of the presence in the film of several hygroscopic substances, the most significant of which is invert sugar. When sufficient moisture has been absorbed from the air the film will support the growth of micro-organisms which promote inversion. It then becomes further diluted and the rate of deterioration of sucrose is compounded. Sherwood & Hines (1951), working in Australia, found that almost all the species of *Aspergillus* (*glaucus, niger, terreus* and *wentii*) isolated from raw sugar were capable of causing inversion, and that *A. glaucus* was by far the most effective. Other fungi and bacteria were relatively unimportant.

The characteristics which affect the keeping and handling quali-

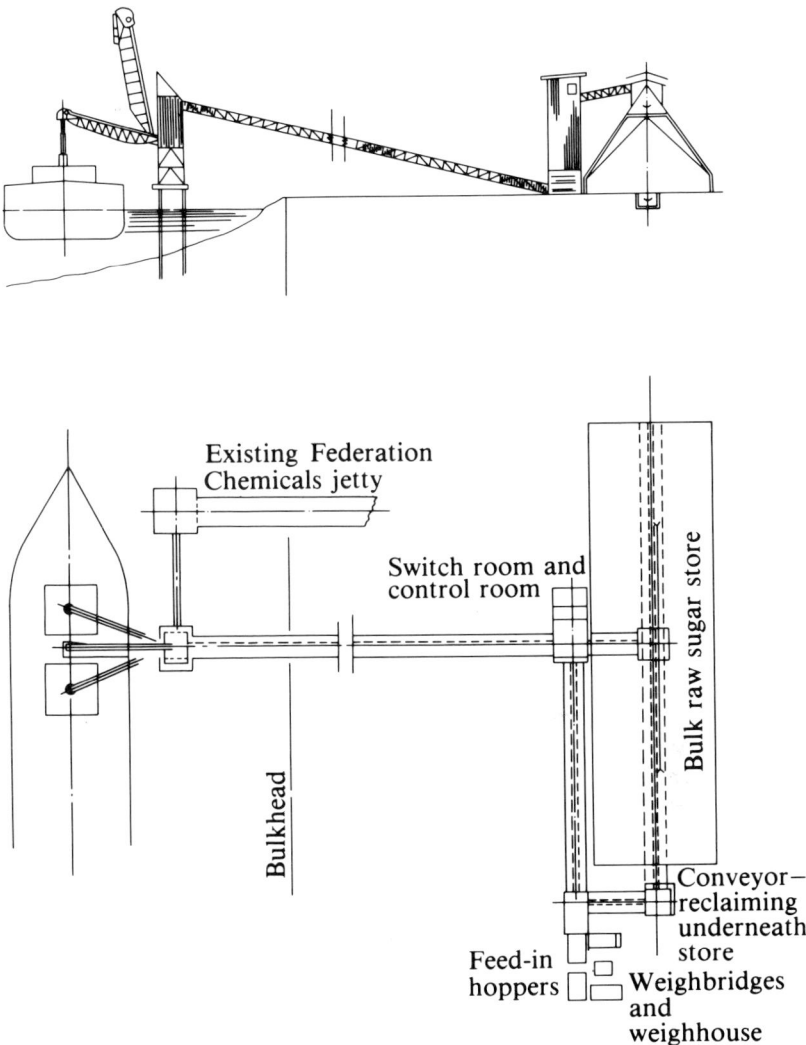

Fig. 10.9 Layout of a typical bulk loading installation

ties of raw sugar were examined in detail by Saint (1960) prior to the building of a bulk store in Barbados. They are:
1. Size and uniformity of crystals.
2. Moisture content and humidity.
3. Temperature.
4. Cleanliness.

Fig. 10.10 Point Lisas bulk loading installation, Trinidad

Size and uniformity of crystals

The smaller the crystal, the larger is the ratio of molasses to sucrose and consequently the danger of deterioration; and the greater the divergence in size, the higher the probability of compaction under pressure. Using the methods recommended by Powers (1948), Saint & Trott (1960) determined the size of crystal (mean aperture) and uniformity (coefficient of variation) of the sugar produced by sixteen factories in Barbados. In the light of experience from other countries in which sugar was handled in bulk, it was decided that the mean aperture should be not less than 0.7 mm and, of course, that every effort should be made to achieve uniformity of size.

These factors are especially important if sugar from several factories, under different management, is being delivered to one store. Where there is central control of manufacture there is much greater likelihood of consistency.

Moisture content and humidity

A quantitative assessment of keeping quality, called the safety factor, was first devised at the Colonial Sugar Refining Company in Australia. The safety factor is a measure of the dilution of the molasses film: the ratio of moisture to all non-sucrose substances.

$$\text{Safety factor} = \frac{\text{moisture per cent sugar}}{100 - \text{pol}}$$

Some questioned the formula, others attempted to elaborate on it.

The most widely accepted alternative, the dilution indicator, is the ratio of moisture to all non-sucrose substances except water (and for good measure, though for no apparent reason, it is also multiplied by 100). Thus:

$$\text{Dilution indicator (DI)} = \frac{100 \times \text{moisture per cent sugar}}{100 - (\text{pol} + \text{moisture per cent sugar})}$$

In Barbados it was laid down that the DI should be not more than 40; penalties were imposed on consignments with DIs between 40 and 50, and those with DIs above 50 were rejected. This is an agreement with the guide commonly accepted throughout the industry that, for good keeping, the safety factor should not be greater than 0.33.

Much thought has been given to means of controlling humidity in the store. At Salt River in Jamaica louvres are opened from 10.00 a.m. to 4.00 p.m. on sunny days, to let in air of low relative humidity; at all other times they are closed. Easton (1968) discussed the possibility of keeping sugar in airtight stores and of reducing the humidity by condensation, using refrigerated coils, or by heating. However, experience has shown that humidity is unimportant, even in wet tropical countries, provided that the other criteria for safe keeping are met.

Temperature

The temperature of sugar entering the store affects its keeping quality and its colour. When considerable difference exists in a heap, moisture migrates from the warmer to the cooler parts. This can lead to concretion and to colour formation where the moisture content of the molasses film has been depleted, and to inversion where it has increased. Colour is critical if exports are sent to refineries in the USA: penalties are imposed when certain levels are exceeded.

In Barbados it was decided that the temperature of sugar received at the store should be not greater than 43 °C (110 °F), a limit which later was found to be satisfactory in practice. Where cooling was necessary, equipment was installed at the factory to blow air on to the sugar as it ascended the elevators to fall into the transport bins.

Cleanliness

It is generally agreed that cleanliness also influences keeping quality; and that cleanliness can be judged by the proportion of insoluble matter in the sugar and by the kind and number of micro-organisms present. The two are probably closely connected because cush-cush (the organic substances, mostly fibrous, remaining in suspension after inadequate clarification), which with mud forms the bulk of insoluble matter, increases the moisture content of the sugar and is a breeding ground for micro-organisms. Spencer & Meade (1945) regarded sugar with less than 40 mg insoluble solids per 100 g as 'extremely clean'. Sugar with more than 100 mg/100 g was 'poor' and a reflection of deficient clarification. In Barbados a tentative limit of 70 mg/100 g was set with the hope that, given time, this might be reduced. It was. The present limits are that cush-cush should not exceed 9 mg/100 g and mud 32 mg/100 g.

Standards having been set with great care, it is gratifying to read in the Barbados Sugar Technology Research Unit, 27th Annual Report, 1976 that in 1976 no load of sugar was rejected at the store and no penalty imposed on any shipment sent to the USA. Statistics concerning the characteristics of sugar received at the store during the period 1972–78, adapted from that report and further data given by Smith (1978), are recorded in Table 10.4.

Sugar in bulk stores is remarkably free from damage by insect pests.

Table 10.4 *Characteristics of raw sugar received at the bulk store, Barbados, 1973–76*

	1973	1974	1975	1976	1977	1978
Grain size						
Mean aperture (mm)	0.70	0.69	0.75	0.73	0.68	0.67
Coefficient of variation	34	34	34	33	35	36
Moisture						
Dilution indicator	30.3	33.3	34.1	31.3	31.6	29.2
Temperature						
Average (°F)	92.9	98.4	99.1	100.3	100.9	98.6
(°C)	34.9	36.9	37.3	37.9	38.3	37.0
Cleanliness						
Cush-cush average (mg/100 g)	9	16	9	7	6	6
Mud average (mg/100 g)	27	40	25	21	16	19

Chapter 11

By-products and sucrochemistry

The by-products of sugar manufacture: molasses, bagasse and filter mud, are used in many ways; and sugar increasingly is regarded not only as food for man but also as the basic raw material for a burgeoning sucrochemical industry. Paturau (1969) has written the standard work on this subject.

Molasses

Final or blackstrap molasses is the dark, viscous liquid discharged from the 'C' centrifugals when all the sucrose which can be recovered from the syrup by economic means has been removed. Estimates on which equipment is installed are based on a yield of 23–28 litres of molasses/tonne cane ground; but, as already described, vastly increased quantities must be handled when stale, burnt cane is processed.

Because molasses contains the greater part of the concentrated non-sugars in cane juice, as well as the commercially unextractable sugars, its qualitative analysis reflects that of the cane juice from which it was made. This, in turn, depends on variety, soil, climatic conditions and also on process treatment. Consequently there is considerable variation in the composition of molasses from different countries, and even sometimes from neighbouring factories. An example is given in Table 11.1. The main constituents are sucrose (roughly 32–36 per cent by weight); glucose and fructose (20–30 per cent); and water (20 per cent). The mineral content also varies but, in general, about 4 per cent by weight is potash (expressed as K_2O). Detailed analyses of the minerals in molasses from Louisiana, Hawaii, Argentina and Java are given by Meade & Chen in their *Cane Sugar Handbook*.

It is the sugars and, in much less degree, the potash in molasses which make it a valuable by-product. The sugars and other organic matter can be used to augment the carbohydrate content of animal fodder; they can also be converted into many chemicals, of which

Table 11.1   *The difference in composition of final molasses from nine factories in Barbados, 1978*

	Per cent by weight			
Range	*Sucrose*	*Glucose and fructose*	*Water*	*Sulphated ash*
High	36.48	30.19	19.95	11.29
Low	31.93	22.34	17.18	9.81

Source: Adapted from Smith (1978).

ethyl alcohol is by far the most important; and potash is essential for efficient plant growth.

Although statistics concerning the production of molasses are not well documented it is reasonable to assume that 20 m. tonnes are produced annually as a by-product of the manufacture of roughly 50 m. tonnes of centrifugal cane sugar. To this must be added an equal or greater quantity, impossible to estimate, from the manufacture of low-grade sugars (p. 313).

A comprehensive description of the composition, properties and uses of molasses, together with trends in world trade, has been given by Baker (1974).

World trade in molasses

Most of the molasses is used locally as animal feed or in the production of alcohol: only 6 m. tonnes per annum enter into international trade. The major importing areas are the USA, Western Europe and Japan, which receive molasses from countries having large sugar-cane industries (Table 11.2)

The use of imported molasses

By and large the pattern of molasses usage is much the same in the USA and in the UK, but is quite different in Continental Europe and in Japan.

In the immediate post-war period 40 per cent of molasses consumed in the USA was for the production of alcohol by fermentation; now only 2 per cent is used for this purpose. The decline was caused by the increasing production of cheaper synthetic alcohol from ethylene. In recent years most of the molasses, 78 per cent, has been fed to animals, and 13 per cent is used in the manufacture of baker's yeast and citric acid. The rising demand of the soft drinks industry for citric acid has been offset by a decline in the market for yeast. Nevertheless, because of the need for animal feed, which has risen sixfold since 1961, there has been a steady overall increase in

Table 11.2 *World trade in sugar-cane molasses*

Major importing areas	Estimated imports 1976/77 (tonnes)
USA	2,200,000
EEC (the 'Six')	1,500,000
Japan	1,000,000
UK	600,000
Scandinavia	300,000
Canada	200,000
South Korea	200,000
Caribbean area	100,000
Taiwan	100,000
	6,200,000

Major exporting areas	Estimated exports 1977/78 (tonnes)
South and Central America	1,200,000
The Far East	1,100,000
Indian Ocean area	900,000
Brazil	900,000
Caribbean area	700,000
Africa	600,000
Australia and Fiji	300,000
Hawaii	300,000
	6,000,000

Source: From Baker (1978).

the consumption of molasses in the USA since 1961, with temporary recessions in 1965, 1969 and 1971.

A similar position in the UK is illustrated by data given in Table 11.3.

In the Continental Europe, on the other hand, the use of molasses for the manufacture of alcohol has increased. This requirement, together with a rapidly developing market for animal feed, caused imports to rise from 1 m. tonnes in 1974 to 1.5 m. tonnes (estimated) in 1976/77.

In Japan, also, large quantities of molasses are used in the manufacture of alcohol, though pollution difficulties in recent years have led to the importation of substantial quantities of alcohol itself. Nevertheless 55 per cent of molasses consumption is still for this purpose and for the production of yeast; 33 per cent is fed to animals, and 5 per cent is used in the manufacture of monosodium glutamate.

The cost of production of molasses is absorbed in that of the manufacture of sugar; therefore an important element in its price is transport, especially ocean freight charges.

Table 11.3   *The average annual usage of molasses in the UK (1967–71)*

	Tonnes	%
Animal feed	628,000*	74
Citric acid and yeast	166,000	20
Alcohol	55,000	6

* Including 226,000 tonnes of beet molasses.
Source: From Baker (1978).

Molasses as a feeding-stuff for animals

Because of its high carbohydrate content and palatability molasses is a valuable animal food and, as already indicated, is used as such not only in sugar-producing countries but also in North America, Europe, Japan and indeed throughout the world. Its starch equivalent is 52 (or 9 mJ/kg), which is roughly 80 per cent of the energy value of barley grain.

By itself molasses is not a balanced ration, being deficient in protein. Taking advantage of the ability of ruminants, through the microflora of their rumen, to convert non-protein nitrogen into compounds which can be assimilated by their digestive systems, attempts have been made to enrich molasses with inorganic nitrogen. It has long been known that ammonia reacts with reducing sugars and that much of its nitrogen is retained in organic combination. Therefore, many studies have been made on the feeding of ruminants with ammoniated inverted molasses (AIM). The literature was reviewed and the results of experiments recorded by Joblin *et al.* (1954 a and b) and Howes *et al.* (1955). It was found that AIM (nitrogen content 4–5 per cent) caused many of the animals to develop toxic symptoms described as 'epileptic hysteria', due probably to the presence in the feedstuff of pyrazines and imidazoles of small molecular weight. Because of this side effect, little further work on AIM has been carried out. Nor is it necessary: to make a more balanced feed it is now known that urea can be mixed with molasses to form a safe food which can provide one-third of the protein requirement of cattle and sheep.

Local use as animal feed

In most tropical and subtropical countries molasses is diluted with small quantities of water (to perhaps 82 ° Brix) and fed directly to draft animals and to dairy and beef herds. However, there is increasing use, especially in South Africa, of an efficient, wholesome feed comprising mixtures of molasses and urea absorbed on bagasse pith in the ratio of three parts of molasses mixture to one of pith. This ration is supplemented in the dry season, when good pasture is scarce but which coincides with the cane harvest, with an abundance of cane tops.

Very little molasses is made into concentrates.

Export use as animal feed

It has been mentioned already that most of the molasses of international trade, 60 per cent or more, is used as animal feed. As in the producing countries, some is fed directly to beef and dairy cattle, sheep, horses, pigs and poultry; but much is mixed with other ingredients and made into more sophisticated feeds. These take the form of cubes, pellets and crumbles which might comprise as much as 15 per cent molasses. Quite apart from its nutritive value, molasses has several desirable qualities which enhance its value when used in this way. It stimulates the microflora of the rumen and allows cattle and sheep to deal effectively with low-grade roughage such as straw; it is palatable to all classes of livestock and by its flavour and smell promotes their appetites. In addition its physical properties cause dust particles to be absorbed, and thereby help to prevent the development of bronchial diseases, particularly those to which ruminants are prone. Molasses also lubricates the dies used in the manufacture of cubes and the like and helps to preserve their structure, especially when they are being stored or transported.

Fermentation products

Commercial products made by the fermentation of molasses are ethyl alcohol (alcohol), carbon dioxide, citric acid, baker's yeast, food yeast, monosodium glutamate, itaconic acid, acetone and butyl alcohol (butanol). Of these alcohol, in the form of rum in producing countries, and as industrial alcohol in both importing and exporting countries, is by far the most important.

Rum

Manufacture in the eighteenth century

For centuries a potable alcoholic liquor, rum, has been made from the by-products of the manufacture of raw sugar. Two hundred years ago the main centres of the rum industry were in the West Indies (especially Jamaica) in Britain and in North America. Because of the different composition of the raw materials used, Jamaican rum was more highly flavoured and better received than that distilled in more temperate climates, and eventually production was confined to countries in which raw sugar was produced. Long (1774) described the reasons for the differences between the two types of rum:

The melasses spirit, distilled in *Britain* and *North America*, is so defective in the *volatile oil*, which is the great corrective, and gives the characteristic to rum, that it is most palpably different from it in taste and flavour, as well as in its most salubrious qualities. For this reason the North American spirit is better than the British; the former being made from the first-drawn melasses, which generally contains a portion of sugar, and a large share of this oil. The French melasses indeed is impoverished very much, by their boiling it over again, to make their *paneel* sugars; but in Jamaica this piece of economy not being practised, the melasses sold here to the North Americans is twice as rich as what they purchase at the French islands; and their distillers probably find it so in the yielding.

In Britain the melasses is proportionably jejune, and deprived of its richness; as the muscovado sugars, by the time they fall into the baker's hands, have been pretty well drained; so that what is drawn in the refining process, and afterwards sold to the distillers, must be very much impoverished.

Some distillers buy up the dark uncured sugars, which yield a spirit of better quality; but it is impossible for them to produce the same spirit as Jamaica rum, where the liquor for distillation is compounded of various mixtures, not to be obtained by the British distiller.

This liquor, for example, consists of:
1-part skimmings,
1-part washings,
1-part cool lees.

To these variously compounded, according to the particular judgement of the manufacturer, and other circumstances, the melasses is added during their fermentation in the cisterns, and in the proportion of about six gallons of melasses to every hundred gallons of liquor.

Sometimes it is made wholly of crude cane-liquor and melasses, run into fermentation together.

So that not only ingredients are various and differently compounded; but the melasses, which is the principal or only substance used in Britain and North-America, bears in Jamaica but a very small proportion to the other ingredients, being only as 6 to 100, or thereabouts.

At this time the daily issue of rum to Royal Navy personnel at the Jamaica station was one half-pint (285 ml). Admiral Vernon, called 'Old Grog' because of the grogram boat cloak which he wore, insisted that the rum be diluted 1 : 3 with water before being dispensed and in 1740 issued a Fleet Order to that effect; hence the name 'grog' for the drink. This use by seamen probably led to a general preference in those days and for many years afterwards for a heavy, highly flavoured rum containing a high content of Long's 'volatile oil'; whereas the market now favours a much lighter blend.

A description of the manufacture of rum in the eighteenth century, in much greater detail, was given by Edwards (1793). In summary it was:
1. The distillery of a large plantation comprised:
A dunder cistern of 3,000 gallons capacity (to provide yeast).
A cistern for the scummings (from sugar manufacture).
Twelve fermenting vats, each of 1,200 gallons capacity.

Two copper stills, one of 1,200 gallons and the other of 600 gallons capacity.

Two pewter worms (the condensers), preferably placed in running water, otherwise in stone tanks.

2. The wash consisted of:

Dunder	50 gallons
Sweets Molasses	6 gallons
Scummings	36 gallons
Water	8 gallons
	100 gallons

(Edwards also used the older spelling 'melasses'.)

3. By the addition of hot or cold water the heat of fermentation was kept within the limits of 32–34 °C (90–94 °F). After approximately seven days the liquor was placed in the large still and heated. Transparent spirit called 'low-wines' was allowed to emerge from the worm until it was no longer inflammable. The low-wines were then transferred to the smaller vessel and redistilled, producing each day 220 gallons, or 2 puncheons, of oil-proof rum, i.e. spirit in which olive oil would sink.

4. The expected proportionate yield was 200 gallons of Jamaica proof rum to 3 hogsheads (of 16 cwt each) of sugar.

Modern manufacture

Pot stills, moonshine and mountain dew Vessels known as pot stills, which are copper retorts not unlike those described by Long and Edwards, are used to produce heavy-bodied, highly aromatic rum from fermented molasses. During distillation little attempt is made to separate alcohol from the other products of fermentation: fusel oil, aldehydes, esters and their congeners. Indeed their concentration often is increased by the bacterial decomposition of pieces of meat and dead animals thrown into the fermenting vats. A dark colour is given to the distillate by the addition of caramel (burnt sugar). The market for this type of rum has diminished greatly during the past twenty to thirty years.

Illicit rum, called 'moonshine' or 'mountain dew', is prepared by the distillation of fermented molasses or cane juice in pot stills of varying degrees of sophistication, yet always located in seclusion. If the raw material is cane juice, it is extracted in simple mills, hand or animal driven, but the bagasse must be hidden from inquisitive excise officers. When molasses is used such problems do not arise. Mountain dew is a colourless liquid, highly and to many (including myself) offensively flavoured. Above all it is extremely potent. As with *gur, panela* and *jaggery*, but also because of the secrecy in which it is made, production statistics are not available.

Column stills Most of the rum of legal commerce is now sepa-rated in column stills, and each stage of its manufacture is subject to rigorous scientific control. The process may be summarized as follows:

molasses + yeast + nutrient + water → rum + carbon dioxide + lees (dunder)

The ingredients Various qualities are sought of the basic ingre-dients. Molasses should be rich in total sugars but have a low mineral content. As the purpose of the factory chemist is to extract as much sucrose as possible from the molasses, his interests often conflict with those of the distiller. The presence of a high proportion of minerals in molasses, reflected in its ash content, inhibits fermen-tation; and besides slowing down the process contributes to the loss of unfermented sugars in the lees.

The yeasts used are strains of *Saccharomyces cerevisiae*. They act quickly and efficiently, even in the presence of high concentrations of alcohol. However, the various strains of *S. cerevisiae* give rise to markedly different end-products. For example one strain produces low-flavoured rum with a high yield of alcohol, while another does the opposite: its alcohol yield is poor but the rum is highly flavoured and rich in fusel oil, aldehydes and esters. Thus the strain used determines the type of rum produced, and is kept in pure culture in the distillery laboratory.

Nitrogen, essential for the rapid propagation of yeast, is supplied as ammonium sulphate, or as the cheaper but equally effective urea.

The water used in rum manufacture must be uncontaminated, and all liquid components must pass through inert conduits. A marked sharpness in the taste of the products of one distillery was eliminated after an artesian well had been drilled to supply pure water, and stainless steel pipes and vessels had been installed to replace those made of cast iron.

Fermentation Molasses is first diluted with water in the ratio 1 : 6; thereby its specific gravity is reduced from 86 °Brix to 17 °Brix. The wash is then pumped into vats called fermenters, which contain cooling coils, and is mixed with a concentrated (2 m cells/ml) inoculum of yeast. Sulphuric acid is used to control the pH of the liquid and the multiplication of yeast cells is encouraged by the addition of urea. Fermentation causes the temperature to rise from 30 °C (86 °F) to 33 °C (92 °F), at which point it is controlled by passing water through the cooling coils. If this were not done the yield of alcohol would be low and that of fusel oil high. Fermen-tation proceeds for thirty-six hours for a light rum; and for forty-eight to sixty hours, at a slightly higher temperature, if a heavy rum is required. The yield should be roughly one unit of proof alcohol from 1.35 of molasses (see p. 336 for the definition of proof spirit).

Distillation Rum is separated from the fermentation products (beer) in column stills. These are vertical vessels, 1–2.5 m in diameter and 6–18 m in height. Placed in the columns at intervals of 150–300 mm are horizontal plates, perforated to allow the passage of gases and fitted with overflow devices to retain on them layers of liquid 25–50 mm in depth. Steam is fed into the bottom of the column and causes the beer on the lowest plate to boil; the vapours travel upwards to the second plate, on which some are condensed and others pass to the third plate, and so on. Vapour from the top of the column is liquefied in a condenser, drawn off and placed in barrels. Manufacture can be continuous if there is a constant flow of beer into the column; whereas a batch process must be operated in pot stills.

Distilleries are of varying degree of complexity: they may comprise a single column or, if several types of rum and high purity alcohol are required, as many as four columns. These columns have different purposes: the more volatile components are removed by stripping columns and the less volatile by rectifying columns. A simple two-column distillery is illustrated in Fig. 11.1.

Maturing Rum matures best in barrels made of seasoned American white oak, charred internally, which previously have been used to hold sherry or spirits. It is in such barrels, marked and placed in tiers on racks for easy inspection and good aeration, that rum is allowed to age. The wood of the casks is sufficiently porous to allow entry of air and the evaporation of liquid. In these circumstances

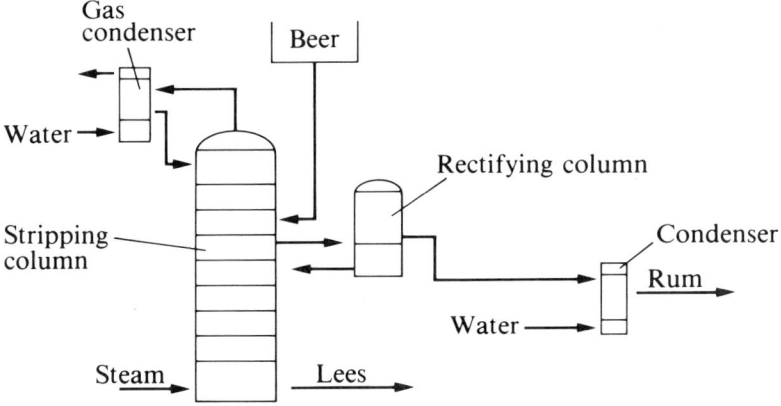

Fig. 11.1 The flow sheet of a simple distillery

free organic acids and alcohols react to form esters, and the alde-
hyde content of the rum is reduced. However, if left too long in the
wood, rum will absorb tannins and furfural and acquire a sharp
taste.

Strength The strength of a spirit is determined by measuring its
specific gravity with a hydrometer. Like other instruments used to
record the physical properties of solutions, for example polarimeters
and Brix spindles, these hydrometers also are calibrated at standard
temperatures; therefore if the liquid tested is cooler or warmer than
the standard the reading must be adjusted by reference to an appropriate table. To meet the excise requirements of different countries
the hydrometers may be graduated in terms of specific gravity, or
with reference to an arbitrary scale, or to give the percentage by
weight or by volume of alcohol, or in relation to proof spirit. To
make confusion worse confounded there is no universally accepted
definition of proof spirit. In the USA it is 50 per cent by volume of
alcohol; but in the UK, where Sikes hydrometers are used, the
corresponding figure is 57 per cent. When rum is placed in casks it
is 40–45 per cent overproof on the USA scale, i.e. it contains
between 70.0 and 72.5 per cent alcohol by volume.

Blending The various types of mature rum, in proportions
decided by the blender, are poured into tanks, diluted with dimi-
neralized water to the strength required and caramel is added.
Insoluble substances are allowed to settle, and after several days the
rum is drawn off, filtered and bottled.
    Blending is an art rather than a science, and the success of a
distillery depends largely on the skill of the blender. By means of
tasting trials he produces a rum which in bouquet, mildness to the
palate and strength, meets the requirements of the market; thereafter his goal is to maintain the consistency of his product.

Distillery effluent After rum has been distilled from the
fermented wash, a foul-smelling suspension of yeast, called 'lees' or
'dunder', remains in the still. The lees contain most of the potassium
removed by cane in its growth and therefore in some countries is
diluted and applied to deficient soils as a fertilizer. This should be
done only in remote areas where the stench is not a public nuisance.
Alternatively, some of the lees might be dried and used as animal
feed, but the cost is usually prohibitive. The disposal of the
remainder is a problem. Because of its organic matter content its
BOD (biological oxygen demand) is high; therefore effluent should
be discharged into rivers and other inland waters only during periods
of wet weather. Even when this is done, the smell is such that lees
become the prime suspect when wildlife is affected. For example,

notice for the closure of a distillery was served on its owners because fish in a nearby river were dying in large numbers. Later it was found that the real cause of the disaster was highly toxic effluent discharged upstream into the same river from a factory in a recently opened industrial estate. In limestone areas lees may be absorbed in 'soaks' without causing offence; and, if the distillery is close to an unfrequented coastline, disposal presents no problem.

Other potable spirits

Gin, whisky, brandy and other potable liquors are now successfully blended and marketed for local consumption in many sugar-producing countries. Concentrates which give their unique flavours to these spirits are imported and mixed with alcohol and water. Provided that the alcohol and water are of high quality, the product is indistinguishable from that made overseas. It is also much cheaper: there is a considerable reduction in freight charges and the excise duty paid on locally produced alcoholic beverages – in most countries the largest item in their cost – usually much less than that levied on imported spirits.

Industrial alcohol

Absolute (anhydrous) alcohol cannot be recovered by simple fractional distillation. At normal atmospheric pressure alcohol and water form an azeotrope, i.e. a liquid which distils without change of composition at a temperature lower than the boiling-points of its components. The azeotrope contains 96 per cent by volume of alcohol but the proportion of alcohol to water varies with the pressure, and the azeotrope ceases to exist at pressures below 70 mm. Consequently, absolute alcohol can be produced by distillation in vacuum. It can also be made by the distillation of a triple azeotrope of water and alcohol with another substance such as benzene or a chlorinated hydrocarbon. First the triple azeotrope, containing the unwanted water, is removed; and then the double azeotrope of alcohol and the added substance. Absolute alcohol remains in the still.

Alcohol is used for a variety of purposes, but in developed countries mainly as solvent for chemicals. There is, however, increasing interest in its ability to replace petrol, at least in part, as fuel for internal combustion engines. Technically it is possible for a blend of 15–20 per cent alcohol with petrol, to an octane rating of 94, to perform satisfactorily without the addition of lead compounds as anti-knock agents; but the alcohol must be anhydrous to overcome corrosion difficulties. Already mixtures containing roughly 12.5 per cent alcohol are being used in Argentina, Taiwan and especially in Brazil; and the car-manufacturing company, Fiat, has an experimental engine which runs on a 4 : 1 mixture of absolute alcohol and petrol. These developments are of special interest because of world-wide unease concerning the future supply of fossil fuels. Coal and oil are wasting assets, and increasingly expensive to recover. Sugarcane, on the other hand, is one of the most efficient agents for the fixation of solar energy. Technological development continues to cause greater productivity in terms of yield per arable hectare per annum; the possibilities of expansion are enormous; and the crop is renewed each year. Nevertheless it must be stressed that, at present, the use of blends of alcohol and petrol is for political and economic reasons rather than for technical efficiency, and therefore fluctuates widely.

Similarly, alcohol more and more is regarded as a basic raw material in the chemical industry, a development beyond the scope of this book.

There are many other small but profitable markets for alcohol, especially in pharmaceutical and cosmetic preparations. For example a refreshing toilet lotion, bay rum, is manufactured in the West Indies by mixing alcohol with an oil distilled from the leaves of the bay tree, *Pimenta racemosa* (Mill.) J. W. Moore (formerly *P. acris* Kostel.). A description of *P. racemosa* with references to experiments on the production of bay oil, rich in eugenol, the chief constituent of clove oil, was given by Williams & Williams (1951).

Carbon dioxide

Carbon dioxide, as well as alcohol, is produced when molasses is fermented by *S. cerevisiae*. The yield of carbon dioxide is 16 per cent by weight of molasses and between 70 and 75 per cent of it can be recovered. The carbon dioxide is then washed, deodorized and either compressed into solid form or liquefied and stored in cylinders. It is used as a cooling agent and in the manufacture of carbonated drinks.

Citric acid

The annual world production of citric acid, roughly 200,000 tonnes, is mostly for the soft drinks and food industries. Formerly extracted from citrus fruits, its main source is now molasses. Although little reliable information about methods of production has been published, it is known that the fungus *Aspergillus niger* is grown in deep culture tanks on a substrate of well-aerated, diluted molasses. Certain nutritional elements are withdrawn in order to encourage the formation of citric acid instead of the proliferation of mycelia. Beet molasses is the preferred substrate, but increasing quantities of cane molasses are being used. Whatever the source, the yield is 1 tonne of citric acid from 3 tonnes of molasses.

Baker's yeast

Baker's yeast, which used to be made from cereals, is now mainly produced by the fermentation of molasses. The yeast is a strain of the same species, *S. cerevisiae*, used in the manufacture of rum and alcohol. Molasses is diluted and sterilized, and after the addition of nitrogen (in the form of ammonium sulphate or urea) and phosphorus (as superphosphate) is further diluted with water until its sugar content is 1.5 per cent. The seed yeast is then introduced, with vigorous aeration, and more diluted, fortified molasses is slowly added to maintain the level of sugar at 1.5 per cent. The temperature is not allowed to exceed 30 °C and the pH is kept at 4.5. Fermentation is completed in ten to twelve hours. The yeast is separated, washed in centrifugals and finally dried on rotary vacuum filters. After filtration it is pressed into fixed weights and wrapped in wax paper for distribution in refrigerated stores and vehicles. The yield is 1 tonne of pressed yeast (29 per cent dry matter) from 1 tonne of molasses. Both beet and cane molasses are used in this process, but beet molasses is deficient in the essential yeast food, biotin. Therefore in the USA and the UK mixtures containing not less than 30 to 40 per cent cane molasses are preferred. In Europe beet molasses is often used by itself and biotin and other nutrients are added.

Food yeast

The diet of many peoples in tropical and subtropical countries is deficient in proteins and vitamins of the B complex. The development of a process to provide these nutrients in the form of dried yeast was therefore of great interest. *Torulopsis untilis* var. *major* was grown, like baker's yeast, in an aerated aqueous solution of molasses with the addition of ammonium sulphate and calcium superphosphate to promote its proliferation; and was then separated, dried and ground. It was mixed with flour and other staples to remedy the dietary deficiency. A large factory to manufacture food yeast was built at Frome in Jamaica in 1946 but soon closed down and was dismantled. There were two reasons for its failure: the difficulty of finding a market for the yeast and the price of molasses, always unstable, which suddenly increased. A smaller factory operates at Durban in South Africa. Floro *et al.* (1947) gave a detailed description of the process used in Jamaica, and the following analysis of the final product:

Protein	45–50%
Aneurin	20 μg/g
Riboflavin	60 μg/g
Nicotinic acid (niacin)	400 μg/g

Monosodium glutamate

Glutamic acid, originally made by the hydrolosis of wheat gluten and similar materials, may also be extracted from the effluent of the

Steffen process in the beet industry. Now it is produced almost entirely by the bacterial fermentation of cane molasses, and mostly in Japan. The yield is roughly 1 tonne of glutamic acid from 4 tonnes of molasses. It is marketed as the sodium salt, in which form it has an intense 'meaty' taste and is used to enhance the flavour of many edible products.

Itaconic acid

Itaconic acid, similar to citric acid in structure, is used as a plasticizer and as a chemical intermediate. It is made by the anaerobic fermentation of glucose or high test (choice) molasses by *Aspergillus terreus*. Little is known of the process, but the yield is 1 tonne of itaconic acid from 2 tonnes of sugar. With a sharp reduction in the supply of high test molasses in recent years it remains to be seen what alternative source of sugar will be used in future for the manufacture of this chemical.

Acetone and butanol (butyl alcohol)

Acetone and butanol can be produced by the bacterial fermentation of molasses, the most effective agents being *Clostridium* spp. However, most acetone and butanol is made synthetically and the bacterial process is little used.

Bagasse

Bagasse, the fibrous residue from crushed cane, is similar in composition to wood except that it has a much higher moisture content (50 per cent compared with as little as 10 per cent in some hardwoods). Most of the water is contained in the pith (the parenchyma). The other components of bagasse are complexes of pentosans, lignin and cellulose. On a dry-weight basis bagasse has more pentosans, less lignin and much the same cellulose content as wood (Table 11.4).

Table 11.4   *A comparison of the approximate composition of bagasse and wood*

Type	Pentosans	Lignin	Cross–Bevan cellulose (% dry matter)	α–cellulose
Softwood (spruce and pine)	13	30	59	34
Hardwood (birch, eucalyptus and oak)	23	25	59	36
Bagasse	27	20	52	39

The Cross–Bevan cellulose, which indicates the potential for paper pulp manufacture, is the residue after extraction by a hot solution of 1 per cent caustic soda, followed by bleaching with chlorine; α-cellulose, on the other hand, is the material required for the production of rayon and is the residue after extraction by a cold aqueous solution of 17.5 per cent caustic soda. Thus both values are empirical and the apparent anomaly in Table 11.4 that the dry matter of hardwoods contains slightly more than 100 per cent of pentosans, lignin and Cross–Bevan cellulose is unimportant. It follows that bagasse can be used instead of wood in many industries; and it has two great advantages: the cost of its collection has been charged to the manufacture of sugar, and the fibres already have been washed and crushed.

By tradition bagasse is burnt to provide the energy required for the operation of sugar factories. However, with constantly improving heat economy, increasing quantities are surplus to that requirement; and if alternative use(s) are economically attractive, more can be provided by replacing bagasse as a boiler fuel, in whole or in part, with oil or natural gas.

Compressed fibre-board

The most extensive use of surplus bagasse is in the production of compressed fibre board. There are two types: panel or insulating board and particle-board.

Panel or insulating board

Panel or insulating board is made by a wet felting process. Bagasse is treated with hot water or steam, under pressure, in rotary digestors. The pulp is then washed and fed into the board machines, from which it emerges as a continuous wet mat. It is then passed through rollers, to remove as much water as possible, cut into sheets and dried. The finished product does not warp and takes paint well. It is also an efficient insulator of heat and can be made resistant to insects, fungi and fire by impregnation with appropriate chemicals. It is marketed under such names as Canec, Canite and Celotex and is used as ceiling material and wall panels.

The fibre-board plant should be located near to the factory(ies) producing bagasse; otherwise considerable freight charges can be incurred.

Particle-board

Particle-board, much denser and harder than insulating board, is made in a dry process by binding (as opposed to felting) the bagasse fibres with a resin. It can be made waterproof and used for most housing needs.

Bagasse is broken in hammer mills, sieved, and particles of suitable size for board production are passed through oscillating screens which separate the fibre from pith and dust. The fibre is then dried to 11 per cent moisture content, sifted again, and mixed with resin (usually a urea–formaldehyde compound). After cold pressing, to give cohesion, the mixture is placed in moulding frames and heated under pressure for perhaps ten minutes at 137 °C to form particleboard. This is then removed, cut into suitable lengths and stored for local use or export. Langreney & Hugot (1969) gave detailed description, with flow diagrams, of the operation of the Bagapan plant in Réunion. There the choice of board size, 153 by 350 cm, with a thickness varying from 4 to 40 mm, was to suit the demand of European furniture manufacturers.

The storage of bagasse and bagassosis

For many years bagasse from the Brechin Castle factory in Trinidad was sent to the United Kingdom to be made into Celotex board. In order to reduce shipping costs, compressed bales were arranged in stacks, allowed to cure for one or two years, and then further compressed in secondary baling machines to much less than their original volume. Meanwhile the stacks were a serious fire hazard. Even more important was the development of bagassosis, a pulmonary hypersensivity disease, by several workers associated with the secondary baling process. This was probably due to the growth of *Thermoactinomyces vulgaris* on old bagasse. Bagassosis, like 'farmer's lung' and 'bird fancier's lung', is caused by the handling of dusty, mouldy organic materials. Protective measures should be taken wherever employees are exposed to such substances. A description of the disease was given by Hearn (1968).

The common practice of using bagasse as deep litter for poultry may also be hazardous in regard to its effect on the health of both the farmer and his flock.

Different means of storing bagasse have been developed in recent years and were described by Atchison (1972). The methods increasingly used are:

1. Wet bulk storage of partially depithed bagasse, with and without pre-treatment.
2. Moist bulk storage of untreated bagasse as delivered from the factory.
3. Dry storage in dense packs, under cover, after depithing and artificial drying.
4. Dry storage as pellets or briquettes.

Paper

Although a patent for the manufacture of pulp and paper from bagasse was issued as long ago as 1838, the first successful mills were started in 1939 in Peru, the Philippines and Taiwan. By 1950 their number had increased only to six, but in 1974 there were at least eighty mills in operation. Production of pulp increased from 100,000 tonnes in 1950 to 500,000 tonnes in 1968 and 925,000 tonnes, roughly 1 per cent of the world's supply, in 1972. This expansion was world-wide and took place in most cane-growing countries. Each of the largest mills was listed by Atchison (1974), who prophesied that production would increase twofold or threefold during the next fifteen to twenty years. Bagasse pulp is now used in the manufacture of bags and cardboard; of corrugated, wrapping and writing paper; and of toilet tissues and towelling. It is probable that in the near future more newspaper will be made from pulp derived, at least in part, from bagasse.

Atchison (1967) described, with flow diagrams, the installation of five new mills:

1. To produce pulp and various grades of bleached and unbleached pulp and paper at Trinidad, Cuba.
2. To produce bleached pulp and paper for writing and printing at Mysore, India.
3. To make the same products as at Mysore, but using bamboo as well as bagasse, at Madras, India.
4. To produce unbleached pulp at Edfu, Egypt.
5. To produce bleached pulp and paper for writing and printing, as at Mysore, at Jujuy, Argentina.

The purpose of pulping is to dissolve lignin and to remove some of the hemicelluloses. Lignin cements the fibres and prevents them from being separated and therefore from forming a loosely felted mat; while hemicelluloses should be present in such proportion (between 8 and 20 per cent) that the pulp may be converted into paper of the desired strength and flexibility.

Bagasse is stored in bales or in bulk, as for the manufacture of board. Pith is first removed, usually in two stages: dry and wet. The remaining fibrous mass is fed into digesters, mixed with a solution of one of several chemicals, depending on the process being used, and heated under pressure for ten to twelve minutes. By this action it is converted into pulp, which later may be bleached. In the manufacture of pulp from wood vigorous digestion is followed by milk bleaching. When bagasse is the raw material, however, it is better to reverse the emphasis: to follow less severe digestion with stronger bleaching.

Furfural

Furfuraldehyde, known as furfural, is a colourless oil of pleasant smell. It has a limited market in highly industrialized countries, mainly for the production of S S Nylon. Furfural can be made from several agricultural waste products: cereal and groundnut hulls, corn cobs and bagasse. These materials contain pentosans which, when treated with dilute acid at a high temperature (150–200 °C) in a digester, are converted by hydrolysis into pentoses. They, in turn, are decomposed to produce furfural. The process may be represented as follows:

$$(C_5H_8O_4)_n + nH_2O \rightarrow nC_5H_{10}O_5 \rightarrow nC_5H_4O_2 + 3nH_2O$$
pentosan   pentose   furfural

Superheated steam is blown through the digester, carries off the furfural and is condensed. The distillate is rectified three times: stripped, dehydrated and refined, and the final product contains 99 per cent furfural.

Most of the world's supply of furfural is made from corn cobs in the USA. However, the Quaker Oats Company is taking part in a joint venture with a sugar company to make furfural from bagasse at La Romana in the Dominican Republic, where a large plant of 15 m. kg annual capacity has been in production for several years. A similar operation was started at Sezela in South Africa. Duffey (1969) laid down the conditions required for the profitable production of furfural. He stressed the importance of access to a market, the size and location of the factory and the availability of technical expertise. The minimum cost of an economic furfural plant in those pre-inflation days was US $10 m. He concluded that these requirements could seldom be met.

Filter mud

Filter mud or filter cake is a useful fertilizer especially when applied to phosphate-deficient soils and to fields in which the topsoil has been removed or redistributed for any reason: for example where drainage patterns have been altered to accommodate mechanical harvesters. It is delivered from factory to field in dump trucks, spread by bulldozers and incorporated with the soil by the passage of chisels or disc ploughs. Heaps of mud become a fire hazard if left undisturbed in dry weather for several days. The smoke is a nuisance, the burning difficult to control and the ability of the mud to improve the physical characteristics of the soil is destroyed.

Sugar-cane wax

The stems of sugar-cane are coated with a thin layer of wax, more noticeable near the nodes than elsewhere. When milling takes place roughly half of the wax is removed in the juice; the other half remains in the bagasse. Most of the portion in the juice is trapped or precipitated when clarification takes place and becomes part of the filter mud; the remainder, mainly waxes with low melting-points, persists in minute quantities throughout the process of manufacture. The wax complex in the filter mud has attracted considerable interest over the years, and at various times has been recovered for commercial use in several countries (for example at Moreton, Queensland, Australia; Natal, South Africa; and Grammercy, Louisiana, USA). The significant physical and chemical properties of refined sugar-cane wax are similar to those of other hard vegetable waxes (Table 11.5). It may therefore be used for the same purposes: the manufacture of polishes and carbon paper.

Filter mud contains 60–80 per cent moisture. The composition of its dry matter is even more variable; in particular the crude wax content may be as low as 6 per cent or as high as 20 per cent. The factors which cause this variation, in order of importance, are variety, pre-harvest burning, climate and intensity of milling.

Table 11.5   *The characteristics of sugar-cane wax and other hard waxes*

	Melting point (°C)	Specific gravity	Acid number	Saponification number	Iodine number
Carnuba wax	85	0.99	10	90	13
Candelilla wax	71	0.98	20	60	38
Sugar-cane wax	78	0.99	15	65	16

Source: Adapted from Wiggins (1949).

Manufacture

The many methods by which cane wax can be extracted and refined were surveyed by Wiggins (1949, 1950). In each of them there are three main stages:
1. The filter mud is dried.
2. Crude wax is extracted from the dried mud.
3. Fatty material is removed from the crude wax.
The yield of refined wax is roughly 60 per cent by weight of crude wax. Perhaps the simplest process, though it also includes demineralization, was developed at Moreton, Queensland. In short, the mud is stored in containers for three weeks, during which time its

temperature rises to 93 °C (200 °F) and its moisture content is reduced to 15 per cent. It is then broken in a hammer mill, sieved and subjected to continuous extraction in heated tanks with a solvent such as benzene. The solvent is recovered by distillation and molten crude wax is run off. Minerals, especially calcium, are removed by heating with hydrochloric acid. The slabs of crude wax are then disintegrated and mixed with absolute alcohol in a vessel equipped with a stirring mechanism. Soft fatty material is dissolved by the alcohol and run off. After the process has been repeated two or three times, refined cane wax remains. Finally the wax is melted and poured into moulds; and alcohol, like acetone, is recovered by distillation for further use. Colour is important and if the refined wax is too dark it can be bleached by boiling with an aqueous solution of potassium chlorate and dilute sulphuric acid.

Markets

Sugar-cane wax is sold in competition with other hard vegetable waxes, especially carnauba, whose supply and price fluctuate widely. Nevertheless, carnauba and similar substances are cheap and simple to produce. Therefore the establishment of cane wax factories to take advantage of temporary high prices is a risky business: many such factories have been closed down after a few years' operation, their products priced out of the market. Examples of this are the demise of wax industries in Java and South Africa when the price of carnauba fell to US $25.00/tonne in 1926–27. Likewise a large plant in Barbados was built in 1949 when the price was £1,200/tonne, and abandoned in 1951 when the price had fallen to £400/tonne. Another factor contributed to its closure: B 37161 was being replaced by newer varieties, such as B 41211, which had thin coatings of low-quality wax.

Carnauba, candelilla and other waxes

The source of carnauba wax is the palm tree, *Copernicia cerifera* Mart., which grows in eastern Brazil. On the underside of its leaves is a thin layer of wax which cracks and can be peeled off in flakes when the leaves are dried. Alternatively the leaves may be cut into sections and boiled in water. The wax then rises to the surface, is skimmed off and separated into three grades according to its colour. Sophisticated refining techniques are not required. The leaves are collected in the dry season, a dozen or so from each palm; and the yield is roughly 1 kg of wax from 100 leaves.

Candelilla wax is extracted from a shrubby plant, *Euphorbia antisiphylitica* Zucc., which grows in the drier parts of Mexico. The wax occurs as a thin coating on the stems and branches. It is separated by boiling in water and yields 3–4 per cent by weight of almost pure wax. Refining is unnecessary. The name candelilla (little candle) describes the smooth, cylindrical, leafless branches of the plant.

Other minor sources of wax include the Colombian palm, *Ceroxylon andicola*, which grows at a higher elevation in a cooler climate and gives an annual yield of 10 kg/tree. They, like carnauba and candelilla, hold two great advantages over filter mud: the product is easily extracted and does not need refinement.

Sucrochemistry
K. J. Parker, Chief Scientist, Tate & Lyle.

The idea that sucrose could be a source of industrial chemical intermediates with considerable potential for development was first conceived by the late Ody H. Lamborn in 1941. His conviction and foresight led to the setting up in 1943 of the Sugar Research Foundation, Inc. by the US sugar producers to sponsor and co-ordinate research into the chemistry of sucrose, for which the name 'sucrochemistry' was coined by the foundation's first President, Dr Henry B. Hass. In 1968 the foundation, recognizing the world-wide interest in promoting non-food markets for sucrose, was re-established as the International Sugar Research Foundation (ISRF), with membership of sugar producers and refiners in Europe, the Americas, South Africa, Australasia and the Far East (Hickson 1977).

During the thirty-five years until the closure in 1978 of the ISRF, when its role was taken over by the World Sugar Research Organization, Ltd based in the UK, an impressive basis of sugar chemistry had been laid down. Numerous applications and uses of sugar derivatives were explored and evaluated under the aegis of the Foundation (Kollonitsch 1970). However, despite the ready availability and low cost of sugar on the world market, the industry has so far been unable to gain a foothold in the chemical markets currently dominated by the petro-chemical industry with its surplus production capacity, and enormous investment in large-scale operations.

Consequently, sucrochemicals are as yet confined to supplying speciality markets with high-value products in which the often unique properties of sucrose derivatives are exploited. This is unlikely to have any noticeable effect in stabilizing the raw sugar price or in reducing the periodic sugar surpluses, one of the principal original objectives of the Sugar Research Foundation.

However, within the last decade oil prices have increased dramatically as the cost of energy has soared. The price of ethylene, which had been falling steadily in real terms until 1976, is now rising despite excess production capacity. The chemical industry is increasingly looking towards regenerable carbon resources for its raw materials. Sugar has unique advantages as an industrial feedstock (Parker 1979).

Unlike other products of biosynthesis, such as cellulose, starch and lignin, sucrose is produced as a pure defined organic compound of low molecular weight. It has multiple chemical functionality, giving the molecule a high potential for versatile application. It is produced as a food in ever-increasing quantity by an established world-wide industry at relatively low cost. Its extraction from sugar-cane is energetically self-sufficient, the energy required to extract, concentrate and purify the sugar being provided by burning the by-product cane fibre or bagasse.

Sugar-cane itself is among the most efficient natural converters of solar energy and atmospheric carbon dioxide into fixed carbon known, achieving a photosynthetic efficiency as high as 2 per cent compared with the average of 0.1 per cent for the world's biomass production. In this respect, it makes optimal use of available arable land, which will ultimately be the limiting resource, in terms of the proportion of recoverable solar energy and carbon immobilized per unit of land area.

Sucrose may be converted into simpler chemical intermediates by thermal degradation or by fermentation. Direct pyrolysis of sugar, yielding mainly carbon, would not be an economic proposition. In chemical degradation of sucrose, despite extensive research, the yields of the isolated products tend to be low, making the processes at present non-competitive with alternative routes from petroleum-derived starting materials. Apart from the inherently low weight yield consequent upon the loss of elements of water, degradative reactions tend not to be readily controllable, giving rise to a mixture of products, and low conversions.

Sucrose is readily fermented and is used for the production of fuel alcohol on a massive scale in several producing countries such as Brazil, Colombia and India. Ethanol itself is an important chemical feedstock, used in the production of acetic acid, esters such as ethyl acetate, acetaldehyde and ethylene. The products are basic raw materials for conventional chemical processing technology.

Numerous other fermentation products (Table 11.6) produced commercially (Demain 1981) use sucrose as the substrate, though other simple sugars, such as glucose derived by hydrolysis of starch or cellulose, may equally well be used. Morever, sucrose is available for fermentation either as molasses or sugar-cane juice, not requiring the need for refining and crystallization.

If sugar is to be regarded as a primary chemical raw material, it is necessary to exploit the synthetic chemistry of sucrose in order to benefit from its unique properties. Chemically sucrose is an octa-hydric alcohol, which will form the expected derivatives, such as

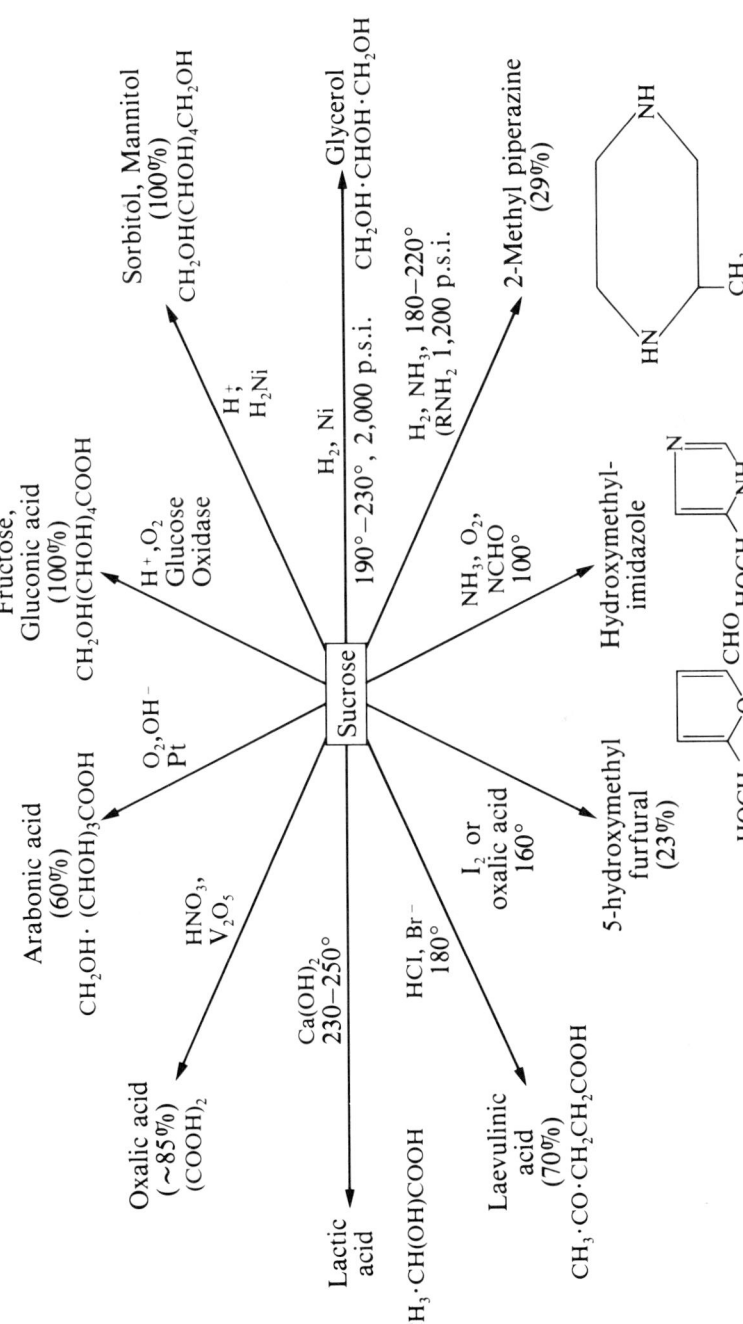

Fig. 11.2 The degradation products of sucrose (From Imrie & Parker 1974)

Table 11.6   *Some products of fermentation of sugars*

Solvents	Acids	Polysaccharides
Ethanol	Citric acid	Alginic acid
Butanol	Lactic acid	Xanthan gum
Acetone	Itaconic acid	Dextran
2 : 3-butandiol	Gluconic acid	Pullulan
Isopropanol	Propionic acid	Poly-γ-Hydroxybutyric acid
	Butyric acid	

Amino acids	Antibiotics	Gases
Glutamic acid	Penicillin	Methane
Lysine	Streptomycin	Hydrogen
Methionine	Cephalosporin	Carbon dioxide
	Tetracycline	

esters, ethers and urethanes. Being polyfunctional, multiple reaction is to be expected, so that selective or partial derivatization may present problems.

The fully substituted esters are readily obtained by reaction with the acid anhydride or chloride. For example, esters such as the octa-acetate, octa-benzoate and hexa-acetate-di-isobutyrate, are commercial products with a wide range of applications and uses as plasticizers, viscosity modifiers, lacquer extenders, etc. Sucrose octa-acetate is also used as a denaturant for sugar and alcohol, being intensely bitter.

Sucrose mono-esters of long-chain fatty acids have surface-active properties and are completely non-toxic. Consequently, they have found numerous applications both in the food and the animal feed industries. They are readily prepared by transesterification of sucrose with the methyl ester of the fatty acid, ideally in the absence of a solvent.

Sucrose esters are readily biodegradable, and can be used in detergents, where environmental problems can be created by persistent detergent residues. For this purpose, however, purified sucrose esters are too costly. The low-cost sucroglyceride mixtures produced by a solventless process (Parker *et al.* 1977), can be economically included in detergent formulations, the detergents being competitive in cost and efficacy with those based on conventional linear alkylsulphonates. Markets include those for domestic and industrial detergents, hand cleansers, hard-surface cleaners, machine cutting lubricants, amongst others.

Sucrose esters of stearic, palmitic, lauric and oleic acid, being completely non-toxic, bland, non-allergenic and non-irritant are suitable for use in foods and are permitted as additives in some countries: for example, in Japan, France, Belgium and Switzerland. They find applications as dispersing agents, emulsifiers and stabilizing agents in baking, confectionery, dairy products, desserts and reconstitutable dried beverages and soups, etc.

An important potentially large market for sucrose and sucrose derivatives is in synthetic resins. Sucrose itself has been evaluated as a filler or substitute for phenol in melamine and novalak-type resins, though there is no evidence that unmodified sucrose will participate in the cross-linking reaction with formaldehyde. The expected hemiacetal link is presumably insufficiently stable at normal curing temperatures to provide any significant cross-linking to sucrose in the final polymer.

Sucrose and its derivatives have been widely explored as components of polyurethane resins, in particular rigid polyurethane foams. Sucrose itself can be used, but tends to lead to brittle products. The polyhydroxyprophyl ether of sucrose is the preferred polyol in practice, having better miscibility with the diisocyanate and fluorocarbon blowing agent used in the preparation of rigid foams. In order to introduce flame-retardant properties into the resin, halogen or phosphorus derivatives of sucrose are frequently included in the reaction. An estimated 65,000 tonnes/annum sucrose is currently used in the manufacture of polyurethane foam resins.

The production of surface coating resins using sucrose esters of drying oil acids, cross-linked with phthalic acid and diepoxides or diisocyanates has been developed in the paint industry, as an alternative to alkyd resins. These products have excellent properties, but have so far proved too expensive to replace conventional resins.

The partial and selective derivatization of sucrose is necessary where the properties of a particular derivative may depend crucially on the position and conformation of the substituent group. This is frequently true of biological or physiological activity where molecular conformation is fundamental to stereo-specific enzyme interaction. For such a synthesis it is necessary to take advantage of the differing relative reactivity of the hydroxyl groups, which will depend not only on inherent steric factors but also on the nature of the reaction, the conditions (temperature, solvent, time, molar proportions of reactants), reagent reactivity, the stability of the activation complex, hydrogen bonding and hydroxyl group acidity. The preparation of the intensely sweet chlorodeoxysucrose derivatives requires a sophisticated synthetic route to obtain a product with the required molecular structure. The position of the chlorine substituent in the sugar molecule is crucial to determining the taste properties of the product (Hough & Khan 1978). Chlorine in the 4,1'- and 6'-positions is associated with high sweetness intensity while substitution in the 2- and 6-positions gives rise to intensely bitter products.

When high-valued products are required, for use for example, as pharmaceuticals or in agriculture, then a costly synthetic process can be justified. The quantity of sucrose used would be insignificant, but the value of a speciality chemical industry to the future sugar economy could be very significant.

Chapter 12

Cane farmers

Reference was made in Chapter 1 to the decision of the CSR to divide its estates in Australia and Fiji into small holdings and to entrust the growing of cane to independent growers (with suitable technical assistance). The growers (farmers) sell their produce to the miller (CSR) and, although there must inevitably be points of disagreement between those engaged in commercial transactions, the operation of this system over the years has been largely responsible for an enviable record of social stability. Such an achievement, and the reason for it, has been widely recognized. For example, long before the sugar industry of Guyana was nationalized in 1978, attempts had been made, with limited success, to establish communities of small growers in that country. In 1966, 2,534 farmers in 15 groups cultivated 2,640 ha and produced 159, 874 tonnes of cane from which 13,512 tonnes of sugar were made. This was but a modest 7 per cent of Guyana's total output, but was twice as much as in 1964. Previously all the cane had been grown by estates which, with the factories, were under common ownership.

Most new projects are so devised that factory and field are separate entities, and that all the cane is produced by independent growers. This is the case, for example, at Central Constancia, which first started grinding operations at Risaralda in Colombia in 1979. However, in a description of the sugar industry in the Sudan, Mathur (1974) emphasized the difference between the ownership of the cane ground at the two factories then in operation. At Guneid all the cane was grown by tenant farmers, each of whom rented a *hosha* of 15 feddans (6.3 ha); whereas at Khasm-el-Girba, where both factory and field were owned and operated by the Government, cane was produced by the old plantation system, i.e. with the help of hired labour. A third factory was to be built at Sennar, and the 13,400 ha to be planted with cane were to be managed in the same manner as at Khasm-el-Girba.

The organization of production in long-established sugar industries is much more complex. In Trinidad roughly one-third of the cane is grown by farmers (37 per cent in 1979) and two-thirds by the estates (63 per cent in 1979). The holdings of about 10,000 farmers vary in size from less than 1 ha to more than 200 ha, but most are small enough to be worked by their owners' families. In the past the company responsible for more than 90 per cent of the country's sugar production (Caroni Ltd) made conscious efforts to increase its farmers' prosperity. When it became possible to control the frog-hopper by the aerial application of insecticides this service was provided, at less than cost price, for the small growers whose crops were most at risk; fertilizer was bought in bulk for the farmers; and a Credit Union, with substantial assets, was managed by the Company's Cane Farmers' Superintendent. Since 1975 the sugar industry of Trinidad has been owned and operated by the Government (by invitation rather than mandatory nationalization), but the pattern of cane production and land tenure has not changed.

The position with regard to the division of cane production between farmers and estates is much the same in Jamaica as it is in Trinidad, though in Jamaica a higher proportion is grown by farmers.

In Barbados, on the other hand, it is difficult to distinguish between an estate and a small holding. As recently as 1939, in this island of 430 sq. km (166 sq. miles) there were 33 central factories engaged in the manufacture of raw sugar, and there were also 72 plants (35 driven by steam and 37 by wind) in which either musco-vado sugar or fancy molasses were made. Since 1939 there has been a slight decrease in the area planted with cane, but only 16 factories of all types operated in 1966, 9 in 1978 and 8 in 1979. Some owners of the 105 estates which had plants of one sort or another in 1939, or their successors, now have interests in the company which owns the eight surviving factories; others do not. In the annual publication *The Yield of Sugar Cane in Barbados*, an arbitrary division of 4 ha (10 acres) between estates and small holdings is made. If more than 4 ha are planted with cane, the cultivation is deemed an estate; if less, it is a small holding. In 1979, 14.5 per cent of the cane ground was produced by small holders from an estimated 3,240 ha (8,000 acres).

Whatever the relationship between the farmer and the manufac-turer it must be based on mutual understanding and respect. A factory cannot operate without cane; and it would be pointless to grow cane if there were no factory in which it could be processed. Therefore it is in their common interest that each should be efficient and profitable. Both parties must remember that the land on which cane is grown could be used to produce other crops. If the price of cane is remunerative, the farmer will get the benefit of the use of machinery which has been bought and erected at considerable cost by the manufacturer. However, if some other crop becomes more rewarding, the farmer can grow it instead of cane, without any other economic consideration, because none of his money is invested in the factory.

Payment for cane

Regardless of his status, the farmer must be paid by the manufacturer for cane delivered. The sum involved and the method of its calculation has been a source of controversy over the years. Many formulae for payment have been devised and, by and large, they have been, and are, well suited to the special circumstances of the industries in which they are used.

Methods of cane purchase

In Barbados, where for many years there was competition between the factories to buy sufficient cane to meet their requirements, it has been unnecessary to devise an all-embracing formula or to pass laws controlling the purchase of cane. Elsewhere the methods used for calculating the price of farmers' cane may be divided into two groups: those in which a flat rate is paid for all cane and those in which the price varies according to cane quality and/or factory efficiency. All take into account the price of sugar, most include the value of molasses, and some also include the value of by-products.

Flat rate systems of payment

Guyana and Trinidad are examples of countries in which flat rate systems are used.

Guyana

The small scale of farmers' involvement in Guyana did not warrant the introduction of a complicated method of price calculation. Instead, a simple formula was devised whereby, for each tonne of cane delivered, the farmer was given two-thirds of the average price received per tonne sugar, divided by the average tonnes of cane required to make 1 tonne of sugar (Table 12.1). In the *Report of a Commission of Inquiry into the Sugar Industry of Guyana* (Persaud *et al.* 1968) it was stated that this ratio for the distribution of sugar proceeds was also used in Queensland, Réunion, Mauritius and Jamaica; and that the more complicated method of calculation employed in Louisiana gave roughly the same result. One of the recommendations of the Persaud Commission was that farmers should receive two-thirds of the value of the molasses, as well as the sugar, derived from their cane.

Table 12.1 *Examples of the calculation of the price of farmers' cane in Guyana*

Average price per tonne sugar (£)	Farmers' share (£) (two-thirds)	Average tonnes cane per tonne sugar	Price of farmers' cane per tonne (£)
198	132	12	11.00
		11	12.00
		10	13.20
264	176	12	14.66
		11	16.00
		10	17.60

Trinidad

Between 1950 and 1977 the cane farming industry of Trinidad was controlled in all its aspects by the Production of Cane Ordinance (PCO) (Revised Ordinances 1950, Ch. 23, No. 12), which succeeded the Production of Cane and Sugar Ordinance 1944. Each year the price to be paid to the farmers for their cane was calculated by the Accountant-General according to the following formula:

$$\frac{(A + B + C + D + E) - (F + H + J)}{K} - G$$

where $A + B + C + D + E$ = the average income from 1 ton of raw sugar (with refined and special sugars included at raw value) and the molasses produced in its manufacture.

F = the average expense of handling cane and of processing and marketing (in respect of 1 ton of sugar).

H = an allowance for depreciation ($804,966 divided by the total production expressed as tons of sugar).

J = an allowance for interest ($4.80/ ton sugar)

K = the average number of tons of cane required to produce 1 ton of sugar.

G (rarely used) = extraordinary expenses incurred in carrying out orders made under the PCO.

All figures referred to 1 long ton (1.02 tonnes) of cane or sugar, to Trinidad and Tobago dollars, and represented averages for the industry. Power was given under the PCO to disregard any costs which were in excess of the average of those of other manufacturers for similar operations.

The benefit of the PCO formula to the manufacturers was that, whatever the price of sugar, they received $4.80 (£1) as interest plus a slightly smaller sum as depreciation for each ton of sugar made from farmers' cane. However, with increasing inflation these rewards became inadequate. The farmers had a guaranteed market and a reasonable price for their crops.

Following the collapse of the 'free' market for sugar in 1976–77 (Ch. 1) the Government of Trinidad and Tobago abandoned the PCO and instead decreed that farmers were to receive a guaranteed minimum price for each ton of cane sold, regardless of the price of sugar or factory efficiency. No doubt social and political considerations influenced those in authority when this decision was made. Be that as it may, the result is that the farmers have been encouraged to continue to plant cane and in 1979, a disastrous financial year for the industry as a whole, they were well paid for their produce.

Advantages and disadvantages of flat rate systems of payment

Flat rate formulae have several commendable features. They are easily calculated and do not require detailed and expensive analyses of juice or cane. They also remove the motive which causes some farmers to contrive to reap all their cane, often by unseemly means, when its sucrose content (and therefore its value) is at its highest. Consequently, cutting orders are issued to farmers in a sequence which is seen to be fair to all, and the crop is reaped in a disciplined manner.

The danger inherent in the use of flat rate formulae is that the farmers, being paid for their cane by weight, might strive for quantity at the expense of quality; they might forget that the purpose of the industry is to make sugar and not to grow cane. Therefore the PCO of Trinidad included the following safeguards:

1. In each cane farmer's contract the manufacturer stipulated the varieties to be cultivated. In practice they were the same as those grown on the estates.
2. The cane delivered by the farmer was to be of good quality, i.e. well topped and free from trash. If a farmer did not meet this requirement, a warning from the cane inspector usually was sufficient remedy.
3. The manufacturer was not required to accept cane which had been burnt more than three days before delivery; and for all burnt cane 60 cents per ton (59 cents per tonne) was to be deducted from the price determined by law. The occasions on which farmers' cane was rejected because it had been burnt for more than three days were rare indeed; and the penalty for burnt cane was more often waived than imposed. Nevertheless the very

existence of these provisos was sufficient to ensure and maintain a high standard of cane quality.

Systems based on cane quality

By and large, however, payment based on cane quality is preferred to a flat rate, and is much more widely used. To assess quality, the juice or cane must be sampled and analysed.

Sampling

The juice required for analysis is that extracted by the first mill (first expressed or crusher juice). Its sampling is not difficult provided that loads delivered by different farmers, or groups of farmers, are roughly separated on the carrier so that they can be readily identified. Signal systems connected with automatic devices cause representative samples of juice to be taken, and their sucrose content (pol) is then determined.

Except in South Africa, the many methods of sampling cane fall into two categories: those in which a stipulated number of stems, or weights of cane, is taken from each load; and those in which a core sample is extracted. In Louisiana, for example, it is required that the size of the sample must be between 22.7 kg (50 lb) and 45.4 kg (100 lb). The choice must be made at random, be free from personal bias and the sample must represent accurately the composition of the whole load, including trash and soil. In practice this is difficult to achieve and differences of opinion concerning the validity of the sample as being representative of the whole can lead to squabbles and acrimony. Indeed a Commission of Enquiry into the price to be paid for farmers' cane in Jamaica, where a similar method of sampling was used, reported that the efforts of the Cane Farmers' Association, on behalf of its members, had been 'directed too intensively to 'policing' at the factories'; but that was because many of the elements in selection were more or less arbitrary (Biggs 1963).

The great merit of the core sampling technique, described by Payne & Rhodes (1967), is that it removes much of the ground for suspicion between farmer and manufacturer. A rotating core tube with a serrated cutting edge, 25 cm (10 in) in diameter, bores through the width of each truck (capacity 18 tonnes) and extracts a sample of cane weighing 14 kg (30 lb). An alternative method is to take the sample by pivoting a core tube 20 cm (8 in) in diameter at an angle of 60° through the top of the load. Subsamples of 0.9 kg (2 lb) are then treated in the manner described by Meade & Chen (1977) and their pol, Brix (refractometer) and fibre content are determined. Core sampling is used in Hawaii, and also in Guadeloupe, where the horizontal method has been adapted to suit local conditions (Lemaire 1972).

Formulae for payment

Having established the quality of the farmers' cane, or the juice extracted from it, means must be found to determine its value. This is done by the application of one of many formulae, either agreed or statutory. Estimates, allowances and averages enter into all of them and none is universally accepted.

Four examples will be given: the Australian CCS (commercial cane sugar) formula, by which payment is made according to the available sugar in the cane, regardless of factory efficiency; the South African DAC (direct analysis of cane) method, which takes into account factory recovery; the Louisiana system, which is based on cane of 'standard' quality; and the complicated (though possibly the fairest) method used in Mauritius.

Payment based solely on the available sugar in the cane (or juice)
The Australian CCS formula

In the 1880s Kottmann, in Australia, derived a formula to measure the sugar which a factory should recover from cane of a given quality. This formula, first applied to cane purchase in 1899, was called POCS (pure obtainable cane sugar). In 1915 the POCS formula was adopted by the Cane Prices Board, renamed the CCS formula, and thereafter has been used throughout Australia (with modifications from time to time), under the Regulation of Sugar Cane Prices Acts, to determine the price to be paid for farmers' cane. The CCS formula is defined as 'that percentage by weight of a quantity of cane which would be recovered as pure sucrose if milling and refining operations were conducted at a prescribed standard of efficiency'. The prescribed standard is that, for every two parts of soluble impurities in the cane, one part of sucrose is lost in the process; and that there are no other losses of sucrose. Therefore:

   CCS = sucrose per cent cane $- \frac{1}{2}$ impurities per cent cane

where CCS is sugar of 94 net titre,* an empirical estimate of the percentage of pure sugar recoverable from a raw sugar. In applying the formula the following assumptions are made:

Brix	= total soluble solids
Sucrose	= pol
Impurities	= Brix − pol
Brix per cent cane	= Brix per cent first expressed juice × $\frac{100 - (F + 3)}{100}$
Pol per cent cane	= pol per cent first expressed juice × 100 $\frac{100 - (F + 5)}{100}$

---

* Net titre = pol − (reducing sugars + 5 times the ash content).

Therefore
CCS = pol per cent cane $-\frac{1}{2}$ (Brix per cent cane − pol per cent cane)

$$=\frac{3}{2}\text{pol per cent cane} - \tfrac{1}{2}\text{Brix per cent cane}$$

$$= \frac{3P}{2}\left(1 - \frac{F+5}{100}\right) - \frac{B}{2}\left(1 - \frac{F+3}{100}\right)$$

where P = pol per cent first expressed juice,
    B = Brix per cent first expressed juice,
    F = fibre per cent cane.

Methods of sampling and analysis are described in *The Laboratory Manual for Queensland Sugar Mills*, the fifth edition of which was published in 1970.

Payment to the farmers, based on the CCS formula, is then determined by local cane price boards. In theory, widely varying proportions might be allotted in different areas; in practice, all receive roughly 70 per cent of the sugar value of their cane.

Although the CCS formula is based on an arbitrary standard of factory efficiency and includes many empirical elements, it has withstood the test of time. Much of the credit for the South Pacific industries having the the highest ratio in the world of sugar to cane is correctly attributed (especially by factory technologists) to its use. Dixon (1956), however, also mentioned other contributory factors: the sweet variety, Badila, brought from New Guinea by Tryon in 1896, set the standard and dominated the cane variety position in Australasia for thirty or forty years; and the climate is ideal for ripening (Ch. 2).

Payment according to the sucrose (pol) in the cane, related to factory recovery
The South African DAC method

In South Africa, as in Australia, payment for cane is determined by quality, but actual rather than theoretical factory recovery is used in the formula.

For many years the sucrose content of each individual consignment of cane was estimated by multiplying the pol of its crusher juice by the weekly Java ratio of the factory, the Java ratio being:

$$\frac{\text{pol per cent cane}}{\text{pol per cent first expressed juice}} \times 100$$

It was wrongly assumed that for any given factory there was little variation in this relationship, regardless of cane quality. The main factor which causes the Java ratio to be an inaccurate measure of the quality of a consignment of cane in regard to all cane ground is that no cognizance is taken of difference in fibre content. Douwes

Dekker (1956) described the disadvantages of the system. He pointed out that some values used in the calculation of the Java ratio were of true sucrose while others were of pol; mentioned the importance of fibre content; and quoted the definition of the Java ratio from the Sugar Act: 'Weight of sucrose in cane crushed during the period divided by the summation of the products of the weight of each consignment multiplied by the polarization of the crusher juice of such consignment in respect of the same period.'

The main advantage of allotting sugar proceeds according to the Java ratio was its simplicity. However, its use as an expression of comparative quality became increasingly suspect, especially when the development of mechanized harvesting caused an increase in the trash content of cane. After an exhaustive enquiry, it was replaced in the 1972–73 season by the DAC method. The basic concept of DAC is that cane is in its most homogeneous condition after it has been shredded. Therefore samples are taken through hatches in the elevator chutes between the shredders and the first mills. To ensure that the samples are representatve, it is essential that the hatches should span the full width of the elevators. The pol of the samples is then determined.

The farmer's payment, for each 100 tonnes of cane, is then calculated as follows:

pol in cane × average overall factory recovery for the year × grower's share × sugar price
(all, except the price of sugar, expressed as percentages).

Details of the DAC method, with special reference to sampling, were given by Buchanan & Brockensha (1974). It is now used in all of the twenty factories which operate in South Africa.

Payment based on cane of 'standard' quality
The Louisiana method

The method of cane payment in Louisiana, prescribed by the US Department of Agriculture, is quite different and highly complex. All farmers' cane is assessed by reference to 'standard' cane, which is defined as having 12 per cent pol and juice with a purity of 75. A premium is added or a penalty deducted when the 'normal' juice (i.e. first-expressed juice, or laboratory mill juice, to which an arbitrary factor is applied) exceeds or fails to reach this quality. For 'standard' cane the farmer is paid US 1.06 per ton (US 1.17 per tonne) for each cent per lb (0.45 kg) of the average price of raw sugar. For example, if a farmer delivers cane of 12 per cent pol and 75 purity 'normal' juice, and the average price of raw sugar during the period of delivery is 17 cents per lb, he will receive 18.02 (17 × 1.06) per ton (19.84 per tonne). The premiums and penalties, within limits, are 5 per cent of the standard price for each 0.5 per cent pol above or below 12 per cent. Those for differences in purity are not so easily defined but are readily available by reference to tables. Payment for molasses is made on the basis of price in excess of 6 cents per gallon (4.55 litres).

SJM and Winter–Carp formulae

Payment for cane in some countries is still founded on the widely known *SJM* and Winter–Carp formulae for the estimation of available sugar.

The *SJM* formula

The *SJM* formula, proposed by Deerr (1911), postulates that if a juice of *J* purity produces a sugar of *S* purity and molasses of *M* purity, the percentage of sucrose (pol) in the juice to go into the sugar will be

$$\frac{100S(J - M)}{J(S - M)}$$

The Winter–Carp formula

Douwes Dekker (1954) stated that Carp's name is wrongly associated with the Winter–Carp formula is, which was put forward by Winter in 1897. It is based on Winter's experience in Java, where he found that for every part of non-sucrose in the juice, 0.4 part of sucrose (pol) was retained in the final molasses. The formula is:

$$x = s\, 1.4 - \frac{40}{p}$$

where x = available sucrose (pol) per cent cane,
  s = per cent sucrose (pol) in terms of weight of cane,
  p = purity of juice.

The Mauritius system

Although most frequently used nowadays as measures of boiling-house efficiency, the *SJM* and Winter–Carp formulae are still the basis for cane payment in some countries. For example Antoine (1961), described the complicated method used in Mauritius under the auspices of the Central Arbitration and Control Board. The following standards of factory efficiency are set:

mill extraction	: 95 per cent at 12.5 fibre per cent cane
boiling-house recovery	: calculated according to the *SJM* formula, assuming a sugar purity of 99 per cent, final molasses purity (Clerget gravity) of 40 and boiling-house efficiency of 99 per cent.

Samples of cane are analysed for fibre, Brix and apparent purity. The figures are then corrected to reflect conditions throughout the whole crop rather than the harvest period of the individual farmer, and the sugar content of the cane is estimated. The farmer is paid two-thirds of the sugar value of his cane plus two-thirds of the average receipts for 'scums and molasses' per tonne cane.

Some reflections on cane farmers, the purchase of cane and price formulae

Cane payment systems

Comment has already been made upon the deficiencies of flat rate systems of cane payment and of those based on arbitrary standards of cane quality or factory efficiency. The one feature which each must possess is ready acceptance by both the farmers and the manufacturers. Mutual trust and goodwill, desirable everywhere, are essential in developing countries of doubtful political stability. Many aspiring politicians have risen to power (and some have failed to do so) by purporting to represent farmers' interests, when in fact their intention was to foment discontent between farmers and manufacturers.

Cane weights

The most important item in all price formulae is the weight of cane. Therefore scales must be accurate; and at each factory there should be a set of standard weights, carefully stored in velvet-lined boxes when not in use. Immediately before the start of crop, each scale at which cane is to be bought should be checked and adjusted to comply with the standard weights. Indeed in some countries it is required by statute that this must be done. For example in Trinidad the scales are tested by police officers, acting on behalf of the Inspector of Weights and Measures.

The scales should not only be accurate, but also be incapable of wrongful manipulation. However, human ingenuity is such that, whatever safeguards are built into their mechanisms, eventually someone will find ways and means of recording false weights. To counter this, each consignment should be reweighed at the factory, and regular comparisons made between the weight of cane bought and cane received. Even when deliveries are sent directly to the factory, they should be weighed twice: on a scale outside the compound and on one inside.

The tares of the vehicles carrying cane (farmers' and estate) should be checked regularly. During one year in Trinidad, appreciably larger quantities of cane were being received at the Brechin Castle factory than had been bought at the scales. If the figures were correct, the farmers were being cheated. Fortunately, they were incorrect. Investigation showed that the transport vehicles had not been tared since the start of crop, after which some had been strengthened by the addition of brackets and other pieces of iron-mongery; and all carried varying quantities of extraneous material (mostly mud). The metal and mud accounted for the discrepancies in weight. Thereafter all vehicles were retared at the beginning of each week.

By-products

When thriving new industries based on the by-products of sugar-cane manufacture are established (Ch. 11), some farmers (or their associations) feel that they should share in the profits. So they should, provided that, from the beginning, they have participated fully in the projects, and have made proportionate contributions to the capital at risk. This rarely happens. Nevertheless, farmers should receive the market value of the raw materials derived from their cane which are used in the new ventures.

For many years farmers in Jamaica shared in the proceeds of the sale of rum. A Commission of Enquiry (Biggs 1963) could find no grounds for this participation and recommended that, instead, the farmers' share of the total income from molasses should be derived from:

(a) the sale of molasses based upon the net export value in respect of molasses which are [*sic*] actually exported, plus

(b) a deduced price (based upon the net export price) for molasses which are [*sic*] retained by the manufacturers for the manufacture of rum and alcohol'.

It is difficult to disagree with these recommendations, and the principle on which they are founded, though traditionally the rum industry of Jamaica has been regarded as part of the sugar industry (Ch. 1).

There is a considerable world trade in molasses, but none in bagasse and filter mud. Therefore it is easy to assess the value of molasses, even where (as in Brazil) none is exported; but difficult to place a realistic value on bagasse used, for example, in the manufacture of a particle-board. However, when such industries become profitable, the farmers should be paid for the raw material which they provided. It is desirable that this be done, even if the figures used are to some extent notional, in the interests of good relations.

REFERENCES

Abbott, E. V. (1938) 'Red rot of sugarcane', *U.S. Dep. Agric. Tech. Bull.*, 641.
Abbott, E. V. (1956) 'Some aspects of testing sugarcane varieties for red rot resistance', *Proc. Int. Soc. Sug. Cane Technol.*, **9**, 1066–78.
Abbott, E. V. (1961) 'Mosaic', in *Sugar-cane Diseases of the World*. Vol. I, Elsevier, Amsterdam.
Abbott, E. V. (1963) 'Problems in sugar cane disease control in Louisiana', *Proc. Int. Soc. Sug. Cane Technol.*, **11**, 739–42.
Abbott, E. V. and Hughes, C. G. (1964) 'Red rot', in *Sugar-cane Diseases of the World*. Vol. II, Elsevier, Amsterdam.
Abbott, E. V. and Martin, J. P. (1952) 'The sugarcane situation in Peru', *Plant Disease Reporter*, **36**, 387–8.
Abbott, E. V. and Summers, E. M. (1951) 'An appraisal of some sugar-cane parent varieties and their progenies with particular reference to disease resistance', *Proc. Int. Soc. Sug. Cane Technol.*, **7**, 97–109.
Abbott, E. V., Zummo, N. and Tippett, R. L. (1967) 'Methods of testing sugarcane varieties for disease resistance at the U.S. sugar cane field station, Houma, Louisiana', *Proc. Int. Soc. Sug. Cane Technol.*, **12**, 1138–43.
Agee, H. P. (1931) 'Fertilization for soil amendment and maintenance', in *Report Ann. Gen. Mtg. H.S.P.A., Honolulu*, 1931.
Agee, H. P. and Dass, U. K. (1933). 'The day degree', *Rep. Hawaiian Sug. Technol.*, **12**, 45–8.
Alam, M. and Hudson, C. (1976) 'Root borer outbreak, January 1976', *Barbados Sug. Indust. Rev.*, **27**, 2–7.
Alexander, A. G. (1973) *Sugarcane Physiology*. Elsevier, Amsterdam, pp. 25–66.
Allison, R. V., Bryan, O. C. and Hunter, J. H. (1927) *Florida Expt. Sta. Bull.*, 190, quoted by H. Paul (1954) in 'The trace element status of peat soils in British Guiana', *Proc. Int. Soc. Sug. Cane Technol.*, **8**, 212.
Allison, W. F. (1974) 'A mechanical harvesting system without burning', *Proc. Int. Soc. Sug. Cane Technol.*, **15**, 1088–95.
Alvarez, L. A. (1954) 'Sugarcane diseases in Paraguay', *Proc. Int. Soc. Sug. Cane Technol.*, **8**, 902–7.
Anderson, J. (1964) 'Control of Childers cane grub', *Cane Grow. Q. Bull.*, **27**, 129–32.
Anon. (1969–70), Ann. Report, Expt. Sta. SASA.
Anon. (1970–71), Ann. Report, Exp. Sta. SASA.
Anon. (1971). *Metodologia para la Seleccion, Propagacion y Desaroolo de las Variedades de Cana*. IICA, Havana, Cuba, 2 vols.
Antoine, R. (1955) 'Cane diseases', *Rep. Maurit. Sug. Ind. Res. Inst.*, 1955, 44–5.
Antoine, R. (1961) 'Smut', in *Sugar-cane Diseases of the World*, Vol. I. Elsevier, Amsterdam.
Antoine, R. (1969) 'The two gumming diseases of sugarcane', *Proc. Int. Soc. Sug. Cane Technol.*, **13**, 1170–9.

Antoine, R. and Ricaud, C. (1961) 'Cane diseases: A method for inoculating leaf scald in field trials', *Rep. Maurit. Sug. Ind. Res. Inst.*, 1961, 55–6.

Antoine, R. *et al.* (1969) 'Sugarcane diseases and their world distribution', *Proc. Int. Soc. Sug. Cane Technol.*, **13**, 1952–77.

Arceneaux, G. (1965) 'Cultivated sugarcanes of the world and their derivation', *Proc. Int. Soc. Sug. Cane Technol.*, **12**, 844–54.

Arceneaux, G. (1968) 'Breeding sugarcane varieties for the northern Caribbean', *Proc. Int. Soc. Sug. Cane Technol.*, **13**, 1034–46.

Arruda, S. C. and Do Amaral, J. F. (1945) 'Leaf scald of sugar cane in Brazil', *Phytopathology*, **35**, 135–7.

Artschwager, E. (1939) 'Illustrated outline for use in taxonomic description of sugarcane varieties', *Proc. Int. Soc. Sug. Cane Technol.*, **6**, 116–28.

Artschwager, E. (1948) 'Vegetative characteristics of some wild forms of Saccharum and related grasses', *US Dep. Agric. Tech. Bull.*, 951.

Artschwager, E. and Brandes, E. W. (1958) 'Sugarcane (*Saccharum officinarum* L.)', *US Dep. Agric., Agriculture Handbook*, No. 122.

Artschwager, E., Brandes, E. W. and Starrett, R. C. (1929) 'Development of flower and seed of some varieties of sugar cane', *J. Agric. Res.*, **39**, 1–30.

Atchison, J. E. (1967) 'Experiences in developing, building and operating bagasse pulp and paper mills', *Proc. Int. Soc. Sug. Cane Technol.*, **12**, 1827–43.

Atchison, J. E. (1972) 'Review of progress with bagasse for use in industry', *Proc. Int. Soc. Sug. Cane Technol.*, **14**, 1189–201.

Atchison, J. E. (1974) 'Present status and future potential for utilization of bagasse in the pulp, paper and paperboard industry', *Proc. Int. Soc. Sug. Cane Technol.*, **15**, 1851–63.

Avasthy, P. N. (1967) 'The problem of white grubs of sugar cane in India', *Proc. Int. Soc. Sug. Cane Technol.*, **12**, 1321–33.

Avasthy, P. N. (1969) 'The top borer of sugar cane, *Scirpophaga nivella* (F.)', in *Pests of Sugar Cane*. Elsevier, Amsterdam.

Avasthy, P. N. and Krishnamurthy, T. N. (1969) 'Distribution and sampling of sugarcane internode borer (*Proceras indicus*) damage', *Proc. Int. Soc. Sug. Cane Technol.*, **13**, 1285–91.

Azizi, H. and Sund, K. A. (1972) 'A study of destructive rodents of Haft Tappeh', *Proc. Int. Soc. Sug. Cane Technol.*, **14**, 571–4.

Bacchi, O. S. *et al.* (1977) 'Minimum threshold temperature for sugar cane growth', *Proc. Int. Soc. Sug. Cane Technol.*, **16**.

Baker, B. P. (1974) *Composition, Properties and Uses of Molasses and Related Products*. United Molasses Trading Co. Ltd, London.

Baker, B. P. (1978) Personal communication.

Baker, R. E. D., Martyn, E. B. and Stevenson, G. C. (1953) 'Sugarcane diseases in the Caribbean', *Proc. Int. Soc. Sug. Cane Technol.*, **8**, 895–902.

Baldwin, A. L. and Fisher, M. W. (1969) 'An industrial approach to manual cane cutting' *Proc. W. Indies Sug. Technol.*, 1969, 166–77.

Baran, R., Bassereau, D. and Gillet, N. (1974) *Proc. Int. Soc. Sug. Cane Technol.*, **15**, 726–35.

Barat, H. (1954) 'Problemes sanitaires de la canne à sucre a Madagascar et en Australie', *Rev. agric. Ile Maurice*, **33**, 207–18.

Barbados SIR (1976) *Barbados Sugar Industry Review*, No. 30.

Barber, C. A. (1918) *Studies in Indian Sugar Canes, No. 3. The classification of Indian canes with special reference to the Saretha and Sunnabile groups*, Mem. Dept. Agric. India, Bot. Ser., 9.

Barnes, A. C. (1974) *The Sugar Cane*. Leonard Hill, London.

Barton, F. M. (1951) 'Benzene hexachloride for the control of *Lepidiota frenchi* grubs in cane', *Proc. Int. Soc. Sug. Cane Technol.*, **7**, 411–14.

Basheer, M. (1959) 'Observations on the usefulness of the large scale operations carried out against sugar cane borers', *Indian J. Ent.*, **20**, 164–6.

Bates, J. F. (1957) 'Developments in weed and pest control in sugarcane in British Guiana', *Proc. Br. W. Indies Sug. Technol.*, 1957, 100–4.

Bates, J. F. (1963) 'The canefield rat in British Guiana and its control', *Proc. Int. Soc. Sug. Cane Technol.*, **11**, 695–704.

Bates, J. F. (1967a) 'Pest control in sugarcane in the Americas', *Proc. Int. Soc. Sug. Cane Technol.*, **12**, 1270–7.

Bates, J. F. (1967b) 'Investigations on moth borers in British Guiana', *Proc. Int. Soc. Sug. Cane Technol.*, **12**, 1349–67.

Bates, J. F. (1969) 'Rodents in sugar cane', in *Pests of Sugar Cane*. Elsevier, Amsterdam.

Beachey, R. W. (1957) *The British West Indies Sugar Industry in the Late 19th Century*. Basil Blackwell, Oxford.

Beg, M. N. and Bennett, F. D. (1973) 'Insects associated with sugarcane on Abaco Island, the Bahamas', *Proc. W. Indies Sug. Technol.*, 1973, 228–34.

Belcher, B. A. (1969) 'List of varieties in the World Reference Collection of Sugarcane at Canal Point, Florida, 1967', *Proc. Int. Soc. Sug. Cane Technol.*, **13**, 876–87.

Bennett, F. D. (1962) 'Outbreaks of *Elasmopalpus lignosellus* (Zell.) (Lepidoptera: Phycitidae) on sugar-cane in Jamaica, Barbados and St. Kitts', *Trop. Agric. (Trin.)*, **39** (2), 153–6.

Bennett, F. D. (1969) 'Tachinid flies as biological control agents for sugar cane moth borers', in *Pests of Sugar Cane*. Elsevier, Amsterdam.

Bennett, F. D. and Pschorn-Walcher (1969) 'Recent investigations on the biological control of *Diatraea* spp. in Trinidad, the lesser Antilles and Trinidad', *Proc. Int. Soc. Sug. Cane Technol.*, **13**, 1321–30.

Bennett, M. C., Gardiner, F. J., Abram, J. C. and Rundell, J. T. (1972) 'The Talofloc decolorization process', *Proc. Int. Soc. Sug. Cane Technol.*, **14**, 1569–88.

Bernstein, L., Clark, R. A., Francois, L. E. and Derderian, M. D. (1966a) 'Salt tolerance of NCo varieties of sugar cane – 2. Effects of soil salinity and sprinkling on chemical composition', *Agron. J.*, **58**, 503–7.

Bernstein, L., Francois, L. E. and Clark, R. A. (1966b). 'Salt tolerance of NCo varieties of sugar cane – 1. Sprouting, growth and yield', *Agron. J.*, **58**, 489–93.

Bhargava, K. S., Joshi, R. D. and Rishi, N. 'Occurrence of strains A and F of sugarcane mosaic virus in Uttar Pradesh (India)', *Proc. Int. Soc. Sug. Cane Technol.*, **14**, 949–54.

Bianchi, F. A. (**1935**) 'Investigations on *Anomala orientalis* Waterhouse on Oahu Sugar Company Ltd', *Hawaii. Plrs' Rec.*, **39**, 234–55.

Bianchi, F. A. (1960a) 'Entomological changes in the sugar cane fields of Hawaii, 1930–1959', *Proc. Int. Soc. Sug. Cane Technol.*, **10**, 989–94.

Bianchi, F. A. (1960b) 'Present status of the rat problem on the sugarcane plantations of Hawaii', *Proc. Int. Soc. Sug. Cane Technol.*, **10**, 1016–18.

Bianchi, F. A. (1963). 'Notes on *Campsomeris marginella* (Klug) *modesta* (Smith) in Hawaii', *Proc. Int. Soc. Sug. Cane Technol.*, **11**, 658–65.

Bieske, G. C. (1965) 'Soil organic matter', in *Manual of Cane-growing*. American Elsevier, New York.

Biggs, H. C. (1963) *Report of a Commission of Enquiry into the Prices to be Paid for Cane Farmers' Cane*, The Government Printer, Kingston, Jamaica.

Bitancourt, J. J. (1939) 'Diseases of the sugarcane in Brazil', *Proc. Int. Soc. Sug. Cane Technol.*, **6**, 187–93.

Blackburn, F. H. B. (1949) 'Some recent developments in froghopper control', *Trop. Agric. (Trin.)*, **26**, 93–102.

Blackburn, F. H. B. (1950) In discussion following 'New horizon in sugar cane breeding' by G. C. Stevenson, *Proc. Br. W. Indies Sug. Technol.*, 1950, 125–31.

Blackburn, F. H. B. (1951). *Report on the Froghopper Outbreak in Jamaica*, Caroni Ltd, Trinidad (private circulation), 3 pp.

Blackburn, F. H. B. (1954) 'Further development in froghopper control', *Proc. Br. W. Indies Sug. Technol.*, 1954, 137–40.

Blackburn, F. H. B. (1967) President's address, *Proc. Sug. Tech. Trin.*, 1967, 132–3.
Blackburn, F. H. B., Hanschell, D. M. and Clarke, M. (1952) 'Some aspects of weed control in Trinidad', *Proc. Br. W. Indies Sug. Technol.*, 1950, 123–34.
Bleszynski, S. (1969) 'The taxonomy of the crambine moth borers of sugar cane', in *Pests of Sugar Cane.* Elsevier, Amsterdam.
Bonnet, J. A. (1953) 'Soil salinity studies as related to sugarcane growing in south-western Puerto Rico', *J. Ag. Univ. Puerto Rico*, **2**, 103–13.
Bonnet, J. A. (1968) 'Sugar cane problems on saline soils', *Sugar y Azucar*, **40** (6), 21–3.
Borden, R. J. (1944) *Hawaii. Plrs' Rec.*, **52**, 113.
Bourne, B. A. (1921) *Rep. Dep. Agric. Barbados*, 1919–20, 10–31.
Bourne, B. A. (1936) 'Methods of selecting cane seedlings in the Everglades as practised at the Florida Agricultural Experimental Station', *Proc. Int. Soc. Sug. Cane Technol.*, **5**, 333–8.
Bourne, B. A. (1954) 'Studies on sugarcane red rot in the Florida Everglades', *Proc. Int. Soc. Sug. Cane Technol.*, **8**, 915–24.
Bourne, B. A. (1961) '*Fusarium* sett or stem rot', in *Sugar-Cane Diseases of the World*, Vol. I. Elsevier, Amsterdam.
Box, H. E. (1939) 'Biological control of *Diatraea saccharalis* (Fabricius) in Saint Lucia, B.W.I.', *Proc. Int. Soc. Sug. Cane Technol.*, **6**, 223–40.
Box, H. E. (1953) *List of Sugar Cane Insects*, Commonw. Inst. Ent., London.
Box, H. E. (1954). 'A preliminary list of the insects affecting sugarcane in the lesser Antilles and Trinidad', *Proc. Int. Soc. Sug. Cane Technol.*, **8**, 549–53.
Box, H. E. and Guagliumi (1954) 'The insects affecting sugarcane in Venezuela', *Proc. Int. Soc. Sug. Cane Technol.*, **8**, 553–9.
Brandes, E. W. (1929) 'Into primeval Papua by seaplane', *Nat. Geog. Mag.*, **56**, 253–332.
Brandes, E. W. and Sartoris, G. B. (1936) 'Sugarcane: its origin and improvement', *US Dept. Agric. Yearbook of Agric.*, 1936, 561–623.
Breemen, J. F. van, Ellis, T. O. and Arceneaux, G. (1965). 'Sugarcane breeding at Central Romana', *Proc. Int. Soc. Sug. Cane Technol.*, **12**, 976–84.
Bremer, G. (1929) 'Short remarks on the cytology of *Saccharum*', *Proc. Int. Soc. Sug. Cane Technol.*, **3**, 403.
Bremer, G. (1932) 'On the somatic chromosome numbers of sugar cane forms and the chromosome numbers of indigenous Indian canes', *Proc. Int. Soc. Sug. Cane Technol.*, **4**, Bull. 20.
Bremer, G. (1961) 'Problems in breeding and cytology of sugar cane. I. A short history of sugar cane breeding. The original forms of *Saccharum*', *Euphytica*, **10**, No. 1.
Brenière, J., Betbeder-Matibet, M., Etienne, J. and Rakotondrahaja, R. (1966) 'Une tentative d'introduction à la Réunion et à Madagascar de *Diatraeophaga striatalis* Townsend pour la lutte contre *Proceras sacchariphagus*, borer ponctúe de la canne à sucre', *Agron. Trop., Nogent*, **21**, 361–84.
Brett, P. G. C. (1950) 'Flowering and pollen fertility in relation to sugarcane breeding in Natal', *Proc. Int. Soc. Sug. Cane Technol.*, **7**, 43–56.
Brett, P. G. C. (1954a) 'Temperature and ovule fertility of sugarcane in Natal', *Proc. Int. Soc. Sug. Cane Technol.*, **8**, 463–5.
Brett, P. G. C. (1954b) 'The first ten years of sugarcane breeding in Natal', *S. Afr. Sug. J.*, **38**, 309–15.
Briton-Jones, H. R. (1927) 'A note on green muscardine (*Metarrhizium anisopliae* Soroklin)', *Minut. Proc. Froghopper Invest. Comm.*, **1**, 293–306.
Browne, C. A. (1939) 'The development of the sugarcane industry in Louisiana and the southern United States', *Proc. Int. Soc. Sug. Cane Technol.*, **6**, 46–70.
Browne, C. A. and Zerban, F. W. (1941) *Sugar Analysis.* Chapman and Hall, London.
Buchanan, E. J. and Brockensha, M. A. (1974) 'The application of direct cane testing

to the South African sugar industry', *Proc. Int. Soc. Sug. Cane Technol.*, **15**, 1456–69.

Bull, T. A. and Glasziou, K. T. (1975) Chapter 3 *Sugar Cane* in *Crop Physiology*, L. T. Evans (ed.), Cambridge University Press.

Burr G. O. *et al.* (1957). *Ann. Rev. Pl. Physiol.*, **8**.

Butani, D. K. (1961) 'Annotated list of insects on sugarcane in India', *Indian J. Sugarcane Res.*, **5** (1 or 2), 126–37.

Butler, E. J. (1918) *Fungi and Diseases in Plants.* Thacker, Spink, Calcutta.

Buxo, D. (1968) 'Insecticides and moth borer damage', *Proc. Sug. Techn. Trin.*, **2**, 52–6.

Buzacott, J. H. (1948) 'The use of benzene hexachloride in north Queensland canefields', *J. Aust. Inst. Agric. Sci.*, **14**, 24–7.

Buzacott, J. H. (1960) 'Methods of cane breeding used by the Queensland Bureau of Experiment Stations', *Proc. Int. Soc. Sug. Cane Technol.*, **10**, 690–4.

Buzacott, J. H. and Hughes, G. C. (1951) 'The 1951 cane collecting expedition to New Guinea', *Cane Grow. Q. Bull.*, **15**.

Bynum, E. K., Ingram, J. W., Charpentier, L. J. and Hayley, W. E. (1951) 'Control of soil insects in Louisiana sugar-cane fields', *Proc. Int. Soc. Sug. Cane Technol.*, **7**, 423–34.

Byther, R. S. and Steiner, G. W. (1974) 'Comparison of inoculation techniques for smut disease testing in Hawaii', *Proc. Int. Soc. Sug. Cane Technol.*, **15**, 280–8.

Byther, R. S., Steiner, G. W. and Wismer, C. A. (1971) 'New sugarcane diseases reported in Hawaii', *Sugarcane Pathol. Newsletter*, **7**, 18–21.

Caresche, L. (1962) 'Les insects nuisibles à la canne à sucre dans l'Isle de la Réunion', *Agron. Trop., Nogent*, **17**, 632–46.

Caresche, L. and Brénière, J. (1962) 'Les insects nuisibles à la canne à sucre à Madagascar. Aspects actuels de la question', *Agron. Trop. Nogent*, **17**, 608–31.

Carl, K. (1962) 'Graminaceous moth borers of West Pakistan', *Tech. Bull. Commonw. Inst. Biol. Contr.*, **2**, 29–76.

Carnegie, A. J. M. (1971) 'Our most important cane insects in the South African sugar industry', *S. Afr. Sug. J.*, **55**, 611–15; **56**, 13–19.

Carnegie, A. J. M. (1974) *Rep. Expt. Sta. S. Afr. Sug. Ass.*, 1973–74, 32.

Carnegie, A. J. M. (1976) *Rept. Expt. Sta. S. Afr. Sug. Ass.*, 1975–76, 36–40.

Charpentier, L. J. and Mathes, R. (1969) 'Cultural practices in relation to stalk moth borer infestations in sugar cane', in *Pests of Sugar Cane*. Elsevier, Amsterdam.

Chen, C. B. and Hung, T. H. (1962) 'Experiment results of introducing parasitic wasps and fly for controlling sugar cane borers from India and Taiwan in 1961–62', *J. Agric. Ass. China* (New Ser.), **40**, 63–71.

Chen, C. B. and Hung, T. H. (1963) 'Experimental results of the introduced *Isotima javensis* Rohw. for the control of the top borer, *Scirpophaga nivella* F., in Taiwan', *Rep. Taiwan Sug. Exp. Sta.*, **31**, 137–46.

Chen, C. B. and Hung, T. H. (1969) 'Experimental results with biological control of sugar cane borers in Taiwan', *Proc. Int. Soc. Sug. Cane Technol.*, 1300–4.

Chilton, St J. P., Paliatseas, E. D. and Perdomo, R. (1965) 'Production of true seed of sugarcane in Louisiana', *Proc. Int. Soc. Sug. Cane Technol.*, **12**, 742–50.

Chinloy, T. S. (1955) 'A survey of the intensity of small moth borer (*Diatraea saccharalis* F.) infestation in the sugar cane of Jamaica Sugar Estates', *Proc. Br. W. Indies Sug. Technol.*, 1955, 73–8.

Chinloy, T. and Hogg, B. M. (1969) 'Rotation of Pangola grass and sugarcane on a Jamaican sugar estate', *Proc. Int. Soc. Sug. Cane Technol.*, **13**, 636–42.

Chona, B. L. (1956) Chairman's address, Pathology section, *Proc. Int. Soc. Sug. Cane Technol.*, **9**, 975–86.

Chu, C. C. and Kong, L. (1971) 'Photorespiration of sugar cane', *Taiwan Sugar Exp. Sta. Ann. Rept.*, 1–14.

Chu, H. T. and Wang, S. C. (1949) 'Effects of several fungicides on the pathogens of some important diseases of sugar cane', *Rep. Taiwan Sug. Exp. Sta.*, **4**, 204–9.

Chu, T. L. (1965) 'Sugarcane breeding in Taiwan', *Proc. Int. Soc. Sug. Cane Technol.*, **12**, 855–63.

Cil, J. A. (1974) 'Pneumatic removal of extraneous matter by sugarcane harvesters', *Proc. Int. Soc. Sug. Cane Technol.*, **15**, 1124–35.

Clarke, G. F. (1959) *Indian Sugar Industry*, BWISA Reports, 1959.

Clarke, G. F. (1962) *Sugar in Mauritius*, BWISA Reports, 1962.

Clarke, I. S. (1960) Internal memorandum, Agric. Dept., Caroni Ltd, Trinidad (unpublished).

Clarke, M. (1979) Personal communication.

Clarke, M., Connell, H. le R. and Elcock, H. L. (1975) 'The yield of sugar cane in Barbados in 1975', *Min. Agric., Sci. and Tech., Barbados Bull.*, 60.

Clarke, M., Reece, N. E. and Elcock, H. L. (1976) 'The yield of sugar cane in Barbados in 1976', *Min. Agric., Sci. and Tech., Barbados Bull.*, 61.

Clarke, M., Reece, N. E. and Elcock, H. L. (1977) 'The yield of sugar cane in Barbados in 1977', *Min. Agric., Food and Consumer Affairs, Barbados Bull.*, 62.

Clayton, J. E. (1969) 'Harvester developments in Florida, Puerto Rico, Louisiana, Australia and Hawaii', *Proc. W. Indies Sug. Technol.*, 1969, 160–5.

Cleare, L. D. (1928) 'A method for the rearing of egg parasites of the sugar cane moth borers', *Bull. Ent. Res.*, **19**, 31–8.

Cleare, L. D. (1939) 'The Amazon fly (*Metagonistylum minense* Towns.) in British Guiana', *Bull. Ent. Res.*, **30**, 85–102.

Clements, H. F. (1948) 'Crop logging sugar cane in Hawaii', *Better Crops with Plant Food*, **32**, 11–18, 45–8.

Clements, H. F. (1972) 'The crop logging system for sugarcane production: 1971 edition', *Proc. Int. Soc. Sug. Cane Technol.*, **14**, 657–72.

Clements, H. F. (1980) *Sugarcane Crop Logging and Crop Control. Principles and practice.* Pitman Publishing, London, 520 pp.

Clements, H. F. and Awada *India J. of Sugar Cane*, Res. & Dev. Jubilee No. VIII, Pt. 2.

Clements, H. F. and Kubota, T. (1942) *Hawaiian Pl. Rec.*, **46**, 17–35.

Cobb, N. A. (1906) 'Fungus maladies of the sugar cane', *Hawaii. Sug. Plrs. Ass., Div. Path. and Phys., Bull.*, 5.

Cobb, N. A. (1909) 'Fungus maladies of the sugar cane', *Hawaii. Sug. Plrs. Ass., Div. Path. and Phys., Bull.*, 6.

Coburn, G. E. and Hensley, S. D. (1971) 'Differential survival of *Diatraea saccharalis* larvae on two varieties of sugarcane', *Proc. Int. Soc. Sug. Cane Technol.*, **14**, 440–4.

Cole, A. C. (1973) 'A locally designed container for transporting chopper harvested cane in St. Kitts', *Proc. West Indies Sug. Technol.*, 1973, 219–25.

Coleman, R. E. *Proc. Int. Soc. Sug. Cane Technol.*, **10**, 805–19.

Coleman, R. E. (1968a) 'Physiology of flowering in sugar cane', *Proc. Int. Soc. Sug. Cane Technol.*, **13**, 795–812.

Coleman, R. E. (1968b) 'Physiology of flowering in sugar cane', *Proc. Int. Soc. Sug. Cane Technol.*, **13**, 992–1000.

Collado, J. C. and Ruano, M. A. (1963) 'The rat problem in the sugar cane plantations of Mexico', *Proc. Int. Soc. Sug. Cane Technol.*, **11**, 705–11.

Cruger, H. (1892) 'Cane diseases and pests' (written in 1863), *Agric. Red., Trinidad*, **7**, 78–83.

Cust, R. J. (1865) *The West Indian Incumbered Estates Acts.* Williams Amer, Lincoln's Inn Gate, London.

Da Eira, A. F., De Carvalho, P. C. T. and Sanguino, A. (1974) 'Studies on aggressiveness of *Fusarium moniliforme* Sheldon, causal agent of pokkah boeng in sugarcane', *Proc. Int. Soc. Sug. Cane Technol.*, **15**, 374–83.

Dale, W. T. (1950) *Report on a visit to Uitvlugt Estate, British Guiana, 23rd–25th November, 1950, in Connection with an Outbreak of Sugar Cane Leaf Scald (X. albilineans).* Report to BGSPA.

Dallas, R. C. (1803) *The History of the Maroons*. Longman and Rees, London.

Daniels, J. (1965) 'Improving sugarcane breeding methods to increase yields', *Proc. Int. Soc. Sug. Cane Technol.*, **12**, 742–50.

Daniels, J., Horsley, D. R., Masilaca, A. S., Miles, K. G., Singh, H., Stevenson, N. D. and Wilson, B. (1971) 'The mass stool population technique of sugarcane selection', *Proc. Int. Soc. Sug. Cane Technol.*, **14**, 163–9.

Daniels, J., Husain, A. A., Hutchinson, P. B. and Wismer, C. A. (1969) 'An insectary method for testing sugarcane varieties for resistance to Fiji disease', *Proc. Int. Soc. Sug. Cane Technol.*, **13**, 1100–6.

Daniels, J., Husain, A. A. and Hutchinson, P. B. (1972) 'The control of sugarcane diseases in Fiji', *Proc. Int Soc. Sug. Cane Technol.*, **14**, 1007–14.

Dantas, B. and Da Silva, A. P. (1956) 'Amelhoria de germinacão de caña de azucar, pelo tratamento fungicida das estacas', *Bo. Techn. Inst. Agron. Nordesto (Brazil)*, **4**, 18–45. Cited by Wismer (1961).

David, H. and Kalra, A. N. (1967) 'A study of the sugar cane internode borer, *Proceras inducus* Kapur', *Proc. Int. Soc. Sug. Cane Technol.*, **12**, 1444–53.

Davies, J. G. (1938) *Experiments using titanium chloride for yellow sugar production*, ICTA, circular.

Davies, J. G. (1942) Opening address, *Proc. W. Indies Sug. Technol.*, 1942, 8.

De Charmoy, D. d'E. (1912) *Report on Phytalus smithi Arrow, and other Beetles injurious to the Sugar Cane in Mauritius*, Govt. Print., Port Louis.

De Charmoy, D. d'E. (1917) 'Notes relative to the importation of *Tiphia parallela* Smith from Barbados to Mauritius for the control of *Phytalus smithi* Arrow', *Bull. Ent. Res.*, **8**, 93–102.

De Charmoy, D. d'E. (1923) 'An attempt to introduce Scoliid wasps from Madagascar to Mauritius', *Bull. Ent. Res.*, **13**, 245–54.

de Gues, J. G. (1967) 'Fertiliser guide for tropical and sub-tropical farming', 1967, Centre d'Étude de l'Azote. Zurich, 115–45.

Deacon, H. F. E. (1969) 'Experience in the use and development of plantation scale fully mechanized harvesting in the Caribbean', *Proc. Int. Soc. Sug. Cane Technol.*, **13**, 1496–512.

Deerr, N. (1911) *Cane Sugar*. Norman Rodger, London.

Deerr, N. (1949) *The History of Sugar Cane*, Vol. I. Chapman and Hall, London.

Demain, A. L. (1981) *Science*, **214**, 987–95.

Dick, J. (1951) 'Sugar-cane entomology in Natal, South Africa', *Proc. Int. Soc. Sug. Cane Technol.*, **7**, 377–97.

Dick, J. (1969a) 'The results and prospects of nematocidal soil treatment in sugar cane fields', in *Pests of Sugar Cane*. Elsevier, Amsterdam.

Dick, J. (1969b) 'The mealybugs of sugarcane', in *Pests of Sugar Cane*. Elsevier, Amsterdam.

Dixon, J. M. (1956) 'Sugar milling by C.S.R.', in *South Pacific Enterprise*. Angus and Robertson, Sydney, Australia.

Dodds, H. H. (1943) 'Sugarcane diseases and insect pests in South Africa', *S. Afr. Sug. J.*, **27**, 340–405.

Donawa, L. P. (1967) 'The use of self-loading trailers', *Proc. Sug. Tech. Ass. Trin.*, 1967, 64–5.

Donawa, L. P. (1973) 'The development of a mechanical cane system in Trinidad 1962–1972', *Proc. W. Indies Sug. Technol.*, 1973, 206–18.

Doty, J. W. (1960) 'Rodent control in commercial cane areas of Florida Everglades', *Proc. Int. Soc. Sug. Cane Technol.*, **10**, 1011–15.

Douwes Dekker, K. (1954) 'Judging boiling house work', *Proc. Int. Soc. Sug. Cane Technol.*, **8**, 671–82.

Douwes Dekker, K. (1956) 'The determination of certain qualities of individual consignments of sugar cane', *Proc. Int. Soc. Sug. Cane Technol.*, **9**, No. 2, 722–33.

Duffey, H. R. (1969) 'Bagasse as a raw material for furfural', *Proc. Int. Soc. Sug. Cane Technol.*, **13**, 1880.

Dutt, N. L. and Rao, J. T. (1950) *Coimbatore canes in Cultivation. Their morphological descriptions and agricultural characteristics.* The Indian Central Sugarcane Committee, New Delhi.

Earle, F. S. (1928) *The Sugarcane and its Culture.* Wiley, New York.

Easton, M. E. G. (1968) 'The deterioration of sugar in storage', *Proc. Sug. Tech. Ass. Trin.*, 1967, 97–105.

Edgerton, C. W. (1951) 'Results of recent investigations in Louisiana on red rot', *Proc. Int. Soc. Cane Technol.*, **7**, 518–23.

Edgerton, C. W. (1955) *Sugarcane and its Diseases*, Louisiana State Univ. Studies, Biol. Sci. Series No. 3, Louisiana State Univ. Press, Baton Rouge.

Edwards, B. (1793) *The History, Civil and Commercial of the British Colonies in the West Indies.* Vol. 2, Luke White, Dublin, pp. 228–37.

Edwards, B. (1801) *An Historical Survey of the Island of Saint Domingo.* John Stockdale, Piccadilly, London.

Egan, B. T. (1972a) 'Leaf-scald disease: introduction', *Proc. Int. Soc. Sug. Cane Technol.*, **14**, 920–4.

Egan, B. T. (1972b) 'Breeding for resistance to leaf scald disease', *Proc. Int. Soc. Sug. Cane Technol.*, **14**, 920–4.

Egan, B. T. *et al.* (1972) 'Report of the standing committee on sugarcane diseases', *Proc. Int. Soc. Sug. Cane Technol.*, **14**, 1655–81.

Egan, B. T. *et al.* (1974) 'Sugarcane diseases and their world distribution', *Proc. Int. Soc. Sug. Cane Technol.*, **15**, xxxii–xxxiv.

Egan, B. T., Kirby, L. K. and Noble, A. G. (1978) 'A review of recent developments concerning the biodeterioration of sugar cane', *Proc. Int. Soc. Sug. Cane Technol.*, **16**.

Empig, L. T. (1974) 'Philippine sugarcane breeding programme: philosophies and strategies', *Proc. Int. Soc. Sug. Cane Technol.*, **15**, 17–23.

Etienne, J. (1969) 'Two years rearing of the Javanese tachinid, *Diatraeophaga striatalis* Towns., a parasite of the cane moth-borer *Proceras sacchariphagus* Boj., in Réunion island', *Proc. Int. Soc. Sug. Cane Technol.*, **13**, 1305–30.

Evans, D. E. and Buxo, D. A. (1972) 'Insecticidal control of the sugarcane froghopper', *Proc. Int. Soc. Sug. Cane Technol.*, **15**, 507–15.

Evans, H. (1935) 'The root system of the sugarcane. I. Methods of study', *Emp. J. Exp. Agr.*, **3**, 351–62.

Evans, H. (1936) 'The root system of the sugar cane. II. Some typical root systems', *Emp. J. Exp. Agr.*, **4**, 208–20.

Evans, H. (1947, 1948) *Annual Reports.* Sugar Cane Research Station, Mauritius.

Evans, H. (1960) 'Elements other than nitrogen, potassium and phosphorus in the nutrition of sugar cane', *Proc. Int. Soc. Sug. Cane Technol.*, **10**, 473–508.

Evans, H. (1962) *Sugar in Mauritius.* BWI Sug. Ass. (inc.).

Evans, H. (1965) Golden Jubilee Souvenir of the Anakapella Research Station, Andra Pradesh, India.

Evans, H. and Wiehe, P. O. (1947) 'Experiments on the treatment of cane setts at planting under Mauritius conditions', *Mauritius Sugarcane Res. Sta. Bull.*, 19.

Fawcett, G. L. (1924) 'Las enfermedales de la caña de azucar en Tucuman', *Estac. Exp. Agric., Tucuman, Bol.*, 1.

Fennah, R. G. (1939) 'A summary of experimental work on varietal resistance of sugar cane to *Tomaspis saccharina* 1936–39', *Trop. Agric., Trin.*, **16**, 223–40.

Fennah, R. G. (1969) 'Damage to sugar cane by Fulgoroidea and related insects in relation to the metabolic state of the host plant', in *Pests of Sugar Cane.* Elsevier, Amsterdam.

Fewkes, D. W. (1966) 'Some observations on the biology of 'jumping borer' (Lepidoptera, Phycitidae) on sugar cane in Trinidad', *Proc. Br. W. Indies Sug. Technol.*, 1966, 233–5.

Fewkes, D. W. (1969a) 'The biology of sugar cane froghoppers', in *Pests of Sugar Cane.* Elsevier, Amsterdam.

Fewkes, D. W. (1969b) 'The control of froghoppers in sugar cane plantations', in *Pests of Sugar Cane*. Elsevier, Amsterdam.

Fewkes, D. W. (1972) 'Notes on a recent outbreak of *Aulacapsis tegalensis* (Homoptera Diaspididae) on sugarcane in Tanzania', *Proc. Int. Soc. Sug. Cane Technol.*, **14**, 527–35.

Fewkes, D. W. and Buxo, D. A. (1965) *Rep. Tate and Lyle Cent. Agric. Res. Sta.*, Trinidad, 1964, 12–20.

Fewkes, D. W. and Buxo, D. A. (1969) 'Chemical control of sugarcane froghopper (Homoptera Cercopidae) infestations from the air in Trinidad', *Proc. Int. Soc. Sug. Cane Technol.*, **13**, 1348–64.

Fitzgerald, J. R. and Lamusse, J. P. (1974) 'Diffusion in South Africa', *Proc. Int. Soc. Sug. Cane Technol.*, **15**, 1486–98.

Flores, C. S. (1972) 'Plagas de la caña de azucar en Mexico', *Boln. Azucarero Mexicano*, April, 4–9.

Flores, S. and Ramirez, A. (1963) 'Eye spot disease in Mexico (*Helminthosporium sacchari* (v. Breda de Haan) Butler)', *Proc. Int. Soc. Sug. Cane Technol.*, **11**, 807–9.

Flores, C. S. and Ruano, M. A. (1961) 'Principales plagas de la caña de azucar en Mexico', *Boln. Divulg. I.M.P.A.*, Mexico, 4.

Floro, M. B., Williams, V., Flook, W. A. and Collier, J. S. (1947) 'Food yeast – some aspects of its development and production', *Proc. Br. W. Indies Sug. Technol.*, 1947, 38–54.

Fogliata, F. A. and Aso, P. J. (1965a) 'Effectos de la salinidad y sodio intercambiable del suelo en el crecimiento de la caña de azucar', *Rev. Ind. Agric. Tucuman*, **43**, 25–45 (Argentina).

Fogliata, F. A. and Aso, P. J. (1965b) 'The effects of soil soluble salts on sucrose yield of sugar cane', *Proc. Int. Soc. Sug. Cane Technol.*, **12**, 682–94.

Follett-Smith, R. R. (1943) 'Natural humus and artificial fertilizers', *Proc. Br. W. Indies Sug. Technol.*, 1943, 21–6.

Fors, A. L. (1960) 'An appraisal and history of Cuban-produced sugar cane varieties', *Proc. Int. Soc. Sug. Cane Technol.*, **10**, 843–50.

Fuchs, T. W., Harding, J. A. and Dupnik, E. (1973) 'Sugar cane borer control on sugar cane in the lower Rio Grande Valley of Texas with aerially applied chemicals', *J. Econ. Ent.*, **66**, 802–3.

Fullaway, D. T. and Krauss, N. L. H. (1945) *Common Insects of Hawaii*, Tongg, Hawaii.

Fuller, R. G. (1976) 'The sugar cane separation process', *Barbados Sug. Ind. Rev.*, **29**, 3–6.

Gard, K. R. (1954) 'The aims of the sugarcane breeding programme at Macknade, Queensland, with special reference to sugar content, vigour and disease resistance', *Proc. Int. Soc. Sug. Cane Technol.*, **8**, 423–9.

Garlough, F. E. (1939) 'Rodents in relation to sugar cane growing', *Proc. Int. Soc. Sug. Cane Technol.*, **6**, 98–105.

George, E. F. (1959) 'Effect of the environment on components of yield in seedlings from five *Saccharum* crosses', *Proc. Int. Soc. Sug. Cane Technol.*, **10**, 755–65.

George, E. F. (1962) 'A further study of *Saccharum* progenies in contrasting environments', *Proc. Int. Soc. Sug. Cane Technol.*, **11**, 488–97.

Ghani, M. A. and Williams, J. R. (1963) 'An attempt to establish the Javanese fly, *Diatraeophaga striatalis* Towns. in Mauritius for control of the cane moth-borer, *Proceras sacchariphagus* Boj., with notes upon parasites of cane moth-borers in Java', *Proc. Int. Soc. Sugar Cane Technol.*, **11**, 626–42.

Ghosh, C. C. (1937) 'The black beetle (*Alissonotum impressicolle* Arr.) a pest of sugarcane in Myitkyina district in northern Burma', *Indian J. Agric. Sci.*, **7**, 907–31.

Gibson, W. (1978) *Int. Sugar J.*, **80**, 362–6.

Gilbert, S. M. (1930) 'The use of legumes on sugar estates', *Minutes and Proc. Sug.*

References

Invest. Cttee., Trin. and Tobago, **3**, 174–92.

Gillaspie, A. G., Davis, R. E. and Worley, J. E. (1974) 'Nature of the ratoon stunting disease agent', *Proc. Int. Soc. Sug. Cane Technol.*, **15**, 218–24.

Girling, D. J. (1972) '*Eldana saccharina* Wlk. (Lepidopterae: Pyralidae), a pest of sugarcane in East Africa', *Proc. Int. Soc. Sug. Cane Technol.*, **14**, 429–34.

Goberdahn, L. P. (1973) 'Twenty years after – the present situation of weed control in the northern areas of Caroni', *Proc. W. Indies Sug. Technol.*, 1973, 82–92.

Gosnell, J. M. and Lonsdale, H. (1977) *Proc. Int. Soc. Sug. Cane Technol.*, **16**, 1565.

Gowing, D. P. and Baniaboassi, N. (1978k 'Observations on cane ripening in the Iranian winter', *Proc. Int. Soc. Sug. Cane Technol.*, **16**.

Gowing, D. P. *et al.* (1977) 'Observations on cane ripening in the Iranian winter', *Proc. Int. Soc. Sug. Cane Technol.*, **16**.

Graham, W. S., Morris, R. M. and Oosthuizen, D. M. (1969) 'Preliminary physio-chemical studies on sugarcane diffusers', *Proc. Int. Soc. Sug. Cane Technol.*, **13**, 122–31.

Grassl, C. O. (1946) '*Saccharum robustum* and other wild relatives of "noble" sugar canes', *J. Arnold Arb.*, **27**, 234–52.

Grassl, C. O. (1967) 'Introgression between *Saccharum* and *Miscanthus* in New Guinea and the Pacific area', *Proc. Int. Soc. Sug. Cane Technol.*, **12**, 995–1003.

Grassl, C. O. (1969) '*Saccharum* names and their interpretation', *Proc. Int. Soc. Sug. Cane Technol.*, **13**, 868–75.

Greene, W. *et al.* (1935) *United Kingdom Sugar Industry Inquiry Committee Report*, Cmd 4871, HMSO, London.

Grist, D. H. (1975) *Rice*, 5th edn. Longman, London.

Guagliumi, P. (1954) 'Nuevas recommendaciones para el combate de la candelilla' *Revta. Fac. Agric. Univ. cent Venez.*, **30**, 17–19.

Guagliumi, P. (1960) 'Actual situation of entomology of sugar cane in Venezuela', *Proc. Int. Soc. Sug. Cane Technol.*, **10**, 1000–11.

Guagliumi, P. (1962) *Las Plagas de la caña de azucar en Venezuela*. Minist. Agric. y Cria., Maracay, Venezuela.

Guppy, P. L. (1914) 'Breeding and colonizing the Syrphid', *Bull. Dep. Agric. Trin. Tobago*, **13**, 217–26.

Gupta, B. D. and Avasthy, P. N. (1957) 'Observations on a new beetle pest of sugarcane crop in Bihar', *Indian Sug.*, **7**, 587–93.

Gupta, B. D. and Avasthy, P. N. (1960) 'Biology and control of the stem borer *Chilo tumidicostalis* Hmpsn.', *Proc. Int. Soc. Sug. Cane Technol.*, **10**, 886–901.

Gupta, M. C. and Gupta, B. D. (1959) 'Sugar cane and its problems: insect pests in India II. The pink borer, *Sesamia inferens* (Walker)', *Indian Sug.*, **9**, 1–4.

Hagley, E. A. C. (1967) 'Studies on the aetiology of froghopper blight in sugarcane. I. Symptom expression and development on sugarcane and other plants. II. Prob-able role of enzymes and amino acids in the salivary secretions of the adult frog-hopper', *Proc. Br. W. Indies Sug. Technol.*, 1966, 183–92.

Handojo, H. and Noordam, D. (1972) 'Purification and serology of sugarcane mosaic virus', *Proc. Int. Soc. Sug. Cane Technol.*, **14**, 973–84.

Hanschell, D. M. (1967) 'Cultivation practices and productivity 1890–1967', *Proc. Sug. Tech. Ass. Trinidad and Tobago*, **1**, 8–14.

Hardy, F. (1926) 'A note on the results of "cyanogas" dusting trials for destroying froghopper nymphs', *Minut. Proc. Froghopper Invest. Comm.*, **1**, 94–7.

Harris, W. V. (1969) 'Termites as pests of sugar cane', in *Pests of Sugar Cane*. Elsevier, Amsterdam.

Harrison, J. B. and Bovell, J. R. (1888) *Report on the Results Obtained upon the Experimental Fields at Dodds Reformatory, Barbados.*

Hazelhoff, E. H. (1932) 'Investigations on the white top borer', *Proc. Int. Soc. Sug. Cane Technol.*, **4**, Bull. 66.

Hearne, C. E. D. (1968) 'Bagassosis: An epidemiological, environmental and clinical survey', *Brit. J. Industr. Med.*, **25**, 267–82.

Hebert, L. P. (1956) 'Effect of seed-piece size and rate of planting on yields of sugarcane and of nitrogen fertilization on yield of seed-cane in Louisiana', *Proc. Int. Soc. Sug. Cane Technol.*, **9**, 301–10.

Hebert, L. P. (1967) 'Row-spacing experiments with sugarcane in Louisiana', *Proc. Int. Soc. Sug. Cane Technol.*, **12**, 96–102.

Hebert, L. P., Matherne, R. J. and Davidson, L. G. (1967) 'Row-spacing experiments with sugarcane in Louisiana', *Proc. Int. Soc. Sug. Cane Technol.*, **12**, 96–102.

Hernandez, O., Jose, V. and Silverio Flores, C. (1956) 'The biology and control of *Aeneolamia postica* (Wlk)', *Proc. Int. Soc. Sug. Cane Technol.*, **9**, 821–36.

Hickson, J. L. (ed) (1977) *Sucrochemistry*. American Chemical Society, Washington, DC.

Hilton, H. W., Robinson, W. H., Teshima, A. H. and Nass, R. D. (1972) 'Zinc phosphide as a rodenticide for rats in Hawaiian sugarcane', *Proc. Int. Soc. Sug. Cane Technol.*, **14**, 561–70.

Hoekstra, R. G. (1978) 'Investigation into the effectiveness of maturity testing for non-irrigated sugarcane fields', *Proc. Int. Soc. Sug. Cane Technol.*, **16**.

Hogarth, D. M. (1978) 'Plant breeding programme (Queensland)', *Producers' Rev. Qd.*, **78** (4), 50–4.

Holguin, F. (1973) Ingenio Mayaguez, Cali, Colombia, personal communication.

Holtzmann, O. V. and Wismer, C. A. (1967) 'Effectiveness of D-D and DBCP against lesion and rootknot nematodes on sugar cane in Hawaii', *Proc. Int. Soc. Sug. Cane Technol.*, **12**, 1413–19.

Hong, H. L. (1956) '*Fusarium*-pokkahboeng resistance trials', *Proc. Int. Soc. Sug. Cane Technol.*, **9**, 1023–9.

Honig, P. (1951) 'Developments in cane sugar production since 1938', *Proc. Int. Soc. Sug. Cane Technol.*, **7**, 750–73.

Horau, M. (1967) 'Sugar cane diseases in Réunion island', *Proc. Int. Soc. Sug. Cane Technol.*, **12**, 1232–5.

Howard, Sir A. 'The writing on the wall', *Int. Sug. J.*

Howes, J. R., Lines, M. G. and Wiggins, L. F. (1955) 'Studies on ammoniated molasses. Part III. Further work on the feeding of ammoniated cane molasses to livestock', *Proc. Br. W. Indies Sug. Technol.*, 1955, 103–8.

HSPA (1957) *Rep. Hawaii Sug. Plrs. Ass.*, 1957, 21–2.

Hudson, J. C (1973). 'Fire water and sugar production in Barbados', *Proc. W. Indies. Sug. Technol.*, 1973, 139–46.

Hudson, J. C. (1974) 'The BSPA cane cutter', *Proc. Int. Soc. Sug. Cane Technol.*, **15**, 1096–1105.

Hudson, J. C. (1978) 'A system for whole stick cane harvesting', *Proc. Int. Soc. Sug. Cane Technol.*, **16**.

Hughes, C. G. (1951) 'The control of cane diseases in Queensland', *Proc. Int. Soc. Sug. Cane Technol.*, **7**, 465–73.

Hughes, C. G. (1954) 'Red rot disease of sugarcane', *Proc. Int. Soc. Sug. Cane Technol.*, **8**, 924–36.

Hughes, C. G. (1961) 'Gumming disease', in *Sugar-cane Diseases of the World*, Vol. I. Elsevier, Amsterdam.

Hughes, C. G. (1974) 'The economic importance of ratoon stunting disease', *Proc. Int. Soc. Sug. Cane Technol.*, **15**, 213–17.

Hughes, C. G. and Robinson, P. E. (1961) 'Fiji disease', in *Sugar-Cane Diseases of the World*, Vol. I. Elsevier, Amsterdam.

Hughes, C. G., Abbott, E. V. and Wismer, C. A. (eds) (1964) *Sugar-Cane Diseases of the World*, Vol. II. Elsevier, Amsterdam.

Hughes, C. G., Steindl, D. R. L. and Egan, B. T. (1968) *Rep. Bur. Sug. Exp. Sta. Qd.*, **70**, 49–54.

Humbert, R. P. (1968) The Growing of Sugar Cane. Elsevier, Amsterdam.

Humbert, R. P. (1971) *Proc. Int. Soc. Sug. Cane Technol.*, **14**, 727.

Humbert, R. P. (1974) 'Improving burns with desiccants as an aid to mechanical harvesting', *Proc. Int. Soc. Sug. Cane Technol.*, **15**, 1065–73.

Humbert, R. P. (1975) 'Harvesting the sun with sugarcane', *World Farming*, Nov., 20.

Humbert, R. P., Zamora, M. and Fraser, T. B. (1967) 'Ripening and maturity control progress at Ingnio Las Mochis, Mexico', *Proc. Int. Soc. Sug. Cane Technol.*, **12**, 446–52.

Husain, A. A. and Hutchinson, P. B. (1972) 'Further experience with the insectary method of testing sugarcane varieties for resistance to Fiji disease', *Proc. Int. Soc. Sug. Cane Technol.*, **14**, 1001–6.

Hutchinson, P. B. (1969) 'A note on disease resistance ratings for sugarcane varieties', *Proc. Int. Soc. Sug. Cane Technol.*, **13**, 1087–9.

Idehara, H. (1966) 'Ring diffusion at Pioneer Mill', *Sug. y Azuc.*, **61**, 46–7.

Illingworth, J. F. and Dodd, A. P. (1921) 'Australian sugar-cane beetles and their allies', *Bull. Bur. Sug. Exp. Sta. Qu. Div. Ent.*, 16.

Imrie, F. K. E. and Parker, K. J. (1974) 'Prospects for the industrial utilisation of sugar', *Proc. Int. Soc. Sug. Cane Technol.*, **15**, 1864–76.

Ingram, J. W., Bynum, E. K. and Mathes, R. (1951a) 'Insect pests of sugar cane in continental United States', *Proc. Int. Soc. Sug. Cane Technol.*, **7**, 395–401.

Ingram, J. W., Bynum, E. K., Haley, W. E. and Charpentier, L. J. (1951b) 'Pests of sugar cane and their control', *Circ. U.S. Dept. Agric.*, No. 878, 1–38.

Ingram, J. W., Jaynes, H. A. and Lobdell, R. N. (1939) 'Sugarcane pests in Florida', *Proc. Int. Soc. Sug. Cane Technol.*, **6**, 89–98.

Innes, R. F. (1954) 'An introduction to the sugarcane agriculture of Jamaica', *Proc. Int. Soc. Sug. Cane Technol.*, **8**, 328–38.

Inniss, B. de L. (1954) 'The methods and policy used in sugarcane breeding and selection at the B.W.I. Central Sugar Cane Breeding Station and the results achieved to date', *Proc. Int. Soc. Sug. Cane Technol.*, **8**, 452–62.

Irvine, J. E. (1967) 'Testing sugarcane varieties for cold tolerance in Louisiana', *Proc. Int. Soc. Sug. Cane Technol.*, **12**, 569–74.

Irvine, J. E. (1969) 'Effects of an early freeze on Louisiana sugarcane', *Proc. Int. Soc. Sug. Cane Technol.*, **13**, 837–9.

Isobe, M. (1969) *Proc. Int. Soc. Sug. Cane Technol.*, **13**, 49.

Iturbe, A. C. and Ruano, M. A. (1963) 'The sugarcane froghopper and its control in Mexico, *Proc. Int. Soc. Sug. Cane Technol.*, **11**, 650–7.

Jagathesan, D. and Sreenivasan, T. V. (1970) 'Induced mutations in sugarcane'. *Indian J. Agric. Sci.*, **40**, 165–72.

James, G. L. (1974) Culmicolous smut of sugarcane, and the effects of its control on yield', *Proc. Int. Soc. Sug. Cane Technol.*, **15**, 292–9.

James, N. I. (1969) 'Delayed flowering and pollen production in male-sterile sugarcane subjected to extended daylength', *Crop Sci.*, **9**, 279–82.

James, N. I. and Miller, J. D. (1975) 'Selection in six crops of sugarcane; II. Efficiency and optimum selection intensities', *Crop Sci.*, **15**, 37–40.

Janaki Amal, E. K. (1936) 'Cyto-genetic analysis of *Saccharum spontaneum* L. I – chromosome studies in some Indian farms', *Indian J. Agric. Sci.*, **6**, 1.

Janes, B. E. (1966) 'Adjustment mechanisms of plants subjected to varied osmotic pressures of nutrient solution', *Soil Science*, **101**, 180–8 (3).

Jepson, W. F. (1954) *A Critical Review of the World Literature on the Lepidopterus Stalk Borers of Tropical Gramineous Crops.* Commonw. Inst. Ent., London.

Jepson, W. F. and Moutia, L. A. (1939) 'The progress of applied entomology in Mauritius during the years 1933 to 1938, with reference to insects of the sugarcane', *Proc. Int. Soc. Sug. Cane Technol.*, **6**, 377–83.

Jeswiet, J. (1925) 'Beschrijving der soorten van het suikerriet. Elfde bijdrage. Bijdrage tot de systematiek von het geslacht *Saccharum*', *Meded. Proefst. Java-Suikerina.*, 1925, 391–404.

Joblin, A. D. H., Howes, J. R. and Wiggins, L. F. (1954a) 'Studies on ammoniated molasses. Part II. Ammoniated inverted cane molasses as feed for ruminants',

Proc. Br. W. Indies Sug. Technol., 1954, 212–17.

Joblin, A. D. H., Thomson, A. F. and Wiggins, L. F. (1954b) 'Studies on ammoniated molasses. Part I. The production of ammoniated molasses'. *Proc. Br. W. Indies Sug. Technol.*, 1954, 208–12.

Johnston, J. R. (1918) 'Diseases of sugar-cane in tropical and subtropical America especially the West Indies', *W. Indian Bull.*, **16**, 275–308.

Johnston, J. R. and Stevenson, J. A. (1917) 'Sugar-cane fungi and diseases of Porto Rico', *J. Dep. Agric. P. Rico*, **1**, 177–251.

Jones, O. A. (1951) 'The early detection of hurricanes', *Proc. Int. Soc. Sug. Cane Technol.*, **7**, 135–49.

Kalra, A. N. and Kulreshta, J. P. (1916) 'Studies on the biology and control of *Lachnosterna consanguinea* (Blanch) a pest of sugarcane in Bihar (India)', *Bull. Ent. Res.*, **52**, 577–87.

Kar, K. and Singh, D. R. (1956). 'Some aspects of controlling red rot disease in Uttar Pradesh', *Proc. Int. Soc. Sug. Cane Technol.*, **9**, 1089–96.

Kar, K., Gupta, S. C. and Kureel, D. C. (1974) 'Screening varieties for red rot resistance', *Proc. Int. Soc. Sug. Cane Technol.*, **15**, 189–93.

Kenning Voss, G. (1967) In discussion following 'The effect of row spacing on sugarcane crops in Natal', *Proc. Int. Soc. Sug. Cane Technol.*, **12**, 111.

Kern, F. (1956) *Plant Prot. Bull.*, **4**, FAO, Rome.

Khan, M. Q. and Krishnamurti Rao, B. H. (1956) 'Assessment of loss due to *Chilotraea infuscatellus* Snell. in sugarcane', *Proc. Int. Soc. Sug. Cane Technol.*, **9**, 870–9.

Khanna, K. L. and Rafay, S. A. (1953) *Annual Report on the Scheme for Investigation and Control of Wilt Disease of Sugarcane for the Year Ending 31st May 1953*. Cent. Sugarcane Res. Sta., Pusa.

King, N. C. (1956) 'The major sugarcane diseases of Natal', *Proc. Int. Soc. Sug. Cane Technol.*, **9**, 1126–34.

King, N. J. (1960) 'The varietal yield decline problem', *Proc. Int. Soc. Sug. Cane Technol.*, **10**, 62–6.

King, N. J., Mungomery, R. W. and Hughes, C. G. (1965) *Manual of Cane-growing*. American Elsevier, New York.

Kirkpatrick, T. W. (1957) *Insect Life in the Tropics*. Longman, London.

Kobus, J. D. (1893) *Historich Overzicht over het Zaaien van Suikerriet*. Archief Java Suikerind., 1.

Koike, H. (1965) 'The aluminium-cap method for testing sugarcane varieties against leaf-scald disease', *Phytopathology*, **55**, 317–19.

Koike, H. (1972) 'Testing sugarcane varieties for leaf-scald resistance', *Proc. Int. Soc. Sug. Cane Technol.*, **14**, 909–19.

Kollonitsch, V. (1970) *Sucrose Chemicals*. The International Sugar Research Foundation Inc., Bethesda, Md.

Kortschak, H. P. (1972) 'Environmental studies', *HSPA Exp. Sta. Ann. Rept.*

Krishnamurti, B. (1977) 'Sugarcane improvement through tissue culture', *Proc. Int. Soc. Sug. Cane Technol.*, **16**, 23–8.

Krishnaswamy, S., Ghosh, M. R., Majumdar, N. and Krishnan, Y. S. (1963) 'An abnormal incidence of *Holototricha serrata* (F) (Melanothidae-Coleoptera)', *Indian J. Ent.*, **25** (4), 381–2.

Kumar, S. and Kalra, A. N. (1965) 'Attack of pink borer *Sesamia inferens* Wlk., as a cane borer in Rajasthan', *Indian Sug. Cane J.*, **9**, 154–6.

Ladd, S. L., Heinz, D. J., Meyer, H. K. and Nishimoto, B. K. (1974) 'Selection studies in sugarcane hybrids. I. Repeatability between selection stages', *Proc. Int. Soc. Sug. Cane Technol.*, **15**, 102–5.

Lalouette, J. A. (1968) 'Breeding of sugarcane varieties in Mauritius', *Proc. Int. Soc. Sug. Cane Technol.*, **13**, 1062–7.

Langreney, F. and Hugot, E. (1969) 'The Bagapan particle board', *Proc. Int. Soc. Sug. Cane Technol.*, **13**, 1891–9.

Lee, C. O. and Van Groeningen (1973) 'Three years of harvesting study in Jamaica', *Proc. W. Indies Sug. Technol.*, 1973, 183–91.

Lee, M. S. and Pao, T. P. (1962) 'A study of ecology and further improvement on the mechanical control of the top borer', *Rep. Taiwan Sugar Exp. Sta.*, **27**, 91–110.

Lee, S. and Loo, Y. S. (1958) 'Report on some experiments on the germination of true sugarcane seed', *Rep. Taiwan Sug. Exp. Sta.*, **18**, 1–13.

Le Grand, F. and Martin, F. G. (1966) 'Maturity testing of sugarcane growing on organic soils of Florida', *Proc. Br. W. Indies Sug. Technol.*, 238–45.

Lemaire, Y. (1966) 'Infestation status of sugarcane plantations in Guadeloupe by *Diatraea saccharalis* – importance of *Lixophaga* and *Metagonistylum* parasitism', *Proc. Br. W. Indies Sug. Technol.*, 304–8.

Lemaire, Y. (1972) 'Cane sampling by coring, hydraulic press, and automatic saccharimetry', *Proc. Int. Soc. Sug. Cane Technol.*, **14**, 1626–36.

Leu, L. S. and Teng, W. S. (1974) 'Culmicolous smut of sugarcane in Taiwan (v): Two pathogenic strains of *Ustilago scitaminea* Sydow', *Proc. Int. Soc. Sug. Cane Technol.*, **15**, 275–9.

Leverington, K. C. (1960) 'The effect of saline conditions on the growth of sugar cane', *Bureau Sugar Exp. Sta. Queensland, Tech. Comm.*, 1–4, 21–5.

Liang, C. J. (1970) 'A survey of borer damage to sugar cane', *Taiwan Sugar*, **17**, 30–1.

Lichtenstein, E. P., Mueller, C. H., Myrdal, G. R. and Schulz, K. R. (1962) 'Vertical distribution and persistence of insecticidal residues in soils as influenced by mode of application and a cover crop', *J. Econ. Ent.*, **55**, 215–19.

Ligon, R. (1657) *A True and Exact History of the Island of Barbados*. London.

Liu, K. (1969a) 'Taiwan and its sugar industry', *Proc. Int. Soc. Sug. Cane Technol.*, **13**, XCIII–CIV.

Liu, L. J. (1969b) 'The effect of temperature on various aspects of the development, occurrence and pathogenicity of *Helminthosporium stenospilum* and *Helminthosporium sacchari* in Puerto Rico', *Proc. Int. Soc. Sug. Cane Technol.*, **13**, 1212–18.

Liu, L. J., Cortes-Monllor, A., Maramorosch, K., Hirumumi, H., Perez, J. E. and Bird, J. (1974) 'Isolation of an organism resembling *Xanthomonas vasculorum* from sugarcane affected by ratoon stunting disease', *Proc. Int. Soc. Sug. Cane Technol.*, **15**, 234–40.

Liu, L. J., Ellis, T. O. and Arcenaux, G. (1967a) 'Diseases of sugar cane and their control at Central Romana', *Proc. Int. Soc. Sug. Cane Technol.*, **13**, 1226–31.

Liu, L. J., Rosario, T. and Roig, F. M. (1967b) 'Diseases of sugar cane in Puerto Rico', *Proc. Int. Soc. Sug. Cane Technol.*, **12**, 1236–40.

Lo, T. T. (1951) 'A report on sugar-cane diseases in Taiwan', *Proc. Int. Soc. Sug. Cane Technol.*, **7**, 452–6.

Loh, C. S. and Wu, W. Y. (1965) 'Developing high yielding varieties of sugarcane in Taiwan', *Proc. Int. Soc. Sug. Cane Technol.*, **12**, 864–70.

Long, E. (1774) *History of Jamaica*, Vol. 2 T. Lowndes, London, pp. 560–1.

Long, W. H. (1969) 'Insecticidal control of moth borers of sugar cane', in *Pests of Sugar Cane*. Elsevier, Amsterdam.

Long, W. H. and Hensley, S. D. (1972) 'Insect pests of sugar cane', *Ann. Rev. Ent.*, **17**, 149–76.

Long, W. H., Concienne, E. J., Hensley, S. D. and McCormick, W. J. (1960) 'Control of the sugarcane borer with insecticides', *Proc. Int. Soc. Sug. Cane Technol.*, **10**, 947–54.

Lopez, A. W. (1931) *Rep. Res. Bur. Philipp. Sugar Ass.*, Manila, 1930–31, pp. 227–73.

McIntosh, A. E. S. (1935) '*An account of the policy and methods employed in the breeding and initial stage of selecting sugar cane seedlings*', *BWI Cent. Sug. Cane Br. Sta., Bull.*, 6.

McIntosh, A. E. S. (1944) 'Nobilisation in cane breeding at the British West Indies Central Sugar Cane Breeding Station and its practical results to date', *Proc. Br. W. Indies Sug. Technol.*, 1944, 26–40.

MacIntyre, P. C. and Keir, W. (1967) 'The use of aircraft in sugar cane areas with special reference to Caroni Limited', *Proc. Sug. Tech. Assoc. Trinidad*, **1**, 41–3.

Maclean, N. R. (1975) 'Long-term effects of sugar cane production on some physical and chemical properties of soils in the Goondi Mill area', *Proc. QSSCT*, 123–6.

McMartin, A. (1945) 'Sugar-cane smut: reappearance in Natal', *S. Afr. Sug. J.*, **29**, 55–7.

McMartin, A. (1948) 'A report on visits to the sugar estates of Southern Rhodesia and Portuguese East Africa, with general observations on the disease', *S. Afr. Sug. J.*, **32**, 737–49.

McMartin, A. (1949) 'Fungicidal treatment for sugarcane cuttings', *S. Afr. Sug. J.*, **33**, 651–5.

McNeal, B. L. (1976) 'Managing salt-affected soils: recent dissolution of some myths', *Crops and Soils*, 22–3.

Mangelsdorf, A. J. (1946, 1953) 'Sugar cane breeding in Hawaii', *Hawaii. Plrs Rec.*, **50**, 141–60; **54**, 101–62.

Mangelsdorf, A. J. (1960) 'Sugar-cane breeding methods', *Proc. Int. Soc. Sug. Cane Technol.*, **10**, 694–701.

Mariotti, J. A. (1972a) 'On the effectiveness of some genetic parameters used in the selection of sugar cane populations', *ISSCT Sugarcane Breeders' Newsl.*, **29**, 8–15.

Mariotti, J. A. (1972b) 'A study on cane yield and its components in the single stool selection stage in sugarcane', *ISSCT Sugarcane Breeders' Newsl.*, **30**, 16–22.

Mariotti, J. A. (1974) 'The effect of environments on the effectiveness of clonal selection in sugarcane', *Proc. Int. Soc. Sug. Cane Technol.*, **15**, 89–95.

Martin, J. P. (1934) 'Symptoms of malnutrition manifested by the sugar cane plant when grown in solutions from which certain elements are omitted', *Hawaii. Plrs. Rec.*, **38**, 21.

Martin, J. P. (1938) *Sugar Cane Diseases in Hawaii.* Advertiser Pub. Co., Honolulu.

Martin, J. P. (1951) 'Sugar-cane diseases and their world distribution', *Proc. Int. Soc. Sug. Cane Technol.*, **7**, 435–52.

Martin, J. P. *et al.* (1956) Committee chairman, 'A revised listing of sugarcane diseases and their world distribution', *Proc. Int. Soc. Sug. Cane Technol.*, **9**, 1174–205.

Martin, J. P., Abbott, E. V. and Hughes, C. G. (eds) (1961a) *Sugar-Cane Diseases of the World*, Vol. I. Elsevier, Amsterdam.

Martin, J. P., Hong, L. H. and Wismer, C. A. (1961b) 'Pokkah Boeng', in *Sugar-Cane Diseases of the World*, Vol. I. Elsevier, Amsterdam.

Martin, J. P. *et al.* (1960) Committee chairman, 'A revised listing of sugar cane diseases and their world distribution', *Proc. Int. Soc. Sug. Cane Technol.*, **10**, 1107–26.

Martin, W. S. (1944) 'Grass covers and their relation to soil structure', *Emp. J. Exp. Agr.*, **12**, 21–32.

Martorell, L. F. and Medina-Gaud, S. (1967) 'Status of important pests of sugar cane in Puerto Rico and their control', *Proc. Int. Soc. Sug. Cane Technol.*, **12**, 1278–86.

Martorell, L. F., Burgos, J. A. and Baiggi, F. M. (1973) 'Preliminary investigations on the sugarcane insects of the Dominican Republic', *Proc. W. Indies Sug. Technol.*, 1973, 247–61.

Martorell, L. F., Medina-Gaud, S. and Cruz-Miret, A. (1967) 'A preliminary report on rat damage in Puerto Rican cane fields', *Proc. Int. Soc. Sug. Cane Technol.*, **12**, 1435–43.

Mathes, R. and Charpentier, L. J. (1969) 'Varietal resistance in sugar cane to stalk moth borers', in *Pests of Sugar Cane.* Elsevier, Amsterdam.

Mathes, R., Ingram, J. W., Questel, D. D., Thames, W. H. and Dugas, A. L. (1954) 'Current status of sugarcane-insect investigations in the United States', *Proc. Int. Soc. Sug. Cane Technol.*, **8**, 560–7.

Mathur, R. B. L. (1974) 'Development of the sugar industry in the Sudan', *Proc. Int. Soc. Sug. Cane Technol.*, **15**, 789–93.

Matsumoto, T. (1952) *Monograph of Sugarcane Diseases in Taiwan.* Taipeh, Taiwan.

Meade, G. P. and Chen, J. C. P. (1977) *Cane Sugar Handbook*. Wiley, New York.

Mehrad, Bahram (1968) 'Effect of soil salinity on sugar cane cultivation at Haft Tappeh, Iran', *Proc. Int. Soc. Sug. Cane Technol.*, **13**, 746–54.

Menon, R. G., Singh, K., Srivastava, N. S. L. and Misra, S. R. (1972) 'The IISR hot-air seed cane treatment plant', *Proc. Int. Soc. Sug. Cane Technol.*, **14**, 991–6.

Metcalfe, J. R. (1964) 'The chemical control of canefly (*Saccharosydne saccharivora* (Westw.) (Hom: Delphacidae) in Jamaica', *Jamaican Assoc. Sug. Cane Technol. J.*, **25**, 39–45.

Metcalfe, J. R. (1966) 'Honey bee poisoning from spraying canefly with malathion', *Proc. Br. W. Indies Sug. Technol.*, 1966, 236–7.

Metcalfe, J. R. (1969) 'The estimation of loss caused by sugar cane borers', in *Pests of Sugar Cane*, Elsevier, Amsterdam.

Metcalfe, J. R. and Brénière, J. (1969) 'Egg parasites (*Trichogramma* spp.) for control of sugar cane moth borers', in *Pests of Sugar Cane*. Elsevier, Amsterdam.

Metcalfe, J. R. and Thomas, G. (1966) 'Preliminary experiments in Jamaica with a method for determining loss of sugar resulting from rat damage to sugar cane', *Proc. Br. W. Indies Sug. Technol.*, 1966, 267–8.

Miller, A. A. (1931) *Climatology*. Methuen, London.

Moir, W. W. G. (1932) 'Hawaiian soils and fertilizer research', *Proc. Int. Soc. Sug. Cane Technol.*, **4**.

Mongelard, J. C. (1973) *Hawaiian Planters Record*, **58** (22), 315–22.

Mongelard, J. C. and Mimura, L. (1971) 'Growth studies on the sugar cane plant. I. Effects of temperature', *Crop Sci.*, II, 795–800.

Moutia, L. A. (1954) 'Notes sur le cycle biologique de trois lépidoptères nuisibles à la canne à sucre à Maurice', *Revue agric. Ile Maurice*, **33**, 116–22.

Moyne, Lord *et al.* (1945) *West India Royal Commission Report*, Cmd. 6607. HMSO, London.

Muir, F. (1917) 'The introduction of *Scolia manilae* Ashm. into the Hawaiian Islands', *Ann. Ent. Soc. Am.*, **10**, 202–10.

Muir, F. (1919) 'The progress of *Scolia manilae* Ashm. in Hawaii', *Ann. Ent. Soc. Am.*, **12**, 171.

Mungomery, R. W. (1949) 'Control of the greyback cane grub pest, *Dermolepida albohirtum* Waterh. by means of "Gammexane" (benzene hexachloride)', *Tech. Commun. Bur. Exp. Sta. Qd.*, 4.

Mungomery, R. W. (1951) 'The status of benzene hexachloride in the control of some Queensland sugar-cane pests', *Proc. Int. Soc. Sug. Cane Technol.*, **7**, 404–10.

Mungomery, R. W. (1954) 'The rise and fall of the sugarcane weevil borer pest in Queensland', *Proc. Int. Soc. Sug. Cane Technol.*, **8**, 586–93.

Mungomery, R. W. (1965) 'Pests', in *Manual of Cane-growing*. Angus and Robertson, Sydney.

Muthusamy, S. (1974) 'Varietal susceptibility to smut (*Ustilago scitaminaea* Sydow.) in relation to bud characters', *Proc. Int. Soc. Sug. Cane Technol.*, **15**, 289–91.

Myers, J. G. (1934). 'The discovery and introduction of the Amazon fly. A new parasite for cane-borers, *Diatraea* spp.', *Trop. Agric., Trin.*, 191–5.

Neate, D. J. H. (1958) 'A note on the use of the Trithion in froghopper control', *Proc. Br. W. Indies Sug. Technol.*, 1957, 190–2.

Nickell, L. G. (1977). *Ecophysiology of Tropical Crops*. Academic Press, New York, pp. 89–111.

Noronha, A. de R. (1972) 'Sugarcane diseases in Mozambique', *Proc. Int. Soc. Sug. Cane Technol.*, **14**, 1059–61.

North, D. S. (1935) 'The gumming disease of the sugarcane, its dissemination and control', *Agr. Rep. (Tech.)*, **10**, CSR, Sydney.

Nowell, W. (1923) *Diseases of Crop Plants in the Lesser Antilles*. The West India Committee, London.

O'Connor, E. (1951) 'Cyclone and drought insurance of sugar crops in Mauritius', *Proc. Int. Soc. Sug. Cane Technol.*, **7**, 150–2.

Ocfemia, G. O. (1939) 'A review of sugarcane diseases in the Philippines', *Proc. Int. Soc. Sug. Cane Technol.*, **6**, 183–7.

Olivier, Lord *et al.* (1930) *Report of the West Indian Sugar Commission.* 1 February 1930, Cmd. 3517. HMSO, London.

Orian, G. (1954) 'The probable origin of the gumming disease of the sugarcane', *Proc. Int. Soc. Sug. Cane Technol.*, **8**, 862–76.

Osada, S. and Flores, S. (1969) 'Varietal resistance trials to eye spot disease (*Helminthosporium sacchari* (v. Breda de Haan) Butler)', *Proc. Int. Soc. Sug. Cane Technol.*, **13**, 1208–11.

Osborne, W. G. T. (1967) 'Bulk handling of sugar at Point Lisas', *Proc. Sug. Tech. Assoc. Trin.*, 1967, 106–9.

Otanes, F. Q. (1924) 'Some observations on root grubs (*Leucopholis irrorata* Chevr.) in the Philippines and suggestions for their control', *Philipp. Agric. Rev.*, **17**, 109–19.

Panje, R. R. (1933) '*Saccharum spontaneum* L. A comparative study of the forms grown at the Imperial Sugar Cane Breeding Station, Coimbatore', *Indian J. Agric. Sci.*, **3**, 1013–44.

Panje, R. R. and Babu, C. N. (1960) 'Studies in *Saccharum spontaneum*. Distribution and geographical association of chromosome numbers', *Cytologia*, **25**, 152–72.

Parker, K. H., James, K. and Hurford, J. (1977) In J.L. Hickson (ed.), *Sucrochemistry*. American Chemical Society, Washington, DC, pp. 97–114.

Parker, K. J. (1979) *Proc. RAPRA Diamond Jubilee Conf.*, July 1979, Rubber and Plastics Res. Assoc., 53–78.

Parris, G. K. (1954) 'James W. Parris: discoverer of sugarcane seedlings', *The Garden Journal*, 1954.

Parthasarathy, S. V. (1950) *Rep. Sug. Res. Scheme, Madras*, 1949–50.

Paturau, J. M. (1963) 'The sugar industry of Mauritius', *Proc. Int. Soc. Sug. Cane Technol.*, **11**, 24–41.

Paturau, J. M. (1969) *By-products of the Sugar Cane Industry.* Elsevier, Amsterdam.

Payne, J. H. (1969) 'Cane diffusion – the displacement process in principle and practice', *Proc. Int. Soc. Sug. Cane Technol.*, **13**, 103–21.

Payne, J. H. and Rhodes, L. J. (1967) 'Assaying cane deliveries by core sampling and direct analysis', *Proc. Int. Soc. Sug. Cane Technol.*, **12**, 1052–50.

Pearson, C. H. O. (1954) 'The general practice and procedure in the cane fields of South Africa', *Proc. Int. Soc. Sug. Cane Technol.*, **8**, 248–53.

Pemberton, C. E. (1936) 'Recent control measures against *Anomala orientalis* Waterh. in Hawaii', *Proc. Int. Soc. Sug. Cane Technol.*, **5**, 591–4.

Pemberton, C. E. (1939) Discussion following paper 'Field rats and their control in Formosa', by Takano and Kondo, *Proc. Int. Soc. Sug. Cane Technol.*, **6**, 112.

Pemberton, C. E. (1951) 'The present status of the insect pests of sugar cane in Hawaii', *Proc. Int. Soc. Sug. Cane Technol.*, **7**, 401–4.

Pemberton, C. E. (1963) 'Insect pests affecting sugar cane plantations within the Pacific', *Proc. Int. Soc. Sug. Cane Technol.*, **11**, 678–89.

Pemberton, C. E. and Williams, J. R. (1969) 'Distribution, origins and spread of sugar cane insect pests', in *Pests of Sugar Cane*. Elsevier, Amsterdam.

Persaud, G. L. B. *et al.* (1968) *Report of a Commission of Inquiry into the Sugar Industry in Guyana.* Government Printer, Georgetown, Guyana.

Pickles, A. (1933) 'Entomological contributions to the study of the sugar-cane froghopper. I. The study of biotic factors of control', *Trop. Agric., Trin.*, **10**, 222–33.

Pickles, A. (1936) *Rep. Dep. Agric. Trin. Tobago*, 57.

Pickles, A. (1942) 'A discussion of researchers on the sugar-cane froghopper (Homop., Cercopidae)', *Trop. Agric. Trin.*, **19**, 116–23.

Pickles, A. (1946a) 'Field trials of insecticides for the control of the sugar-cane froghopper', *Trop. Agric., Trin.*, **23**, 9–11.

Pickles, A. (1946b) 'Pathological problems of sugar cane in St. Kitts', *Proc. Br. W. Indies Sug. Technol.*, 1946, 104–11.

382 *References*

5**Pickles, A.** (1947) *Rep. Dep. Agric. Trin. Tobago,* 15.
Pitman, F. (1917) *The development of the British West Indies, 1700–1763.* Yale University Press, New Haven, Conn.
Potes, A. F. (1954) 'The sugarcane entomology of the Cauca Valley of Colombia', *Proc. Int. Soc. Sug. Cane Technol.,* **8**, 568–9.
Potter, T. E. K. and Carrington, A. J. (1947) *Report on Preliminary Investigation of the Effect of Soil Application of Gammexane formulations on Sugar Cane Froghopper Incidence.* T. Geddes Grant Ltd, Trinidad (private circulation).
Pradhan, S. and Bhatia, S. K. (1956) 'The effect of temperature and humidity on the development of the sugarcane stem borer *Chilotraea infuscatellus* Snell', *Proc. Int. Soc. Sug. Cane Technol.,* **9**, 856–69.
Price, S. (1956) 'A leaf squash technique for studies of somatic chromosomes in *Saccharum*', *Proc. Int. Soc. Sug. Cane Technol.,* **9**, 780–3.
Price, S. (1957) 'Cytological studies in *Saccharum* and allied genera. II. Geographical distribution and chromosome numbers in *S. robustum*', *Cytologia,* **22**, No. 1.
Price, S. (1960) 'Cytological studies in *Saccharum* and allied genera. VI. Chromosome numbers in *S. officinarum* and other noble sugar canes', *Hawaii Pltrs. Rec.,* **56**, 183–94.
Purseglove, J. W. (1972) *Tropical Crops: Monocotyledons.* Longman, London.
Raja Rao, S. A. (1964) 'Biological control of top borer *Sciropophaga nivella* F. on sugar cane at Pugalur (Madras State)', *Proc. Conf. Sug. Dev. Wkrs. India,* **5**, 588–96.
Rao, J. T. (1965) 'Current breeding techniques and selection procedures adopted at Coimbatore', *Proc. Int. Soc. Sug. Cane Technol.,* **12**, 840–3.
Rao, J. T., Krishnamurthi, T. W. and Natarajan, B. V. (1967) 'Heritability of economic characters in sugarcane and its implications in selection', *Indian Sug.,* **17**, 153–61.
Rao, J. T., Srinvasan, K. V. and Alexander, K. C. (1966) 'A red rot resistant mutant of sugarcane induced by gamma irradiation', *Proc. Indian Acad. Sci.,* **64**, 224–30.
Rao, P. S. (1972) 'Radiosensitivity and non-flowering mutants in sugarcane', *Proc. Int. Soc. Sug. Cane Technol.,* **14**, 408–12.
Rao, P. S. (1977) 'Effects of flowering on yield and quality of sugarcane', *Exp. Agric.,* **13**, 381–7.
Rao, P. S. (1980) 'Fertility, seed storage and seed viability in sugar cane', *Proc. Int. Soc. Sug. Cane Technol.,* **16**.
Rao, V. P. and Nagaraja, H. (1969) '*Sesamia* species as pests of sugar cane', in *Pests of Sugar Cane.* Elsevier, Amsterdam.
Rao, V. P. and Sankaran, T. (1969) 'The scale insects of sugar cane', in *Pests of Sugar Cane.* Elsevier, Amsterdam.
Rajabalee, M. A. (1969) 'Rats as cane pests', *Rep. Maurit. Sug. Ind. Res. Inst.,* 1969, 81–2.
Rajabalee, M. A. (1970) 'Rats as cane pests', *Rep. Maurit. Sug. Ind. Res. Inst.,* 1970, 114–16.
RAM (1948) 'Câmbate ao carvão' nos canaviais de São Paulo', *Brasil Azucar,* **31**, 278–80, in *Rev. Appl. Mycol.,* **28**, 83.
Rands, R. D. (1961) 'Root rot', in *Sugar-cane Diseases of the World,* Vol. I. Elsevier, Amsterdam.
Rands, R. D. and Abbott, E. V. (1939) 'Sugarcane diseases in the United States', *Proc. Int. Soc. Sug. Cane Technol.,* **6**, 202–12.
Rands, R. D. and Dopp, E. (1938) '*Pythium root rot of sugar cane, U.S. Dep. Agric. Tech. Bull.,* 666.
Raynal, Abbé (1777) *A Philosophical and Political History of the Settlements and Trade of the Europeans in the East and West Indies,* translated by J. Justamond, 3rd edn. T. Cadell, London.

Redhead, T. D. (1972) 'Dynamics of a sparse population of the rat *Melomys littoralis* Lönnberg (Muridae) in sugarcane and natural vegetation', *Proc. Int. Soc. Sug. Cane Technol.*, **14**, 548–60.

Reece, R. (1857) In *Hints to Young Barbados Planters*. Israel Bowen, Bridgetown, Barbados.

Reyes, G. M. (1954) 'Post-war observations on sugarcane diseases in the Philippines', *Proc. Int. Soc. Sug. Cane Technol.*, **8**, 907–10.

Ricaud, C. (1974a) *Rep. Maurit. Sug. Ind. Res. Inst.*, 1974, 45.

Ricaud, C. (1974b) 'Problems in the diagnosis of ratoon stunting disease', *Proc. Int. Soc. Sug. Cane Technol.*, **15**, 241–9.

Ricaud, C. and Sullivan, S. (1974) 'Further evidence of population shift in the gumming disease pathogen in Mauritius', *Proc. Int. Soc. Sug. Cane Technol.*, **15**, 204–9.

Ricaud, R. (1977) *Proc. Int. Soc. Sug. Cane Technol.*, **16**, 1039–47.

Ridge, D. R. and Kelly, G. J. (1975) 'Water quality and watering problems', *Qld. Soc. Sugarcane Tech. Proc.*, **42**, 77–80.

Risco, S. H. (1960) 'Combating the borer in Peru. Success of the campaign of biological control', *Proc. Int. Soc. Sug. Cane Technol.*, **10**, 877–86.

Roach, B. T. (1971) 'Nobilisation of sugarcane', *Proc. Int. Soc. Sug. Cane Technol.*, **14**, 206–16.

Robinson, F. E. (1963) 'Soil moisture tension, sugarcane stalk elongation and irrigation interval control', *Agron. J.*, **55**, 481–3.

Robinson, P. E. and Martin, J. P. (1956) 'Testing sugar cane varieties against Fiji disease and downy mildew in Fiji', *Proc. Int. Soc. Sug. Cane Technol.*, **9**, 986–1011.

Robinson, R. A. (1959) 'Sugar cane smut', *E. Afr. Agric. J.*, 240–3.

Rorer, J. B. (1910) 'The froghopper fungus', *Bull. Dep. Agric. Trin. Tobago*, **9**, 182–4.

Rorer, J. B. (1911) 'The green muscardine of froghoppers', *Bull. Dep. Agric. Trin. Tobago*, **10**, 50–8.

Rorer, J. B. (1913) 'The green muscardine fungus and its use in cane fields', *Circ. Dep. Agric. Trin. Tobago*, **8**, 1–14.

Rosenfeld, A. H. (1939) 'Minor sugarcane diseases in Egypt', *Proc. Int. Soc. Sug. Cane Technol.*, **6**, 194–8.

Roxburgh, W. (1819) *Plants of the Coast of Coromandel*. W. Bulmer, London.

Ruschel, R. (1977) 'Phenotypic stability of some sugarcane varieties (*Saccharum* sp.) in Brazil', *Proc. Int. Soc. Sug. Cane Technol.*, **16**, 275–81.

Russell, E. W. (1973) *Soil Conditions and Plant Growth*, 10th edn. Longman, 849 pp.

SAC (1975) *Report by the Chairman for 1974–5*. The Sugar Association of the Caribbean (Incorporated), Trinidad.

Saint, S. J. (1953) 'Sugar production in Barbados during the past 100 years', *Proc. Int. Soc. Sug. Cane Technol.*, 966–77.

Saint, S. J. (1960) 'Provisional standards for Barbados bulk storage sugars', *Proc. Br. W. Indies Sug. Technol.*, 228–33.

Saint, S. J. and Trott, R. R. (1960) 'The average size, weight and number of crystals in massecuites', *Sug. J.*, **23** (Dec.), 23–7.

Samol, H. H. (1972) 'Rot damage and control in the Florida sugarcane industry', *Proc. Int. Soc. Sug. Cane Technol.*, **14**, 503–6.

Samuels, G. and Capo, B. G. (1954) 'A survey of the agronomic practices in growing sugarcane in Puerto Rico in 1952', *Proc. Int. Soc. Sug. Cane Technol.*, **8**, 240–7.

Saplala, V. L. (1959). 'White grub of sugarcane and its control', *Philipp. Sug. Inst. Q.*, **5**, 85–9.

Sartoris, G. B. (1947) 'New kinds of sugarcane', in *The Yearbook of Agriculture 1943–1947, Science in Farming*. United States Department of Agriculture, Washington.

Sayed, G. el-K., Ahmed, A. el-B. and Mohammed, M. S. (1974) 'Effect of method of extraction of cane juice quality', *Proc. Int. Soc. Sug. Cane Technol.*, **15**, 1215–23.

Scaramuzza, L. C. (1960a) 'Damage by the sugarcane borer in Louisiana and Cuba. The importance of biological control', *Proc. Int. Soc. Sug. Cane Technol.*, **10**, 938–42.

Scaramuzza, L. C. (1960b) In discussion following 'A list of insects and animals affecting sugarcane in Cuba', *Proc. Int. Soc. Sug. Cane Technol.*, **10**, 1000.

Scaramuzza, L. C. and Barry, F. V. (1960) 'A list of insects and animals affecting sugarcane in Cuba', *Proc. Int. Soc. Sug. Cane Technol.*, **10**, 994–1000.

Schenxnayder, C. A. (1956) 'The ratoon stunting disease of sugar cane in Louisiana with notes on its control', *Proc. Int. Soc. Sug. Cane Technol.*, **9**, 1058–65.

Schmidt, N. O. and Wise, W. S. (1956) 'Some aspects of the theory of cane diffusion', *Proc. Int. Soc. Sug. Cane Technol.*, **9**, 370–81.

Scott, P. (1971) *Proc. Int. Soc. Sug. Cane Technol.*, **14**, 859.

Seaton, O. M. (1969) 'Field mechanization in British Honduras', *Proc. W. Indies Sug. Technol.*, 1969, 143–8.

Shaw, H. R. (1954) 'An international glance at sucrose content of cane', *Proc. Int. Soc. Sug. Cane Technol.*, **8**, 283–91.

Shen, I-Sun and Tung, H. L. (1964) 'Study on salt tolerance in sugar cane', *Taiwan Sugar Exp. Sta. Report*, **35**, 1–24.

Shepherd, E. F. S. (1926) *Diseases of Sugar Cane in Mauritius*. Dept. Agric. Mauritius, Gen. Ser., Bull. 32.

Shepherd, E. F. S. (1931) *Diseases of Sugar Cane in Mauritius*. Dept. Agric. Mauritius, Gen. Ser., Bull. 41.

Sherwood, I. R. and Hines, W. J. (1951) 'Microbiological aspects of the deterioration of raw sugar', *Proc. Int. Soc. Sug. Cane Technol.*, **7**, 591–607.

Shukla, U. C. and Prasad, K. G. (1974) 'Amelioration role of zinc on maize growth under alkali soil condition', *Agro. J.*, **66**, 804–6.

Siddiqui, Z. A., Rajani, V. G. and Singh, O. P. (1959) 'Simultaneous control of sugar cane termite and shoot borers through soil application of *gamma* BHC liquid and its boosting effect on crop yield', *Indian J. Sug. Cane Res. Dev.*, **3**, 227–32.

Simmonds, F. J. (1951) 'The small moth borers of sugar cane, *Diatraea* spp., in Trinidad', *Trop. Agric., Trin.*, **28**, 80–108.

Simmonds, F. J. (1960) 'The successful control of the sugarcane moth borer, *Diatraea saccharalis* F. (Lepidoptera, Pyralidae) in Guadeloupe, F. W.I.', *Proc. Int. Soc. Sug. Cane Technol.*, **10**, 914–19.

Simmonds, N. W. (1979) *Principles of Crop Improvement*. Longman, London.

Singh, H., Singh, S. and Singh, N. (1956) 'Varietal resistance to red rot disease of sugarcane in the Punjab', *Proc. Int. Soc. Sug. Cane Technol.*, **9**, 1071–8.

Singh, Y. (1976) *Proc. W.I. Sug. Techn. Conf.*, 91–9.

Skinner, H. M. (1929) 'The giant moth borer of sugar cane (*Castnia licus* Drury)', supplement to *Trop. Agric., Trin.*, **7**.

Skinner, J. C. (1961) 'Sugarcane selection experiments. 2. Competition between varieties', *Tech. Commun. Bur. Exp. Stns. Qd.*, 1961, No. 1.

Skinner, J. C. (1965) 'Grading varieties for selection', *Proc. Int. Soc. Sug. Cane Technol.*, **12**, 938–49.

Skinner, J. C. (1971) 'Selection in sugarcane: a review', *Proc. Int. Soc. Sug. Cane Technol.*, **14**, 149–62.

Smith, C. E. M. (1965) 'The variety selection programme in Jamaica', *Proc. Int. Soc. Sug. Cane Technol.*, **12**, 956–64.

Smith, P. D. (1978) 'Dark crystal sugar delivered to the bulk store in 1978', *Barbados Sug. Tech. Res. Unit, Tech. Rep.*, 69.

Smith, P. D. and Brooks, S. A. (1978) 'Barbados fancy molasses, 1978', *Barbados Sug. Tech. Res. Unit, Tech. Rep.*, 65.

Smyth, E. G. (1917) 'The white-grubs injuring the sugar cane in Porto Rico: the common white grub, *Phyllophaga portoricensis* n. sp.', *J. Dept. Agric. P. Rico*, **1**, 145–52.

South Pacific Enterprise (1956) Angus and Robertson, Sydney.

Steindl, D. R. L. (1961) 'Ratoon stunting disease', in *Sugar-cane Diseases of the World*. Elsevier, Amsterdam.

Steindl, D. R. L. (1972) 'The elimination of leaf scald from infected planting material', *Proc. Int. Soc. Sug. Cane Technol.*, **14**, 925–9.

Steindl, D. R. L. (1974) 'Ratoon stunting disease history, distribution and control', *Proc. Int. Soc. Sug. Cane Technol.*, **15**, 210–12.

Stevenson, G. C. (1957) 'The British West Indies Sugar Cane Breeding Station – twenty-five years' progress', *Proc. Br. W. Indies Sug. Technol.*, 1957, 24–33.

Stevenson, G. C. (1965) *Genetics and Breeding of Sugar Cane.* Longman, London.

Stevenson, R. A. and Rands, R. D. (1938) 'An annotated list of fungi and bacteria associated with sugarcane and its products', *Hawaii. Plrs. Rec.*, **42**, 247–313.

Subramaniam, L. S. (1936) 'Diseases of sugar cane and methods for their control', *Imp. Council of Agric. Res. India, Misc. Bull. 10.*

Summers, E. M., Brandes, E. W. and Rands, R. D. (1948) 'Mosaic of sugar cane in the United States, with special reference to strains of the virus', *Tech. Bull. U.S. Dept. Agric.*, 955.

Sund, K. A. (1967) 'The effect of freezing temperatures on the 1963–1964 sugar cane crop, Haft Tapeh, Iran', *Proc. Int. Soc. Sug. Cane Technol.*, **12**, 561–8.

Symposium (various authors) (1962) 'Defects for which the majority of seedlings are discarded during selection (in Australia, Mauritius, Louisiana, Hawaii, S. Africa)', *Proc. Int. Soc. Sug. Cane Technol.*, **11**, 410–35.

Szent-Ivany, J. J. H. and Ardley, J. H. (1963) 'Insects of *Saccharum* spp. in the territory of Papua and New Guinea', *Proc. Int. Soc. Sug. Cane Technol.*, **11**, 690–4.

Takano, S. and Kondo, T. (1939) 'The field rats and their control in Formosa', *Proc. Int. Soc. Sug. Cane Technol.*, **6**, 106–12.

Takara, T. and Azuma, S. (1969) 'Important insect pests affecting sugarcane and problems on their control in the Ryukyu Islands', *Proc. Int. Soc. Sug. Cane Technol.*, **13**, 1424–32.

Tan, S. W. and Johnson (1974) 'Yellowing of sugarcane leaves caused by *Phaenacantha saccharicida* Karsch (Hemiptera: Lygaeidae)', *Proc. Int. Soc. Sug. Cane Technol.*, **15**, 453–6.

Tantera, D. M. and Steib, R. J. (1972) 'Temperatures and treatment durations with hot air for control of ratoon stunting disease in Louisiana', *Proc. Int. Soc. Sug. Cane Technol.*, **14**, 997–1000.

Tate and Lyle (1969) *Sugar Production Costs and Profits 1969.* Tate and Lyle, London (private circulation).

Tate and Lyle (1975) *The TAFLOC Process for Sugar Refining.* Tate and Lyle Enterprises, Bromley, England.

Taussig, C. W. (1940) *Some Notes on Sugar and Molasses.* Charles William Taussig, New York.

Teakle, D. S. (1974) 'The causal agent of ratoon stunting disease (RSD)', *Proc. Int. Soc. Sug. Cane Technol.*, **15**, 225–33.

Thompson, G. D. (1976) *S.A. Sugar J.*, **61**, 161–74.

Thompson, G. D. (1981) Introduction, *Experiment Station Annual Report 1979–80.*

Thompson, G. D. and Boyce, J. P. (1971) *Proc. Int. Soc. Sug. Cane Technol.*, **14**, 813–26.

Thompson, G. D. and Du Toit (1967) 'The effects of row spacing on sugarcane crops in Natal', *Proc. Int. Soc. Sug. Cane Technol.*, **12**, 103–12.

Thompson, H. A. (1976) 'Developments in cane harvesting and transport in Africa', *Sugar y Azucar*, **71**, (II), 27–34.

Thompson, H. A. (1977) TLTS, Bromley (personal communication).

Trouse, A. C. and Humbert, R. P. (1961) 'Some effects of soil compaction on the development of sugar cane roots', *Soil Sci.*, **91**, 208–17.

Tucker, R. W. E. (1951) 'A twenty-year record of the biological control of one sugarcane pest', *Proc. Int. Soc. Sug. Cane Technol.*, **7**, 343–54.

Turner, P. E. (1945) 'A grouping of soils according to their tillage and drainage requirements', *Jam. Assoc. Sug. Technol. Quart.*, **9**, 37–47.

Urata, R. and Warner, J. N. (1959) 'Criteria for sugarcane selection in Hawaii', *Proc. Int. Soc. Sug. Cane Technol.*, **10**, 702–14.

Urich, F. W. (1912) *Bull. Dept. Agric. Trin. Tobago*, **11**, 298.

Urich, F. W. (1914) 'Plant diseases and pests', *Bull. Dept. Agric. Trin. Tobago*, **13**, 101–3.

Valdivia, S. V. and Pinna, J. C. (1974) 'A theoretical salt effect limit for sugarcane considering soil physical properties', *Proc. Int. Soc. Sug. Cane Technol.*, **15**, 736–42.

Van Dillewijn, C. (1952) *Botany of Sugar Cane*. Chronica Botanica, Waltham, Mass.

Van Dillewijn, J. (1951) '*Fusarium* pokkahboeng', *Proc. Int. Soc. Sug. Cane Technol.*, **7**, 473–99.

Van Zwaluwenburg, R. H. (1937) 'Summary of laboratory studies of *Anomala* 1933–35', *Hawaii. Plrs. Rec.*, **41**, 25–32.

Van Zwaluwenburg, R. H. (1951) 'The insects affecting sugar cane in Mexico', *Proc. Int. Soc. Sug. Cane Technol.*, **7**, 373–7.

Venkatraman, T. S. and Rao, U. V. (1928) 'Coimbatore seedling canes Co's 205, 210, 213, 214 and 223 described and illustrated', *Agric. J. India*, **23**, Part 1.

Venkatraman, T. S. and Thomas, R. (1931) 'Coimbatore seedling canes Co. 281 and Co. 290 described and illustrated', *Agriculture and Live Stock in India*, **1**, Part 2.

Venkatraman, T. and Vasudeva Menon, P. P. (1964) 'Occurrence of *Sesamia inferens* (Walker) in South India on grown up sugar cane', *Indian Sug. Cane J.*, **9**, 53–4.

Wagenaar, G. A. W. (1959) 'Selection schedule used at the Java Sugar Experiment Station, Pasuruan', *Proc. Int. Soc. Sug. Cane Technol.*, **10**, 721–32.

Wahid, M. A., Steib, R. J. and Chilton, S. J. P. (1953) 'The use of fungicides to reduce the occurrence of red rot infection in stalks of standing cane', *Proc. Int. Soc. Sug. Cane Technol.*, **8**, 936–9.

Waiyaki, J. N. (1974) 'The ecology of *Eldana saccharina* Walker and associated loss in cane yield at Arusha-chini, Moshi, Tanzania', *Proc. Int. Soc. Sug. Cane Technol.*, **15**, 457–62.

Wakker, J. H. and Went, F. A. F. C. (1898) *Ze siekten van het suikeriet op Java*. E. J. Brill, Leiden.

Walker, D. I. T. (1962) 'Family performance at early selection stages as a guide to the breeding programme', *Proc. Int. Soc. Sug. Cane Technol.*, **11**, 469–83.

Walker, D. I. T. (1980) Personal communication.

Walker, D. I. T. and Simmonds, N. W. (1980) 'A comparison of the performance of sugarcane varieties in trials and in agriculture', *Exp. Agric.*

Walter, M. (1954) 'Extraction process for bagasse diffusion', *Proc. Int. Soc. Sug. Cane Technol.*, **8**, 766–74.

Warner, J. N. (1954a) 'The philosophy of sugarcane breeding in Hawaii', *Proc. Int. Soc. Sug. Cane Technol.*, **8**, 410–14.

Warner, J. N. (1954b) 'Techniques of breeding and testing sugarcane in Hawaii', *Proc. Int. Soc. Sug. Cane Technol.*, **8**, 415–22.

Warner, J. N. and Grassl, C. (1960) 'The 1957 New Guinea expedition for collection of sugarcane', *Proc. Int. Soc. Sug. Cane Technol.*, **10**, 792–4.

Warren, G. T. (1945) 'Observations on the sugar cane moth borer (*Diatraea saccharalis*) in Antigua and the control of same by parasites', *Proc. Br. W. Indies Sug. Technol.*, 1945, 78–82.

Wells, S. D. and James, G. D. (1976) 'Rapid dextrans formation in stale cane and its processing consequences', *Proc. Qld. Soc. Sug. Cane Technol.*, **43**, 287–93.

WICSCBS (1969) *Ann. Rept. W.I. Cent. Sug. Cane Br. Stn.*, **36**, 15–17.

Wiehe, P. O. (1944) *Rep. Dept. Agric. Maurit.*, **1944**, 11–12.

Wiehe, P. O. (1963) 'The control of sugar cane diseases in Mauritius', *Proc. Int. Soc.*

Sug. Cane Technol., **11**, 743–8.

Wiggins, L. F. (1949) 'Sugar cane wax, Part I', *Proc. Br. W. Indies Sug. Technol.*, 1949, 24–8.

Wiggins, L. F. (1950) 'Sugar cane wax, Part II', *Proc. Br. W. Indies Sug. Technol.*, 1950, 16–21.

Wiggins, L. F. and Davison, B. K. (1951) 'Sugar cane wax, Part III – The refining of sugar cane wax', *Proc. Br. W. Indies Sug. Technol.*, 1951, 50–5.

Williams, C. B. (1921) 'Report of the froghopper-blight of sugar-cane in Trinidad', *Mem. Dept. Agric. Trin. and Tobago*, **1**, 179 pp.

Williams, E. (1964) *Capitalism and Slavery*. Andre Deutsch, London.

Williams, J. R. (1963) 'Army worm', *Ann. Rep. Maurit. Sug. Ind. Res. Inst.*, 1963, 92.

Williams, J. R. (1969) 'Nematodes as pests of sugar cane', in *Pests of Sugar Cane*. Elsevier, Amsterdam.

Williams, J. R. and Mamet, J. R. (1962) 'The insects and other invertebrates of sugar cane in Mauritius and Réunion', *Occ. Pap. Maurit. Sug. Ind. Res. Inst.*, **8**.

Williams, J. R. and Moutia, L. A. (1954) 'Some aspects of sugarcane entomology in Mauritius', *Proc. Int. Soc. Sug. Cane Technol.*, **8**, 570–3.

Williams, R. O. and Williams, R. O., Jnr. (1951) *The Useful and Ornamental Plants of Trinidad and Tobago*. Guardian Commercial Printery, Trinidad, pp. 256–7.

Wilson, G. (1951) 'Frenchi grub control in North Queensland by benzene hexachloride', *Proc. Int. Soc. Sug. Cane Technol.*, **7**, 414–16.

Wilson, G. (1956) 'Control of the cane grub pests *Dermolepida albohirtum* Waterh. and *Lepidiota frenchi* Blkb. by benzene hexachloride', *Tech. Commun. Bur. Sug. Expt. Stns. Qd.*, **2**.

Wilson, G. (1969a) 'White grubs as pests of sugar cane', in *Pests of Sugar Cane*. Elsevier, Amsterdam.

Wilson, G. (1969b) 'Insecticides for the control of soil inhabiting pests of sugar cane', in *Pests of Sugar Cane*. Elsevier, Amsterdam.

Wilson, G., Hitchcock, B. E., Woods, J. A., Cook, I. M. and Moller, R. B. (1965) *Rep. Bur. Sug. Exp. Stns. Qd.*, **65**, 47–56.

Winstanley, J. (1954) 'A general survey of the possible uses of bagasse', *Proc. Br. W. Indies Sug. Technol.*, 1954, 198–208.

WISA (1972) *Annual Report by the Chairman for 1971–72*. West Indies Sugar Association (Inc.), Barbados.

WISA (1973) *Annual Report by the Chairman for 1972–73*. West Indies Sugar Association (Inc.), Barbados.

WISA (1974) *Annual Report by the Chairman for 1973–74*. West Indies Sugar Association (Inc.), Barbados.

Wismer, C. A. (1951) 'Controlling pineapple disease of sugarcane', *Hawaii. Plrs. Rec.*, **54**, 23–53.

Wismer, C. A. (1961) 'Pineapple disease', in *Sugar-cane Diseases of the world*. Vol. I, Elsevier, Amsterdam.

Wismer, C. A. (1971) 'A sugar-cane clone apparently immune to RSD', *Sugar-cane Path. Newsl.*, **6**, 46.

Withycombe, C. L. (1926) 'Studies on the aetiology of sugar cane froghopper blight in Trinidad. I. Introduction and general survey', *Ann. App. Biol.*, **13**, 64–108.

Wolcott, G. N. (1936) 'The chemical control of white grubs in Puerto Rico', *Proc. Int. Soc. Sug. Cane Technol.*, **7**, 417–23.

Wood, D. (1968) *Trinidad in Transition*. Oxford University Press, London.

Wood, E. F. L. (1922) *Report by the Hon. E. F. L. Wood, M.P. (Parliamentary Under-Secretary of State for the Colonies), on his Visit to the West Indies and British Guiana, December 1921–February 1922* Cmd. 1679. HMSO, London.

Wu, H. S. (1969) 'A brief review of recent diffuser development', *Proc. Int. Soc. Sug. Cane Technol.*, **13**, 165–8.

Yadav, R. P., Anderson, H. L. and Long, W. H. (1965) 'Sugar cane borer resistance to insecticides', *J. Econ. Ent.*, **58**, 1122–4.

Yates, R. A. (1967) *Austr. J. Agric. Res.*, **18**, 903–20.

Yearwood, R. D. E. (1940) *Further Factory Trials in the Manufacture of Y.C. Sugar without using Stannous Chloride – the Development of a New Process.* ICTA circular.

Young-Kong, V. M. (1976) 'The potential impact of sugar cane smut (*Ustilago scitaminea*) on the Caribbean sugar industry', *Proc. W. Indies Sug. Technol.*, 1976, 125–9.

Young-Kong, V. M. and Jackson, N. E. (1976) 'The effect of conversion of cambered beds to a ridge and furrow layout on sugarcane yields in Guyana', *Proc. W. Indies Sug. Technol.*, 1976, 87–90.

Zummo, N. (1974) 'Sugarcane mosaic virus strain L: a new virulent strain of sugarcane mosaic virus from Meigs, Georgia', *Proc. Int. Soc. Sug. Cane Technol.*, **15**, 305–9.

Index